Pure Adulteration

AT BAY. Eighth Cartoon of the Series. Fig. 193.

August Berghaus's hydra-headed monster of adulteration, from *Rural New Yorker*, May 1887.

Pure Adulteration

CHEATING ON NATURE IN THE AGE OF MANUFACTURED FOOD

Benjamin R. Cohen

The University of Chicago Press CHICAGO AND LONDON

The University of Chicago Press, Chicago 60637
The University of Chicago Press, Ltd., London

Published 2019
Paperback edition 2021
Printed in the United States of America

30 29 28 27 26 25 24 23 22 21   1 2 3 4 5

ISBN-13: 978-0-226-37792-6 (cloth)
ISBN-13: 978-0-226-81674-6 (paper)
ISBN-13: 978-0-226-66709-6 (e-book)
DOI: https://doi.org/10.7208/chicago/9780226667096.001.0001

Library of Congress Cataloging-in-Publication Data

Names: Cohen, Benjamin R., author.
Title: Pure adulteration : cheating on nature in the age of manufactured food / Benjamin R. Cohen.
Description: Chicago ; London : The University of Chicago Press, 2019. | Includes bibliographical references and index.
Identifiers: LCCN 2019017188 | ISBN 9780226377926 (cloth : alk. paper) | ISBN 9780226667096 (e-book)
Subjects: LCSH: Food adulteration and inspection—United States—History—19th century. | Food adulteration and inspection—United States—History—20th century.
Classification: LCC TX515 .C65 2019 | DDC 363.19/2640973—dc23
LC record available at https://lccn.loc.gov/2019017188

♾ This paper meets the requirements of ANSI/NISO Z39.48-1992 (Permanence of Paper).

FOR MY PARENTS

Contents

A Note on the Digital Companion to This Book

A number of maps included throughout this book refer readers to https://purefood.lafayette.edu/. The maps included here are static versions of a larger series of dynamic maps tracing changes in various features of the three main cases in this book between the 1870s and 1910s. Readers should refer to that site for further maps and images from the book.

Table of Maps

(Full sets of maps are available at the digital companion site https://purefood.lafayette.edu.)

Prologue

When humans cheat on each other, we call it adultery. When they cheat on nature, we call it adulteration. Both senses of *adulter* come from a deeper moral calculus of what is right and proper and what is not. Both carry within them a notion of dishonesty, impropriety, and lack of faith. As an environmental writer, I'm not entirely comfortable personifying nature with the word *cheating*, as if humanity were separate from nature and akin to a partner, but people in the nineteenth century didn't much share my concern as they laid out the main questions of the pure food and adulteration debates: how could you tell the food you bought was the food you thought you bought? How could you verify whether it was authentic or pure or from nature? Are pure and natural synonymous? And what if the food was made in a factory instead of harvested from a field, as so many new late-1800s products were? Is it okay to manipulate nature far enough to produce new foods but not so far that you question the safety of the foods and the health of those who consume them? How do you know where that line is? And who decides?

The concerns weren't entirely new in the 1800s, as people have worried about their food for as long as people have bought and traded it. But manufactured foods in the later nineteenth century changed the dimensions of worry by confusing stable notions of nature and artifice. By 1906, the people asking the above questions in the United States, pure food agitators, helped justify legislation that would lead to the Food and Drug Administration (FDA), an agency thereafter cast as the culmination of the food-safety movement of this era.

Typical textbook accounts, with the pitfalls of simplification, explain the agency's founding with a few quick strokes. There was Upton Sinclair's *The Jungle*, also 1906, and to this day usually the first thing people think about when the topic of pure foods and the origins of the FDA comes up. There was Harvey Wiley's decades-long work at the Bureau of Chemistry in the US Department of Agriculture (USDA). Even if Wiley's name is not as recognizable as Sinclair's, he is well known to readers of food and Progressive Era history as a central figure in the fight against adulteration, and he's also known for absorbing too much of the credit, overshadowing other advocates in the domestic-science world, civic associations, women's clubs, grocers' forums, and offices of public health. The quick accounts also tend to disregard widespread public food-health precedents from European states that had been well under way for decades. And such clipped accounts almost always leave readers with a comment that after all of this adulterated mess, everything was better. Yet even today, we are still wondering whether some new product or some "natural" food or preservative or additive is a sham, a cheat, an adulteration. In the face of daunting questions about genetic modification, highly processed foods, and the veracity of health claims, we are still struggling to answer the eternally braided questions of what to eat, who says so, and how do they know.

There is a remarkable history here, one that places continuing concerns about food onto a longer and more complicated trajectory. I've been looking into that nineteenth-century history while working in the local food scene in the twenty-first. My motivation in our time is to envision a future that goes beyond the limitations of the industrial food system of the past century. This attention to the future sparked my interest in understanding where manufactured foods came from in the past and what they left us. Studying the origins of agricultural science in my first book led me to the beginnings of the USDA in the 1860s. I started the research for this book thinking about the next phase in agricultural history, which, as it happens, leaned toward food history and the beginnings of the FDA after 1906. What I found in finishing this project was that the story to be told was less about the regulatory pinnacle of the FDA and more about the shifting understanding—or definition, or conception, or idea—of pure food itself. Across the latter nineteenth century, the very concept of a pure food changed from one grounded in the environmental work of agriculture, community, and cooking to one that was outsourced, so to speak, to agents with certified analyses working at the end of the food life cycle at the storefronts and grocery shelves of labeled items. The transition from an agriculturally anchored agency, the USDA, to a consumer-anchored one, the FDA, followed the shifts from field to kitchen, from farm to city, from producer- to consumer-oriented viewpoints. Those trends all grew substantially throughout the 1900s, but their foundations were established early in the century.

Debates over purity and adulteration provide a captivating window onto those origins of manufactured foods and the perceived problems they wrought. Indeed, disputes of what was then called the era of adulteration revealed confusing new tensions in the ways people understood, procured, trusted, and ate their food. Cultural factors help explain why anyone cared. They show that a prevailing suspicion of cheats, deceivers, hucksters, and con men—and a corollary fervor for sincerity, authenticity, and honesty—provided the foundation from which the era of adulteration was born. Environmental factors help explain why the prevailing cultural concerns were exacerbated. New supply chains, complex commodity flows, far-flung land-use patterns, and theretofore unknown ingredients all formed a new infrastructure of food and agriculture that made the view from farm to fork thick and opaque. Together, that host of cultural and environmental conditions shaped the pure-food crusades and the set of chemists, analysts, and public health officials bent on commandeering and policing the concept of purity. Whereas into the mid-nineteenth century it was still common to understand food based on its origins, its provenance, and its purveyors, by the early twentieth century the concept of purity had moved from its agricultural setting and relocated into the hands of the analyst and lab. Purity had become a scientific concept policed by government agencies and backed by certified analysis. It would be the analysts and scientists who drew the line between pure and adulterated.

Pure Adulteration looks to the pure food crusades to tell a story about cultures of authenticity and manufactured foods of questionable purity. Along the way, it shows the ways farmers, manufacturers, grocers, housewives, government officials, scientific analysts, public health officers, customers, and consumers sought to demarcate and patrol the ever-contingent, always contested border between purity and adulteration. This book about the pure food crusades is about how people drew the line between pure and adulterated. There is (and was) no natural, prehuman distinction that we can simply uncover and enforce; we have to decide where to draw it and how to police it. Today's world is different from that of our nineteenth-century forbears in so many ways, but the challenge of policing the difference between acceptable and unacceptable practices remains central to daily decisions about the foods we eat, how we produce them, and what choices we make when buying them.

they wrought. Indeed, disputes over what was then called the era of adulteration revealed confusing new tensions in the ways people understood, procured, trusted, and ate their food. Cultural factors help explain what anxieties existed. They show that a pervasive suspicion of cheats, chemists, hucksters, and counterfeits—and a corollary fervor for sincerity, authenticity, and honesty—provided the foundation from which the era of adulteration was born. Environmental factors help explain why the prevailing cultural concerns were so pressing. New supply chains, complex commodity flows, the filling land use patterns, and the adoption of unknown ingredients all played a role in the consumer's food and agriculture that made the view from the larder look a bit different. Together, the history of cultural and environmental conditions shaped the pure food crusades and the rise of chemists, analysts, and public health officials bent on commanding and policing the concept of purity. Whereas into the mid-nineteenth century it was still common to understand food based on its origin and provenance, and its purveyors, by the early twentieth century the concept of purity had moved from the agricultural setting and entered into the hands of the analyst and laboratory. It had become a scientific concept policed by government agencies and backed by certified analysts. It would be the analysts and scientists who drew the line between pure and adulterated.

Pure Adulteration looks to the pure food crusades to tell a story about cultures of authenticity and manufactured foods of questionable purity. Along the way, it shows the ways farmers, manufacturers, grocers, housewives, government officials, scientific analysts, public health officers, reformers, and consumers sought to demarcate and patrol the ever-changing, always contested border between purity and adulteration. This book about the pure food crusades is about how people drew the line between pure and adulterated. There is (and was) no natural, prehuman distinction that we can simply uncover and enforce; we have to decide where to draw it and how to police it. Today's world is different from that of our nineteenth-century forebears in so many ways, but the challenge of policing the difference between acceptable and unacceptable practices remains central to daily decisions about the foods we eat, how we produce them, and what choices we make when buying them.

1

The Appearance of Being Earnest

It's 1879. The courtroom in Santiago is full. The tables and benches and sidelines hold a defendant, his accomplice, the lawyers for all sides, the justice of the Chilean supreme court, and onlookers. The trial had dragged on for two years. The defendant was incarcerated all the while at the nearby Des Hotel Ingles. This autumn afternoon was the end of a very long journey.

Up to that point in life, the accused had "engaged the most elegant suite of rooms in the most fashionable hotels," charming investors with his "large, eloquent eyes." Having spent the prior decade crisscrossing half the globe from Europe to North America to South America, he was the man papers from the United States to New Zealand called "foremost in the ranks of the world's swindlers," the man who they said had "the black heart of a conscienceless scoundrel," the one the *New York Times* devoted ten long paragraphs to in his obituary six years later as the "king of swindlers."[1]

He was the Chevalier Alfred Paraf.

Paraf was a Frenchman. He was born on June 10, 1844, in the Alsace region near the Rhine. He's not part of historical memory anymore, though the image of his deceptive capacity was a stock reference in stories about swindlers for decades. He had, such sources say, a big personality and a winning smile. He had, they said, "the suave address of a gentleman." And of course he had a wax-tipped moustache. One paper described him as a man with "the form and features of an Apollo" to match "the polished manners of a citizen of the world." Another called him "handsome, polished, well educated, known for his keen intelligence and ready wit." In an age of confidence men—con men—a Brooklyn daily said of Paraf that he stood "above and beyond his fellows, who, compared with him, were mere bunglers in operations which he had reduced to an exact science and of which he was the greatest... exponent."[2] From Alsace to Scotland to New York to Rhode Island to San Francisco to Nevada to Chile,

Paraf made his name as "handsome, refined, clever, brilliant, extravagant, immoral and audacious" (fig. 1.1).

His cheat? His crime? His unconscionable deed? What prompted the *New York Times* to anoint him the "king of swindlers"?

Fake butter. Oleomargarine. The scourge of dairy natures.

Paraf was an adulterator. He made artificial versions of natural things and sought to pass them off as equivalent. *Adulteration* is the term long used to describe the contamination, deception, or false substitution of one product for another. In the nineteenth century, the term was increasingly used to refer to suspicious foods. Some of the Chevalier's milder accusers found it inappropriate. Others—the ones resorting to black-hearted, conscienceless-scoundrel-level rhetoric—clearly didn't mince words, framing his crimes as an assault on moral propriety.[3]

Such critics slammed oleomargarine as a purported adulterant of butter. It was a manufactured product from the factory, not the result of a natural agricultural process. It was wrong, contemporaries thought, because it went against nature. In the century to come, many people would think of it as a cheap substitute for butter. Some would see it as a sign of progress while others would continue to see it as a corruption of the good and the right. Granted, in the particular case of margarine the product's evolution in public consciousness has been so great that rather than concealing the deception, by the later twentieth century advertisers took it as a point of pride that consumers could be tricked—we couldn't believe it's *not* butter. Yet that eventuality was hardly a foregone conclusion. In the decades after its invention in 1869, margarine was cast by some as "the most gigantic swindle of our time." The charges against it were serious and severe. "The Cow Superseded," said the *San Francisco Chronicle*. "That atrocious insult to modern civilization," if we follow the *Washington Post*.[4]

The handsome Chevalier may thus have become famous as the proprietor of the oleo swindle, but his crimes and stature far exceeded that singular product. They were tapping into a larger crisis of confidence and trust by highlighting the contested ethics of purity, nature, and artifice. Similarly, for some, margarine itself may have been a particularly egregious affront to the nature of the cow—the nature of an agrarian world—but its presence was accompanied by a laundry list of artificial products and suspected contaminants entering the new marketplace of the Victorian world. As the frontispiece has it, margarine was one head of a three-headed hydra of adulteration, with cottonseed oil (fake olive oil and fake lard) and glucose (fake sugar), and even those were but a sampling of the wider

FIGURE 1.1. Chevalier Alfred Paraf. He had this portrait taken in 1873 while in San Francisco at the studios of Bradley and Rulofson. Courtesy of Columbia University.

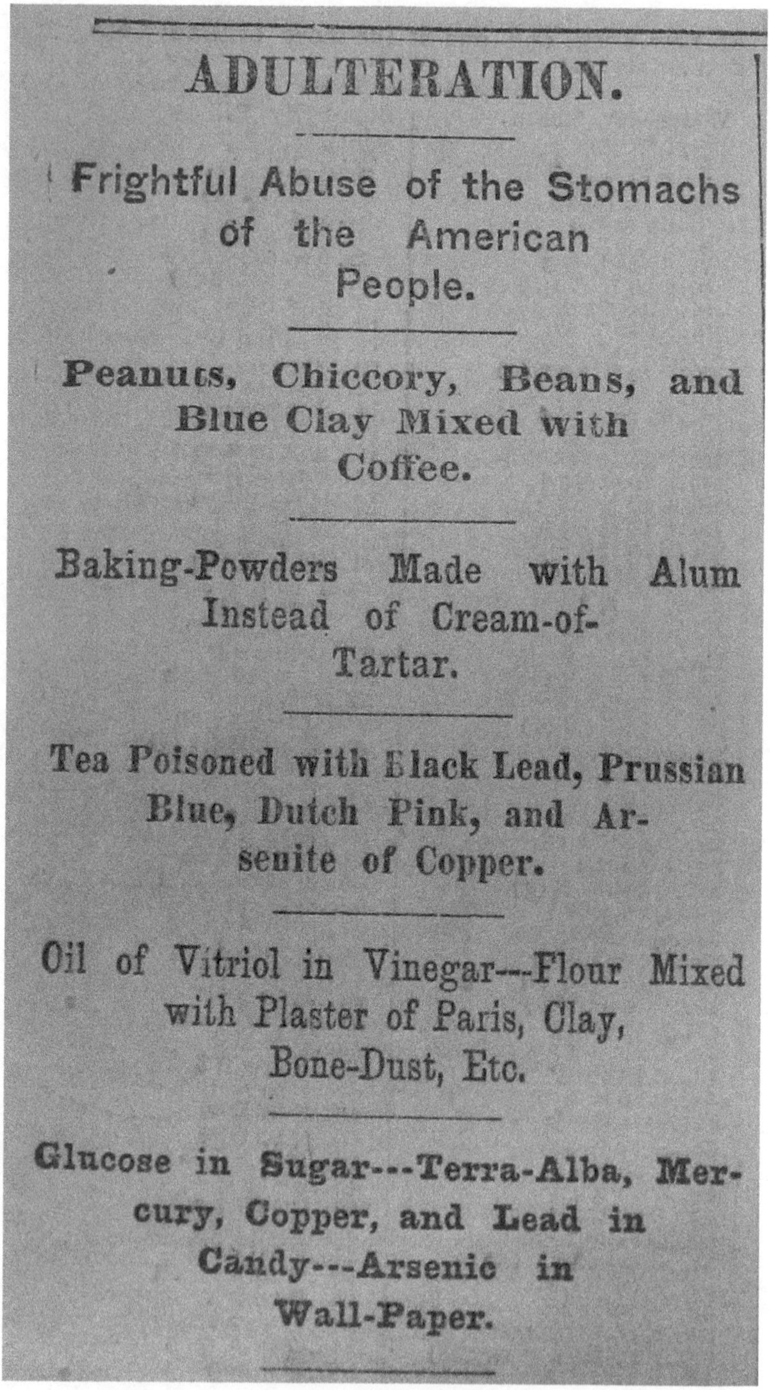

ADULTERATION.

Frightful Abuse of the Stomachs of the American People.

Peanuts, Chiccory, Beans, and Blue Clay Mixed with Coffee.

Baking-Powders Made with Alum Instead of Cream-of-Tartar.

Tea Poisoned with Black Lead, Prussian Blue, Dutch Pink, and Arsenite of Copper.

Oil of Vitriol in Vinegar---Flour Mixed with Plaster of Paris, Clay, Bone-Dust, Etc.

Glucose in Sugar---Terra-Alba, Mercury, Copper, and Lead in Candy---Arsenic in Wall-Paper.

FIGURE 1.2. *Chicago Tribune* reports on antiadulteration legislation, February 9, 1879.

concerns afoot. Milk and sugar, coffee and tea, mustard and ketchup, baking powder, bread, butter, cheese, flour, olive oil, honey, candy, spices, vinegar, ice, beef, pork, lard, fertilizer, beer, wine, canned vegetables—these were all under suspicion of having been contaminated or debased, the blame falling variously on farmers, manufacturers, distributors, grocers, even potentially paid-off inspectors. The entire panoply of the food system was under suspicion at one point or another from one view or another.

How did these suspicions come to be? Where did the era of adulteration come from? Where did it leave us?

The questions bear a few kinds of answers. Paraf's background offers a specific one. The broader background of culture, environment, and science in the nineteenth century adds another. This book, as a whole, takes those together to provide a third.

As for specifics, for his part Paraf was the son of a successful dye maker in Mulhouse, a town near the Rhine River about halfway between Strasbourg and Basel.[5] The area was a hotbed of manufacturing development by midcentury and the seedbed for a chemical revolution over the next half century. To speak of chemistry at the time was to speak of a craft trade devoted to animals, vegetables, and minerals. In the history of science, agricultural chemistry was coming into its own, animal chemistry was thriving, and plant chemistry was an active focus for scientists, dye makers, and druggists alike. Agricultural and manufacturing efforts were often of a piece, all common stages in the exchange of organic material. Fabrics and textiles were woven in mills from the fibers of the land; colors were mixed from extracts of plants; the bulk of new foods drew from combinations of plant and animal oils and fats; and chemists served as aids to the processes.

The young Paraf's chemical prowess began with training from his father. At age fourteen, he went to study in Paris at the Collège de France. His skills gave him access to a profitable trade. He would have over a dozen patents before he was thirty. In his early twenties, though, he was bored and restless.

With traveling money from his generous father, he set out in the early 1860s for Glasgow. It was there that his chicanery began. Having quickly lost his cash on ostentatious living, he devised and sold a new calico dye from a fragrant plant called weld (*Reseda lutea*). The color, it happens, wore off after minimal use. That wearing off was long enough after its application for Paraf to hightail it back to France before he was caught. He was a spry twenty-three-year-old, there just long enough to bilk his own uncle of thousands in another scam before crossing the Atlantic to New York in 1866 or 1867 (reports differ).[6]

"Asunto Paraff"—The Paraf Situation, to quote later Peruvian sources—blossomed in Manhattan. He met, courted, and wed Leila Smith, the daughter of a high-profile New York lawyer, C. Bainbridge Smith. He took side trips to New England to advance his dye-making funds, at one point fleecing the former Governor of Rhode Island out of tens of thousands. He lived lavishly in midtown Manhattan and dined at only the most posh houses, Delmonico's among them. He launched a new manufacturing facility with a pilfered patent. Paraf had a lot going on: a lady to woo and marry, a new industry to start, patents to file, appearances to keep up, investors to dupe, creditors to flee. Here was a man on the move.[7]

He moved quickly not just geographically but in business dealings too. Because he was a learned and wealthy man and a notably suave chemist, Paraf identified and then collapsed the cultural distance between the unknown and the trustworthy. He filled that space with metrics of confidence and the veneer of respectability and allure. In fact, had he been "only a brilliant swindler, obtaining gold and silver from old boots and creating fortunes out of the refuse of tar barrels," wrote the *Brooklyn Eagle*, "the world would have been content to let him squander the money he acquired so easily without grudging him his meteoric prosperity or demanding his punishment." But no, while the American people may have borne "forgery, defalcation [, and] hypocrisy," what he did was "an act too base for characterization." It was, his advocates would write, "the crowning work of his active intellectual life."[8] "What he did," his detractors would shout, "was to introduce oleomargarine, a vicious substitute for butter."[9]

What he did, in short, was to compound worries over the artificial. His character was questionable and insincere, a fake, just as his product was viciously replacing an icon of the agrarian dairy. What is more, he claimed the patent for oleomargarine as his own. But he had nabbed it from his fellow Frenchman, Hippolyte Mège-Mouriès. The Mège Patent, as it was later called, had been motivated by the coming Franco-Prussian War of 1870. As early as the 1867 World Exposition in Paris, Napoleon III was encouraging animal chemists to develop a cheaper substitute for butter, all the better to save important milk and dairy stores for the troops. Mège mixed milk with heated and refined animal fats—tallow from rendered beef suet for its stearin and olein—to reduce the dairy content of butter.[10] A new product was born. Paraf thus misrepresented his relationship to the invention of a product that people feared was a misrepresentation of natural butter (fig. 1.3).

Stolen recipe in hand, Paraf charmed investors in New York into establishing the Oleomargarine Manufacturing Company in 1873. While maneuvering to start the company, he had meetings with the eminent

ALFRED PARAF, OF NEW YORK, N. Y., ASSIGNOR TO THE OLEOMARGARINE MANUFACTURING COMPANY.

IMPROVEMENT IN PURIFYING AND SEPARATING FATS.

Specification forming part of Letters Patent No. **137,564**, dated April 8, 1873; application filed February 11, 1873.

FIGURE 1.3. Header of Paraf's original ill-gotten patent for "Improvement in Purifying and Separating Fats," that is, oleomargarine, approved on April 8, 1873, two months after filing it and three years after stealing the recipe.

though undoubtedly less suave chemist Charles Chandler. Chandler—a professor at Columbia's School of Mines (1864), the first chemist of the newly chartered New York Board of Health and its Anti-Adulteration Division (1866), and soon a founding member of the American Chemical Society (1876)—dismissed the young man as a curiosity without much comment. Some later reports claimed that Paraf had "studied with Chandler" while in New York. If so, the things he studied were not academic.

Paraf stayed a step ahead of those who would question him by heading west later in 1873. Ostensibly at the invitation of capitalists in California, he helped build the new California Oleomargarine Company in San Francisco. In a show meant to build public trust, the company hosted a demonstration of the process in October. The public display was a hallmark maneuver of con men; it was not a big step from Paraf's margarine to Barnum's circus to the carnival of historical tricksters engendering trust through purportedly transparent demonstrations. Supporters published in local papers the invitation to see "the matter explained lucidly by Professor Alfred Paraf" and sent formal letters to prominent figures, including Professor Chandler in New York. "Advancement and invention are seen everywhere," the authors wrote, "and the latest and most important discovery is one that has enabled the inventor to achieve a victory over animal matter." Included with the invitation was a glossy biography of the handsome Chevalier (fig. 1.1). The text matched the image. The biography painted a picture of unsullied character with inventive prowess.[11]

While in California, Paraf entertained an audience with past governor and soon to be senator Leland Stanford. Here was the president of the Central Pacific and Southern Pacific Railroads, the man who hit the spike in Promontory, Utah, in 1869 connecting the transcontinental line, the later benefactor of the university founded in his name. Stanford the robber baron likewise dismissed Paraf without losing any money, and in this case of a swindler's meeting you have to think that perhaps it takes one to know one.

As he traveled in San Francisco, his investors back east had found that he stole—"filched," they said—Mège-Mourié's invention. They dispatched representatives to Paris to secure the rightful use of the Mège patent. Upon return, they kept making the oleo while renaming and reorganizing the venture as the United States Dairy Company.[12] A new name and constitution would distance themselves from their ill-gotten founding.

News traveled quickly along the telegraph lines strung beside those same new rail lines that brought Paraf west. The Chevalier was arrested in California with an associate on August 4, 1874, "under an indictment for forgery."[13] Bail was set at $10,000.

It didn't dissuade him. He of the eloquent eyes somehow posted bail and spent two more years in California dodging his potential jailors before eventually heading south, leaving no record that the trouble back east was a matter of concern or that the forgery arrest was something to take seriously. He presented himself as a patent-wielding inventor of the new manufacturing age, maintaining his self-presentation and the knockoff butter as legitimate, even as the public began to cast him as the disingenuous oleo man and a challenge to sincerity with a jury still out on the matter of legitimacy.

Paraf bounded onto a Gilded Age stage that was prepared to see him. His audience—his mark—was already wrestling with the problem of whether novelty was a change for the better or for the worse. He plied the uncertainty over fabrication and antagonized the search for authenticity.

Admittedly, the adulteration of foods, drinks, and drugs goes back at least as far as notices from the ancient Hittites (ca. 1500 BCE) and Roman legal codes and references in the *Oxford English Dictionary* well before Shakespeare ("adulterate: to render spurious or counterfeit; to falsify, corrupt, debase esp. by admixture of baser ingredients"). Guild-based codes of conduct defined the contours of proper food production and distribution for centuries, where local producers were also local consumers, where the moral standard for providing safe food was policed by community-based traditions and structures. But adulteration became a more pronounced public problem with the breakdown of those guild-based structures and the rise of early industrial artificial foods from manufacturing sites. The later nineteenth century debates over these long-standing problems became known as the Pure Food Crusades. In some circles, they would call it the era of adulteration.[14]

This was an extended period of enormous environmental upheaval in many notable ways. Paraf stood at the head of it, looking on a scene of impressive change. Historians commonly point to a host of rapid changes in the ways people treated the environment during the later nineteenth century: rural-urban shifts, railway-led expansion, colonization, imperialism, the birth of national parks, and a nascent conservationist ethos with which to respond to those changes. But debates over purity and adulteration show that one of the more remarkable transitions was taking place in the agricultural environments that held fields, factories, and kitchens together. "A new generation of industrially produced foods left consumers wondering about what had been added or lost inside the factory," writes Susanne Freidberg in her study of perishability in the same era. In just half a century, conventional agricultural production and food identity were radically upended with new factory systems, new manufactured products, and new ways of buying, cooking, and, ultimately, knowing food.[15]

In Europe and the United States, debates over adulteration provided a primary forum for tackling the changes. The problems were framed in a number of ways. When it concerned contaminated food, officials treated the issue as a public health concern. The anxieties evident in figure 1.2 point to this. The fear was that contrivances like those from Paraf would make you sick. In one of the archives consulted in researching this book, a box held a plastic-covered container with a letter to the Board of Health. Taped to this 1879 letter was a half-bitten piece of chocolate revealing a cockroach (still preserved to this day) inside it. When it was about undermining weights and measures with false ingredients, consumers and nascent regulators treated it as an economic, taxation, and marketplace issue. Think of water in milk or sawdust in flour, two prominent problems of the time. And when it concerned the problems of distant food makers and agriculturalists, as the following chapters show, journalists, novelists, politicians, farmers, and civic groups cast it as a consequence of the increasing complexity of global trade networks and their new food-production and agrarian practices.

The broader background of culture, environment, and politics in the nineteenth century helps give substance to the source of such multiheaded angst. For one thing, prevailing cultural concerns about character were at play. Upended measures of trust, confidence, and sincerity that characterized an age of growth and social flux aggravated anxiety in the nineteenth century. Paraf's questionable character makes that clear enough. The larger issue that his example touched on, though, stemmed from a serious challenge to the very concept of purity and its relationship to nature. In the case of adulteration, the crux of the matter was a prevailing belief that sincere people begot sincere food. Those who were worrying over artificial foods were also worrying over insincere food providers. The concepts of pure and natural were synonymous. Authenticity, nature, and purity formed a constellation that provided a reference point for right living. Deception, artifice, and adulteration stood as a series of counterreferents on the side of impropriety. When the horticulturalist Liberty Hyde Bailey worried in the 1910s that "we find ourselves [in] a staggering infidelity in the use of good raw materials," he was binding together the sense of artifice as false and adulteration as cheating.[16]

In an environmental sense, the lengthening distance between the producer and the consumer of foods and other agriculturally based products (things like fertilizers and cotton, for example) exacerbated adulteration. "As soon as there emerged a consuming public, distinct and separate from the producers of food," writes food historian John Burnett, "opportunities for organized commercial fraud arose."[17] A chemist like Paraf was in a prime position to take advantage of this increasing distance between people and their food. The new foodways both followed and helped foster the rural-urban migrations and immigration patterns that are a stock chapter in surveys of the period between the American Civil War and the first World War. Old familiarities were disappearing; new mechanisms to verify the identity of a product were not yet institutionalized (though they would be).

The new foodways were not so schematically drawn, though, and the environmental context of adulteration was about more than simply pulling point A and point B farther apart. As many historians have shown, the geographical reconfigurations led to complex and expanding networks, not just the thin chain and easy metric of distance and separation. A host of new processors, distributors, wholesalers, and retailers mucked up the visible connections between farms and forks. It was about thickness, opacity, and confusing sightlines as much as distance.[18] As opposed to the nearly one-stop shop of butter from a creamery, margarine was a product of stockyards, slaughterhouses, animal-fat processors, chemists, oil traders, mechanical pressers and heaters and boilers, coloring matter (usually new chemical derivatives), and other ingredient sources.

The satire in figure 1.4, below, exaggerated the distinction to good effect. In the past, *Puck* suggested, butter came from the pastoral calm of the dairymaid, whose pokey cow peeked at the transformation of milk into butter through an open door in summer. In the present, by contrast, a malicious churner works a cauldron of hot chemical ingredients sourced from a disheveled stack of oil barrels and questionable materials. The image is indeed exaggerated; margarine was a product of the agrarian world just as butter was (differing by the degree of human intervention in the

FIGURE 1.4. "Progress and Butter," a satirical take on the difference between real and fake butter accentuating a contrast between rural dairymaids and ill-begotten chemical devilry. From *Puck* 7 (March 31, 1880): 55.

nonhuman world). But as the cartoon accentuates, new foodways of industrial modernity structured the path between farmers and eaters in ways that made the path from one to the other murky. Note, too, that the image captures a gendered dimension to the perceived changes. Where a fair and feminine dairymaid sat at the center of the practice of the past, an animalistic man controls the process of the present. Taken together, the changes were (and remain) disorienting.[19]

What is more, Paraf made his way while political debates took shape through the rise of the modern state, which led to legal statutes demanding standards for goods and services. Rather than caveat emptor—buyer beware—those seeking stronger governance adopted more deliberate approaches. French regional administrators were forerunners, requiring analyses of water, medicine, foods, and even colored candy wrappers by the 1830s.[20] These were the consequence of a public health movement instigated by new Napoleonic Codes of public administration. In Britain, full-bore analyses took root by the 1840s and 1850s, the result of analytical work published in the *Lancet* and authored in large part by the public health pioneer Arthur Hassall. This even spawned a new profession. By the 1870s, writes historian Chris Hamlin, towns were "required to employ a public analyst, an office established to fight food and drug adulteration."[21] There, water quality led the charge for broader public health overtures, with food safety and air measurements to follow. The latter half of the century was host to a bevy of new scientific trades angling to tackle the adulteration problem. English-style public analysts were but one; analytical chemists, food scientists, nutritionists, and other public health officers followed suit.

New forms of public administration resulted from these worries, but the concerns themselves were produced by an evolving sense of the role of government in handling civic affairs. New ways of quantifying and standardizing the food trade provided new forms of measurement. Those new codified measurements replaced the previous dominance of cultural norms based on acquaintance and personal experience. The bureaucratized measures stood in for the trust and familiarity of community-based sales and services, the ones, it is worth noting, that modern local food advocates are working to rebuild.[22]

There were changes in administering the public square from the private sector, too, especially through the development of what we now recognize as modern advertising. Before the twentieth century, advertisements and brands were a scattered affair; they were yet to be organized through professionalized strategies of identifying and appealing to target audiences of willing consumers. The food industry and grocers' empire were some of the most vibrant spheres of activity for a new age of admen to assure customers that their store-bought goods were wholesome and healthy. Procter & Gamble's work with soap and household goods offered an early, visible example, with a "stars and the man in the moon" logo; ads for patent medicines, agricultural supplies, and new foods were also prominent by the Gilded Age. A vibrant print culture aided the cause, as weekly magazines, the penny press, and trade journals created space for such advertisements. The cultural values of trust and authenticity played central roles in terms of new advertising techniques, brand development, and sales strategies across the Gilded Age and Progressive Era. This was a new politics, especially in the ways in which people governed the marketplace through nascent regulation if it was public government and advertising if private. Thus did a broader nineteenth-century set of changes in culture, environment, and politics create and illuminate the angst over food identity.[23]

In the face of those changes, questions about purity were everywhere. During the era of adulteration, the general public would know about the issues of purity from personal experience growing, buying, and preparing food. The reading populace knew even more about it from a new genre of literature, the antiadulteration treatise. The genre began its modern incarnation in 1820 with Fredric Accum's *A Treatise on Adulterations of Food, and Culinary Poisons* (fig. 1.5). This was the first significant work to present a full account of adulteration and its detection as a public and scientific problem. Accum's work was motivated by a moral charge to protect the

A TREATISE

ON

ADULTERATIONS OF FOOD,

AND

Culinary Poisons,

EXHIBITING

THE FRAUDULENT SOPHISTICATIONS

OF

BREAD, BEER, WINE, SPIRITUOUS LIQUORS, TEA, COFFEE,

Cream, Confectionery, Vinegar, Mustard, Pepper, Cheese, Olive Oil, Pickles,

AND OTHER ARTICLES EMPLOYED IN DOMESTIC ECONOMY,

AND

Methods of detecting them.

THERE IS
DEATH
IN THE POT
2 Kings C.IV. V.40

THE FOURTH EDITION.

BY FREDRICK ACCUM,

Operative Chemist, Lecturer on Practical Chemistry, Mineralogy, and on Chemistry applied to the Arts and Manufactures; Member of the Royal Irish Academy; Fellow of the Linnæan Society; Member of the Royal Academy of Sciences, and of the Royal Society of Arts of Berlin, &c. &c.

London:

SOLD BY LONGMAN, HURST, REES, ORME, AND BROWN,

PATERNOSTER ROW.

1822.

FIGURE 1.5. Front cover of Accum's treatise on adulterations. The text went through four editions in just its first two years in print. This cover is from 1822, the fourth edition. Courtesy of the Science History Institute.

community. The chemist's analytical results revealed the contents, as his title indicated, of clearly distinguishable "culinary poisons." He believed that adulteration was willful deceit by the producer.

Accum was at the front of the new catalog of books and "treatises" on food, fraud, chemistry, and adulteration in the century to come. Some of these works aimed at the general problem of adulteration; some took aim at specific products. All of them were part of a body of work that offered a troubling perspective on food identity. It was a body of literature that came to include grocers' papers, treatises on domestic economy, newspaper commentaries, and a range of pamphlets, satires, civic association meeting minutes, and ephemeral public notices. Even when some volumes were, let's say, a bit hyperbolic—it was an unsigned 1886 publication about oleomargarine that called it "the Most Gigantic Swindle of Modern Times"—they were speaking to an available sentiment among a weary public.[24]

Paraf's presence made real the fears stoked by adulteration texts. In 1887, a decade after his splash across the continent, the USDA's Bureau of Chemistry began compiling its monumental Bulletin 13: *Food and Food Adulterants*, bringing antiadulteration attention to a new political level in the United States to match the European attention it was already receiving. The bureau's chief agent, Harvey Wiley, led the ten-part, fourteen-hundred-page, sixteen-year study. His role as a chemist, a bureaucrat, and a staunch pure food crusader represented the thick combination of the antiadulteration movement by the end of the century. When Congress passed the Pure Food and Drug Act of 1906, they vested authority for policing adulteration in Wiley's Bureau of Chemistry. His role as a chemist highlighted the view that the sciences may be the cause of but were also the solution to the problems.[25]

Cultural anxiety, environmental reconfigurations, publishing and bureaucratic attention. It's possible Paraf earned the royal moniker "king of swindlers" for having combined many of these elements so deftly. He capitalized on the uncertainty of a changing agricultural geography by manipulating modes of trust. He added to the confusion by presenting a false front with an appearance as fake as his product. The force of his crime was not lost on a suspicious public: to them, even his con was a con.

Paraf also revealed a conflict about knowing nature that the broader debate over adulteration then exposed further and that forms a great part of my own interest in the matter now. That conflict came from two different scales of environmental attention that were melted onto each other. At the bigger scale, it included a spatial history involving new pathways, new networks, and new patterns of production, distribution, and consumption. These were large-scale, to the point of growing from regional to continental to global. But there was also a smaller-scale, product-specific level: the appearance of a food item, the substance of the product itself, its surfaces and interiors and identities right there on your plate. Things were not always as they seemed. It would thus be a mistake to consider adulteration only a consequence of distance and separation. Recognizing the confidence games at play shows that debates over pure foods were also about the appearance of honesty and fidelity to nature right in front of the eater.

Here we are then, at the crux of two monumental historical features. The second half of the nineteenth century witnessed both the dawn of the age of manufactured food and a moment of rising tension in a century of mobility, invention, and industrial development. The Chevalier was not there to see it (more on that below), but in recognition of this tension and as a spur to the USDA's forthcoming Bulletin 13 the US Congress would pass the Oleomargarine Act in 1886 mandating the taxing and labeling of the artificial butter.[26] This was the first large-scale, federal law in the United States speaking directly to the question of adulteration. It was a sign of the times.

In the almost quarter century between 1879 and 1906, legislators introduced and debated 190 separate pure food laws in the house and senate, averaging a new one about every seven weeks for those twenty-seven years. State-level and municipal ordinances during the time are too numerous to catalog, especially since many of them, such as the antiadulteration treatises, were aimed at particularly offensive foods and not adulteration in general. (Margarine alone was the subject of over seventy pieces of state-based legislation from 1877 to 1899, as shown in the maps accompanying this text at the digital companion website.[27]) The public health precedents from European nations influenced the United States. So did legions of civic groups, chemists, public health officials, agrarian interests, and texts in that new antiadulteration genre.[28]

It would be hard to discount the legislative zeal, civic attention, and analytical publishing efforts as frivolous. Similarly, it would be disingenuous to say, as numerous observers during and since that time have, that such worries were merely the price of progress, the common response to novelty. Saying so gives up the larger historical import of problems about food and health uncertainty. The very questions of progress and novelty were at issue. Taking them seriously allows for a view of the era faithful to its contested ethics of purity, nature, and artifice.

Paraf and his fraternity of charlatans help spotlight that ethical context, though to be fair they can't take all the blame for this. Confidence men and swindlers did not spawn or explain adulteration, as there were any number of reasons a product might change or be different from what it claimed to be: accidents, spoilage, decay, storage issues, even simply losing track of ingredients.[29] But here is the key: the same questions of trust and confidence that framed the world of the charlatan set the stage where concerns and fears over deceptive or contaminated foods played out. In a personal sense, the shady Chevalier crossed a line, legally and morally: his acts were criminal. In the historical sense, he stepped over a line marking a transition from the world of harvested and home-cooked foods to a new world of manufactured food.

Pure Adulteration charts that transition from the mid-nineteenth to the early twentieth century. I show how cultural (part 1) and environmental (part 2) conditions shaped the pure food crusades. I then explain how the cultural and environmental circumstances influenced scientific, chemical responses (part 3). I place the brunt of attention on the new ways farmers, chemists, grocers, housewives, and politicians worked with and came to understand nature and food. Putting the environment back into the pure food crusades doesn't just open up a view of the changing geography of the era; it also shows that the line drawn between pure and adulterated served as a proxy for distinguishing between natural and artificial. And the line between natural and artificial was not a given. It was, rather, an outcome of the ongoing cultural process of distinguishing between acceptable and unacceptable environmental behavior. The era of adulteration that was molded by these considerations led to new legislation and oversight, new bodies of civic action, new professions, and a modern century of scientifically, analytically certified purity.

The two sides of this transition offer a vision of changing forms of confidence and trust. A dominant agrarian world into the mid-nineteenth century led people to understand the identity of their food based on provenance—they knew what it was because they knew or thought they knew where it was from. (74 percent of the US population was rural in 1870 and 51 percent of the labor force was agricultural.) By the early twentieth century, a new industrial world of manufactured food led people to understand food based on labels, scientific analysis, and marketed brands. This was a shift from a preindustrial world where lived agrarian experience framed the ways one understood food-health claims to one where the responsibility was held by the label at the store, not the farm. It was a change from a producer-oriented world to a consumer-based one. Adulteration happened in small rural communities, too, don't get me wrong.

The interesting difference is the ways people policed food identity in those communities, not that they were idylls of purity and nobility. In the history of foods and farms, the move to industrial modernity was a move from trust in knowledge embodied by agrarian life to trust sanctioned by scientifically verified analytical knowledge.[30]

As one instigator of nineteenth-century pure food arguments among many, the audacious Frenchman Paraf challenged an offended public by presenting the end product as legitimate, asking consumers to ignore or not even know about the process by which it got there. His success followed from presenting a genuine front. In subsequent years, those defending the new artificial product (and other new manufactured products) argued that it was no less healthy than butter, that the problem was about transparency and full disclosure, not the devilish poisoning that so many cover images from antiadulteration commentaries showed. Most of the reputable scientific work at the time confirmed oleo's relative healthfulness. Even the esteemed Charles Chandler endorsed it.[31]

Those arguing against his product and manufactured food in general, however, thought there was more to it. They contended that the environmental process of producing the manufactured article was problematic, that the legitimacy of the artificial food was questionable because it violated norms of agricultural activity. In the larger sense, the point of tension was the development of new agricultural networks and the reconfiguration of relationships between producers, consumers, men, women, and the crops, animals, and lands that tied them together.

A wary public confronted the dashing Chevalier in the midst of those changes. He manipulated codes of conduct, he charmed and cajoled, he presented a false front, he used his chemical skills to look under the surface of a product and make it appear similar by sight. Some were taken in. Others came to wonder, is this guy for real?

The answer finally caught up with him, as it were. His itinerary had one more stop.

Paraf hung around California for two years after his 1874 indictment, trying to make a name for himself that escaped the food fraud phase of his con games. His devices still circled out from animal chemistry in oils and fats, but now they focused on new ventures in metalworking too. A trip to Nevada had him using animal fats to transform copper ore into more profitable metals. Then, as San Francisco's *Daily Morning Call* reported in February 1876, the "Oleaginous Paraf" left ahead of his detractors for "more congenial climes." He was off to South America, landing near Santiago with his personal servant in tow.[32]

Paraf was detained and questioned upon landing in Chile in early 1876.

His reputation had preceded him. A Peruvian paper later reported that he, "a notable chemist," had arrived in the country with "a brother and six family members" and "the supposed discovery he has made."[33] His family was certainly not with him. Paraf, it seemed, had a little entourage. It also isn't clear which "supposed discovery" the story referred to, though we do know he would soon apply to the minister of the interior for patent protection for transforming animal fats into olein and stearin. The olein would help his margarine cause.[34] Stearin led him to soap and candle making and, more compellingly for investors, the glyceride that served as an ingredient in smelting, bullet casing manufacturing, and gunpowder production. (Stearin is a glycerin. Nitroglycerine was produced from glycerin and nitric acid.) Always the dignified con man—poised, confident, dashing as ever, appearances first, please—he made it past the dockside interrogation to settle in for that set of new projects.

This time, though, finally, inevitably, came the end. "At last we are revenged," read one headline, for "that inventor of the most detestable stuff, oleomargarine, has met with a fate he well deserved." It was true. For a confidence man, it appears that a year after his landing in Chile, "overconfidence was his ruin."[35] He had accrued too many false fronts in too many places. A retrospective observer might say, Mr. Paraf, you have to know when to stop.

Yet still he stayed busy in the year before his trial ended. In his time at Des Hotel Ingles on the Pacific Coast of South America, he oversaw a pamphlet circulated to highlight his biography and dismiss the purportedly scurrilous charges from the United States and Europe. At one point he sought to delay the proceedings to await the arrival of his family. He said his wife, then in Bordeaux, was on her way. There was no word whether his young son was in tow. He had also somehow persuaded a Chilean to support his mining venture to turn tin into gold with some of those new animal fat inventions. Locals erected the Paraf Smelting Works, an edifice that upon his conviction, a Panamanian newspaper reported—no friend to the Chileans—stood as "a monument of the folly and gullibility of the Santiago public."[36]

And incidentally, that servant who followed in tow? He was actually another French chemist in disguise, hunkered down at the side of the wily Paraf and, reports had it, affecting the image of a peasant.

The court would put the Chevalier away for good in October of 1879, guilty of violating articles 167 and 468 of the Chilean penal code. He was exiled from Santiago and sentenced to five years under guard in a new penal colony a thousand kilometers south at Valdivia. Although his sentence for violating laws of public trust, forgery, and falsification was for

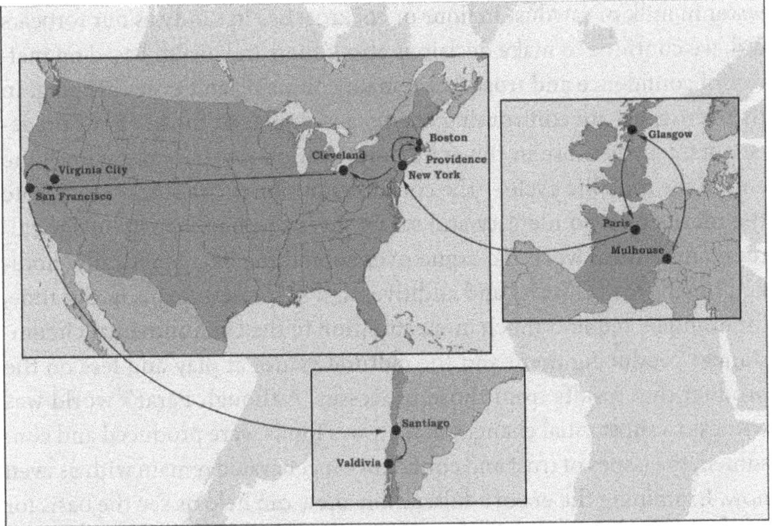

FIGURE 1.6. Overview map of Alfred Paraf's travels from his birth in Mulhouse, France, to his death in Valdivia, Chile. See https://purefood.lafayette.edu for an interactive version of these travels. Map created by John Clarke.

five years, he remained incarcerated for six until his death of pneumonia in 1885.

Paraf's story of transatlantic and Pacific Coast movement opens up the three main parts of this book (fig. 1.6). From Mulhouse to Paris to Glasgow to New York to San Francisco to his final stop in Valdivia, the self-titled Chevalier traced the contours of a new world. His questionable though dashing character and ill-gotten social credibility point to part 1's attention to a culture of adulteration that explores the debates over adulteration and purity that arose from literature, papers, domestic economists, and more. His globe-trotting schemes trace the actual development of new foods from factories that moved across the ocean, across the continental United States, and back and forth between the Americas, intimating the geography that part 2 of the book explores with attention to margarine and two other cumbersome adulterants of the time, cottonseed oil and glucose. His training in chemistry and remarkable chemical skills point both to the often-suspected material source of adulteration—chemistry—and its purported solution—scientific analysis. This chemical aspect frames part 3 of the book.

I return in the epilogue to acknowledge that people today are still confused over the larger and more enduring category of novel and manufactured foods. While developed economies may no longer fear (I hope) water in milk or sawdust in flour or cockroaches in candy as our forbears did, we continue to make decisions about food and health based on metrics of confidence and trust that remain confusing and ever changing. In that sense, we are confronting challenges similar to those of our predecessors. Just as those in the era of adulteration saw it, waiting until the end of the food life cycle—the consumer option at the store—to debate the merits of food identity and safety gives up the majority of agricultural and ethical work. To argue over something like genetically modified organisms (GMOs) and additives and suspected toxins in our time, for example, requires much more attention to the environmental circumstances producing them and the cultural values at play and less on the product that results from those processes. Although Paraf's world was witness to substantial changes in the ways foods were produced and consumed, the issues of trust and confidence it generated remain with us even now. Examining the era of adulteration anew can help us see the basis for the age of manufactured food in which we still live.

How did that era come to be? Where did it leave us? The chapters ahead explore those questions.

1

The Culture of Adulteration

✷ 2 ✷

Surfaces & Interiors

In every walk of life, men assume the expression and manner which they believe will make them as they would wish to be regarded. Thus one might say that our world consists entirely of face values.

FRANÇOIS VI, Duc de la Rochefoucauld, 1665

We that consume are daily in the consumption of lies—we drink *lying* coffee—we eat *lying* food—we patch *lying* clothes with cheating thread—we perfume ourselves with *lying* essence—we wet our feet in *lying* boots—catch cold, however, truly enough—[and] are tormented with adulterated drugs.

HENRY WARD BEECHER, 1851[1]

Appearances matter. I think about this almost daily when talking to people, wondering what their real story is, who they really are beyond surface pleasantries. Their face values may or may not provide valid windows onto their interior truths.

I was once introducing a film screening at a local food forum, and, afterward, a gentleman in a Stetson introduced himself, asking for my contact information. He wanted to talk about the work my campus group was doing in the local food community. I agreed to meet. We set a date. He chose a café I knew well because I read and wrote there often. In fact, that week I was there reading Herman Melville's *The Confidence-Man* (1857) for this very book. Talking about food as an environmental issue has a lot of people—farmers, gardeners, environmentalists, neighbors—getting in touch. His request wasn't unusual.

He'd arrived before me and was sitting in the same chair I was reading Melville in the last time. His hat was set over a stack of four books on the low, wood-planked coffee table. Something about his posture, the "expression and manner" Rochefoucauld pointed out, gave me a sense of unease, so I clarified that I only had maybe twenty minutes. He was gracious, seemed by all rights kind, and told me that was not a lot of time to convey the concepts he wanted to convey, but we are all on a journey, so we must start somewhere.

If I was not suspicious before, his invocation of a journey, his demeanor and presentation, made me so. Melville was on my mind. His novel speaks to mid-nineteenth-century concerns of purity, trust, and authenticity that seemed real and present to me. It was clear from Melville that questions of trusting food were tied up with questions about trusting people, a point that Alfred Paraf's travails already suggested and a view Henry Ward Beecher cast in the epigraph above with his *lying* food. It was also clear that there was more to knowing my behatted coffee partner than his appearance and outer presentation alone.

Published six years after *Moby-Dick*, Melville's *The Confidence-Man* explores public mores, private virtues, and the difficulties of trust and confidence when strangers stop strangers to interact. Its underpinnings are the tension between surfaces and interiors. The novel is a satire set on a steamboat traveling the Mississippi River between New Orleans and St. Louis. Most of the forty-five chapters are episodes on the journey where one passenger tries to gain someone's confidence for the purpose of taking his money. A central plot point has one person approaching another to ask, "Would you have enough confidence to lend me a hundred dollars?" If the answer is no, he will ask, "But what could I do to gain your confidence?" The novel is a study of authenticity and sincerity when challenged by unclear codes of conduct. *What does it take to trust me, to trust anyone? What do I have to do to gain your confidence?*

I was drawn to it because it's a great and underread book. I was also drawn to it because it makes a dual point that scholars often advance. It's one that drives the chapters of this book: the attempt to point to the material world, the things and physical environment of nature, and the ways we come to make claims about what that material world means, our ideas and concepts and methods. Philosophers would call those ontological and epistemological questions about *what* we know and *how* we know it. I was happy to avoid an abstract approach to those issues in the coffee shop. I was fascinated by the coincidence of thinking about appearances and truth while sitting there talking to a guy who may have been trying to sell me something.

It soon became clear my café friend was neither a farmer nor a gardener, and I started to figure he probably saw in me a chance to gain support, financial or otherwise. He told me about a ministry he was either part of or starting, it wasn't clear which, and then brought out the books from under the hat, each of them illustrating a big concept about how we ride as passengers on Spaceship Earth and we need to be better environmental citizens. I was respectful, I wanted to be respectful, I agreed with his points, even as Melvillian suspicion was growing.

After the books and their concepts got their due, our time was up. The café was bustling, doors opening and closing, active as a steamboat at its port of call. Here was what seemed to be a sincere man, earnest and plain-spoken. Our interaction was an effort to gain my trust. Although I hadn't been skeptical about his character beforehand, and he certainly *appeared* forthright, on my drive home I started to worry. Had I made any promises? I resolved to Google him again, maybe try different keywords this time.

The next day, before I had a chance to do that, I checked my email to find a lengthy follow-up ending with a pitch about the character of his ministry (extant or envisioned, it still was not clear) that put me back on the Mississippi, back inside the tangle of trust, faith, and knowledge. And money. "I am willing to take the leap of faith," he wrote, "and, in order to do this, I need funding."

If I was uneasy before, I was full-on skeptical now. This was the next phase of a possible con. Melville satirized and dramatized these interactions, as a thousand books and movies have done since. He didn't coin the term *con man*, the shorthand version of confidence man, but drew from it to set the deeper slipperiness of trust, belief, and faith into a sticky world of cultural relations. His was the age of Phineas T. Barnum, when new codes of public conduct met with everyday challenges in understanding one another. Barnum, for his part, seized the opportunities of flux and confusion to make a career and build a legacy of tricks and artifice. He shared the same nineteenth-century awakening of public displays as Melville, those punctuated by "the carnivalesque, the sleight-of-hand, the tromp l'oeil, the manipulation," in an era that "involved a calculated intermixing of the genuine and the fake, enchantment and disenchantment."[2] His contemporary, the preacher Henry Ward Beecher, saw these things too. He wrote the epigraph above the year Melville published *Moby-Dick*, recognizing that the larger category of fraud and deception could involve anything from coffee and food to thread, perfume, boots, and drugs. Foods and drinks offered the same intermixing of the genuine and the fake, enchantment and disenchantment.[3]

Melville and Beecher wrote and spoke at the head of the historical period I introduced in chapter 1 as the era of adulteration, when farmers, manufacturers, housewives, grocers, cooks, writers, and politicians waged a "War on Adulteration." Their contemporaries of the midcentury were beginning to feel the unease and distrust I had at the café. The public was being cheated. People were worried. They didn't want to drink lying coffee.

Their period marked a turning point in a long time line of adulteration. A few features help explain what was new. One was the breakdown of community-based norms, a trend Melville was already onto in the 1850s. That breakdown followed from a loss of familiarity exacerbated by distance and the increasing scope of mobility.[4] Another was the loss of trust in direct sensory evidence, a loss of confidence brought on by fears of false appearances and untrained eyes. Rochefoucauld's face values. Holding the two kinds of losses together were fears of knowledge lost somewhere between surfaces and interiors, between the way things looked and the way they really were.

To that first point, people indeed moved around a lot. Several patterns of dramatic mobility all overlapped during the later nineteenth century. One was country-to-city traffic, a phenomenon of notable significance in the United States but shared by the broader Western world. Taking the British experience as an instance of fuller Western trends finds that "one in five people in Britain lived in a town in 1801, one in two by 1851, and two out of three in 1881."[5] The United States was behind chronologically—the one-in-two ratio didn't take hold until 1920, with two out of three by 1950—but experienced the trend far more dramatically. Across the nineteenth century, the British population experienced a threefold increase in urban population. The US population experienced a sixfold increase in the same period. Even if the total population was still more rural in the United States, the degree of change toward urban centers was more intense.[6] While mobility was a global phenomenon, it had particular resonance for the American experience.

A second kind of geographical mobility came from terrestrial expansion across North America, often justified as Manifest Destiny. The year 1869, for example, saw the completion of the first transcontinental railroad. Only twenty-one years later, analysts noted from the 1890 census that the frontier was finally closed. Coincident to both of those trends was the third pattern of mobility, that of immigration, most notably from northern, southern, and eastern Europe and Asia. Over twenty-five million people immigrated between the 1860s and 1910s, accounting for over half of the total population increase across that time span.[7]

It all added up. The Melvillian take on this was more about mobility than distance. My own reading of the era aligns with that view. A number of historians have noted the same turbulent expansion, transition, and growth and have suggested a distance-equals-adulteration equation.[8] That relationship's simple schematic doesn't entirely hold up to historical scrutiny, but that's because there was far more to it, not that distance played no role.

Faced with remarkable and seemingly swift demographic revolution, what changed in the era of adulteration was less the *fact* of impure foods and more the ways they came about and the ways people responded to them. Part 1 of this book explores how changes influenced by mobility, confidence and trust laid the foundation from which the era's concerns grew. The basic point is that a culture of concern for sincerity came to frame the ways people understood food identity and health. Trusting food meant trusting people. It's worth stepping back to see how the various cultural elements of the fight for purity—sincere behavior, genuine presentation—set the stage for the war on adulteration of the later nineteenth century. Two broad questions—how did people expect to trust their food, and how did they respond to fears over its identity—provide the basic foundation on which the rest of this book sits.

The Moral and Material Life of Adulteration

Questions about a fractured age of sincerity, truth, and authenticity sounded a rising chorus in the nineteenth century, but the problems of adulterated food, a subcategory of inauthenticity, were not new. Cultural codes, food laws, and trade practices have suggested the ways people developed ideas about what counted as pure and what as adulterated for millennia. An extensive lineage of commentary follows the search for true and genuine behavior, a search that often held together the virtues of people with the qualities of the food they made or sold.

The Bible itself spoke to food purity. The Old Testament prescribed distinctions between clean and unclean foods first in Leviticus 11:3 and later in passages about the crime of unhealthy food. Kosher laws to this day derive from those first articulations of a line between what counted as acceptable and what did not. In ancient Greece, no less a voice than Plato recorded the proper protocol for dealing with adulteration in his *Laws*, circa 360 BCE: "The wardens of the market and the guardian of the law shall obtain information from experienced persons concerning the rogueries and adulterations of the sellers and shall write up what the seller ought and ought not to do in each case." Harvey Wiley would later copy out Plato's work as part of his own research for *Foods and Their Adulteration* (1907), writing parenthetically and sarcastically that Plato's claim two millennia earlier preceded Wiley's Board of Food and Drug Inspection by "several years." The *Laws* did not indicate how to know that a food was adulterated beyond "guardian" testimony, but the text makes it clear that concern for pure food was a feature of the polis. The perpetrator, wrote Plato, "shall be beaten in the market-place with stripes—one stripe for every drachma in the price he asks for the article."[9]

Chinese authorities had long prescribed codes of proper municipal conduct concerning food. It is not surprising that their worries were devoted more to rice, wine, and medicine than to non-Asian staples of bread or milk. In Europe, the worries were more related to bread, meat, and drinks. A 1266 Assize of Bread framed English common law for centuries around standards for size and weight of bread. The Assize became the customary tenet in food-health laws and main legal precedent in the English-speaking world. Its enactment was fit to guild structures, where managing the operation of a trade followed from organizing community life. It was aimed at regulating the health of the food but also at maintaining honest market activity. Centuries of butcher laws in France similarly date from the thirteenth century as efforts in managing civic food standards for both health and market honesty. A fourteenth-century text, *Piers Plowman*, understood the long history of contaminated food, "for they poison the people secretly and often." A fifteenth-century incident in Nuremberg found a man caught adulterating saffron. He was burned at the stake. His accomplice was buried alive. The famous Reinheitsgebot in Bavaria, the German Beer Purity law dating from the sixteenth century, proposed a positive sense of purity with less deadly punishments for violation, describing what to do rather than what not to do. With reference to three ingredients—water, hops, and barley—it dictated (and dictates still) how to meet the definition of "pure" beer.[10]

Historical laws make it clear that adulterated foods have long been part of public life. Be it Athens, China, England, France, Germany, bread, beer, or butchers, these codes of conduct were circumscribed by regional and national bounds. Communities policed the violators through visible and sometimes Draconian measures. Uncovering secrets, avoiding poisons, striving for biblical fidelity, they all led to an accumulation of issues in a lineage of antiadulteration activity marking a consistent path across the centuries because of a basic fact: people have forever wanted to know what was in their food.

In the modern era beginning in the seventeenth century and securing itself by the nineteenth century, the tenor, force, and shape of debates about adulteration shifted. Well before the late nineteenth century's version of it, an evolving set of trade practices marked the inklings of that new phase in the history of adulteration. The first age of globalization—the one marked by Columbus, Cortez, and Hudson and friends' exploration and colonization—laid the pattern for these far-ranging trade routes.

Goods began moving at a larger global scale even if people did not yet (as they would in force in the coming century). This could be good or bad, depending on your perspective and experience—sugar, coffee, spices, and the like were circulating around the world to the delight of merchants and Western eaters but not to those enslaved or colonized in service to the commodity trades. More to the point of food identity, with new patterns of trade the difference between a pure and adulterated product was becoming judged through cross-regional and global trade practices. This led to antiadulteration and antismuggling laws for policing purity. The laws were intended to monitor, tax, and ensure the veracity of imported coffee, spices, sugar, grains, and spirits. Were the imports really what the shipper said they were? Were they the genuine article?

A 1725 antiadulteration law in England made the point. It was intended to verify that bricks of tea trafficked through the "French, Swedes, Dutch, or Danes" from China were valid. The law packed in a host of key terms, punishing "any dealer who should counterfeit, or adulterate, or alter, or fabricate, or manufacture tea [with] sugar, molasses, clay and logwood, as well as terra japonica."[11] The British imported over twelve million pounds of tea annually by the 1770s (a very small portion of which colonial rebels dumped into Boston Harbor). Concern for the tea's authenticity was thus a matter of financial significance.

Other new food- and drink-related laws and texts, like those for corn or alcohol, pushed the same themes about proper conduct. A book of *Advice to the Unwary* from 1780 noted the hundreds of thousands of "quarters" of corn (a quarter is eight bushels) shipped from England to Sweden, France, Holland, and Denmark, where it was used to distill spirits. Smuggling brandies and "corn-spirits" back to England under false names led to tax evasion and contaminated products. Such rules of exchange attended to ensuring truth and trust. Knowing they were getting what they traded for was a matter of specific financial importance and a point of safety to avoid poisoning people either secretly or often.

Another aspect of the changing adulteration debate involved attention to the ways people understood the moral bases of its origins. To that end, treatises and tracts in the eighteenth century increasingly noted that people understood the complex *reasons* for impurity, not just the fact of it, by associating moral virtues of people with moral claims attached to foods and other products. A series of keywords developed in parallel across the eighteenth century to anchor the moral argument from which people understood adulteration. *Sincerity*, *genuineness*, and *artifice* were high among them.

Truthful action in public, at least in Western cultures, had long been tied up in religious expressions about honest presentations of the self in the face of God. Who were you before a deity? What was your best, true self? Sincerity was key in that. As the cultural critic Lionel Trilling observed, "to speak of the sincere doctrine, or the sincere religion, or the sincere Gospel, was to say that it had not been tampered with, or falsified, or corrupted."[12]

A new alignment during the Enlightenment pushed discussions of purity beyond their theological anchor. Namely, the concept of purity moved from outside humanity and in the house of God to one embodied by individuals acting as moral agents with the responsibility to act properly.[13] By the turn of the nineteenth century, people were not only seeking sincerity "to please God" but as part of a more secular vision of authenticity. In that way, sincerity hewed as much to moral conduct in the public sphere as to the church nave. It was becoming "a way to earnestly express feelings, to trumpet the truth of oneself."[14] That mattered when the question of food identity was centered in a new kind of public exchange either at the market or on the docks of imports and trade.

Being "genuine" gave meaning to this newer cultural framework. In arguments over food identity and honest behavior alike, advocates pitched *genuine* as the opposite of *adulterated*. Identifying genuine action was part of a modern move to align *true* and *pure* (qualities that attached to the genuine) against a set of antonyms, such as *adulterated* and *corrupt*. It also came with that useful overlap between a human virtue—how one should act—and a material quality—what a thing should be. King Louis XIV's "chief druggist" authored a 1712 tract on food, drinks, and drugs, the *Histoire générale des drogues* (Complete history of drugs), that held together human and material values. His goal was to hash out "methods of distinguishing the genuine and perfect, from the adulterated, sophisticated, and decayed." A 1740 British "alarm to all persons" sought similarly to laud the "genuine" and warn against the "most shocking, pernicious, and destructive practices" by those "who make and sell divers kinds of eatables and drinkables."[15] An English dictionary in 1764 defined adulteration as "the act of debasing . . . something that was pure and genuine." English and then American works revealed a steady dialogue about the common category of genuine, true, and natural products. Writers aligned *natural*, in this context, with *genuine* and *sincere* and against *sophisticated* and *adulterated*.[16] Adulteration happened, the argument went, because disingenuous people produced disingenuous food.

If one slow change in the debate over adulteration was thus to elevate the virtue of genuineness over that of sophistication, another was to pit the value of purity against a new concern, artifice. Artifice as a concept was age old, but artificial as a physical property was gaining greater visibility with the onset of industrial processes. For Enlightenment-era observers experiencing the first industrial revolution in cities across England, there were ever more examples of products resulting from the artifice of the craft shop. Contrivances everywhere—not just from the dusty spaces of small craft shop floors but from larger factories churning out fabrications at scale. Contrivances, those artificial things, offered benefits and signs of progress to so many but also added to the unease over shifting norms about truth and authenticity.

Artifice was like *genuine* in one sense: it had long been a human characteristic and a description of material things. An Irish author spoke to that duality in 1709 as a way to explain where adulteration came from. He blamed "artifice" and "pride" while seeking to expose "the false arts of life, to pull off the disguises of cunning vanity, and affectation, and recommend a general simplicity in our dress, our discourse, and our behavior."[17] Flourishes and excessive affectations were signs of artifice. Purity by contrast was plain, simple, and unadorned. Samuel Johnson's 1755 dictionary defined *genuineness* in opposition to artifice, as "freedom from any thing counterfeit; freedom from adulteration; purity; natural state."[18] He was less troubled than readers today about assuming a singular, a-cultural concept of nature, instead articulating a contrast of artifice and nature that presented adulteration as a problem caused by deviations from a proper natural order. An author on bread and flour adulteration, Peter Markham, threw around loaded terms that placed the blame on the lapsed morality of the time, an era marked by the artificial and insincere. "It will be said by many people," he wrote of adulteration, "the immoralities and luxuries of the age produced this effect."[19]

That was a common view. It created a conflict where the moral and material conditions of the first industrial revolution influenced purity debates. That set the precedent for the second industrial revolution a century later, the one undergirding pure food crusades.[20] Early experiences with manufacturing and factory conditions led to a frequent theme, namely, the understanding of artifice, insincerity, and impurity as a constellation that was set, as Samuel Johnson captured it, against purity and nature.

Thomas Jefferson carried forward Markham's point about the immoralities of the age. He made the connections to manufacturing explicit in the 1780s when he bemoaned the loss in moral integrity brought by factory life and material artifice. His famous plea telling young America to "leave our workshops in Europe" was about more than food purveyors specifically. It spoke to the process at large around which agricultural life was organized. But it also drew from a critique about the proper way to conduct oneself in the public sphere in ways that reflected the new secular argument against artifice. By the early nineteenth century the lines were drawn. An influential argument was that to be pure and genuine as a person was to be unpolluted by artifice.[21]

Jefferson offers a useful reference to the age in two ways. One is his view on artifice. The other is because he straddled the legacy of initial British industrialization and the forthcoming development of more global industrialism. As the nineteenth century dawned, more changes were afoot with the moral and material worries over sincere and genuine action. Farmers and bakers, politicians and housewives, inspectors and gentlemen alike weren't naive, as if there were no techniques to decipher the truth of food claims or the means to address and police the authenticity of their food. In an agrarian society of relative self-sufficiency or regional reliability, though, questions about food identity were internal to the management of household and community life. Attention to food identity was most applicable and most concerning when food came from beyond the household, as had happened in earlier centuries with antismuggling laws and trade rules. Milk adulteration isn't a problem until you rely on someone else to provide your milk; mislabeled meat isn't contentious until you get it from the butcher in town; bread has long been a source of adulteration angst because the baker, like the butcher and the brewer, has for so long been a village trade, a common source of food beyond the home. In Emerson's words, "Every man alone is sincere. At the entrance of a second person hypocrisy begins."[22]

Community standards regulated codes of proper conduct within the jurisdiction of guilds for centuries. "Traditional norms governing face-to-face conduct," historian Karen Halttunen writes, "had operated in a world where men and women came to know one another gradually... within a well-defined social context of family and community."[23] For those on the production side of the equation—the bakers and butchers and brewers—guild-based training dictated the correct manner of food preparation and sales and operated as a means to produce knowledge about food and health. Surely foods were still contaminated or falsely presented; this "before" picture, alas, is not one of a placid paradise. But for those on the consumer side—the customers and market goers and household eaters—it meant that when foods *were* contaminated, the ways to adjudicate the issue came from within community contexts.

Before the nineteenth century the great majority of colonial Americans sourced their own food, a connection that influenced the ways they knew food. As with their forbears, "they killed their own game, caught their own fish, brewed their own beer, slaughtered their own livestock, sowed their own wheat, tilled their own fields, chopped down and burned their own fuel, grew their own vegetables, and reaped their own harvests." It is striking how intimately those in colonial America knew their food because they lived it. "The most fundamental aim of all farmers in all regions and at all levels was self-provisioning," writes historian Richard Bushman.[24] Theirs was a place-based knowledge born of environmental understandings and agricultural intimacy.

Cookbooks provide insight on how this mattered for understanding food. Amelia Simmons, the author of the nation's first notable cookbook, *American Cookery* (1796), wrote her recipes and knew her food from "the close familiarity . . . a woman who cooked could have with animals' lives." Knowing the cow's age, disposition, and daily habits influenced how Simmons understood "the consistency, flavor, and color of its milk."[25] Having intimate knowledge of terrain, weather, and neighbors structured familiarity with food quality. Advising readers on pure "sweet butter," for instance, Simmons suggested sending "stone pots to honest, neat, and trusty dairy people." Those honest neighbors would bring it back "in the night, or cool rainy morning, covered with a clean cloth wet in cold water, and partake of no heat from the horse, and set the pots in the coldest part of your cellar" to ensure its unadulterated condition.[26] The sense of pure, in other words, derived from the familiar interactions of local life.

The tight relationship between agrarian livelihoods and what historian Ann Vileisis captures as "kitchen literacy" shaped the ways people came to recognize or suspect the identity of foods from outside the home. Compared with a twentieth century of more urbanized settings, those managing the household and its land knew to recognize quality and identity from daily life. Europeans of the later eighteenth and early nineteenth centuries were beginning to interact through urban markets, developing regulated slaughterhouses, and continuing in earnest to legislate consumer practices beyond the household; the early American experience was more fully indebted to understanding food from the experience of harvesting and preparing it well into the nineteenth century.[27]

At the same time, saying that people understood their food from lived experience is fair but also too general. More specifically, in the craft-based societies before the industrial age, people came to know their food through common sensory measures, what scientists would now call *organoleptic metrics*. Sight, touch, taste, and smell served as profound measures of food identity. Purity and adulteration were understood through the lens of sensory experiences that didn't yet involve complicated tools or instruments.

The literature on antiadulteration often recommended avoiding adulteration by diligent sensory tests. A guide from 1832 advised readers to choose butter "based on taste and smell," to ascertain fraud in "the quality of flour [by] colour and feel," and to detect adulterated oil by "smell" and "the dullness of the colour." Cookbooks for centuries called on cooks to gauge quality, identity, and safety of food through sensory interactions. "Olfactory vigilance," to quote the French historian Madeleine Ferrières' lovely phrase, paired with the decisiveness of sight and taste in securing the confidence of the customer.[28]

The color palette of foods served as a complex means for such security. Yellow could be valuable or cautionary, for example. White was ideal for the common household ingredient lard, with butchers frequently using candles and lighting to showcase the color and hide any tint of yellow. But yellow was preferable for butter or other dishes that could benefit from the addition of saffron, or "parsley juice for green" or "sunflower for purple."[29] This is to say that appearance helped inform identity. Sellers knew this, butchers knew it, bakers and spice dealers and dairymen knew it.

These weren't abstract principles of theory or speculation, how people used the surface to guide what they thought of the substance. The precepts guided the ways people made their way through daily life and the food marketplace of the nineteenth century. The flaws exposed in those principles by the time of Melville and friends also pushed on assumptions about seeing and knowing, raising the question, did to see mean to know? The question dogged customers who encountered products and sellers outside the "well-defined social context of family and community."

Scholars have talked about "character" to illustrate that form of confusion. Historian James Salazar has gone so far as to call character "*the* most coveted object of nineteenth-century American culture." I like character, too, since it's another trait—like genuine and artifice—that is both a human virtue and a material property. It's a pronouncement on the inner moral compass of a person, though the term itself comes from printing technology. It referred to the mark left when a sheet is pressed upon. It is the seal imprinted in the wax. We still use the term, adopting it to virtual space to refer to letters on the screen, a vestige of print history as the imprint of a key striking the page. The related concept of the "brand" shows this, too: something pressed into a surface or a body. As the age of advertising developed, manufacturers would press the brand on their product to prove its identity and secure the customer's trust. Brands were impressions. You want to make the right impression. Character traits, material and moral, helped community members build a degree of comfort about how to trust one another.

Barring personal knowledge of the seller, appearances, smells, tastes, and textures provided the basic means to gauge the identity of milk, bread, coffee, butter, and the like. The question of seeing and knowing thus had a practical element. It is a classic problem in economic history, where the asymmetry in knowledge between seller and buyer creates a situation that demands trust and respect between the two. Sellers know more than buyers—thus the need for third-party or independent vouchers such as brands, trademarks, certification, and reputation.

New marketing and advertising practices in the postbellum decades grew from attention to and often in reaction to just that impulse to present the truth of a product on its surface, what philosophers call a mimetic impulse, from the Greek *mimesis,* to imitate.[30] Advertisers identified the flawed assumptions underlying that impulse: under contention was just that modernist assumption "that reality always resides beneath the layers of appearances."[31] The faith of the food purchaser rested on the assumption that appearances matched realities.

Adulterators preyed on that faith. Unsettled norms of genuine selves and true character shook it. Dramatic demographic changes exacerbated the confusion over it. The sixfold increase in the United States' urban population; the swift expansion across the North American continent; the increase in raw population from thirty-eight million to ninety-two million people, half of which were accounted for by immigration: the degree of demographic change in the nineteenth century was hardly trivial.[32]

Amid those statistical realities, cultural observers spoke to the felt experience wrought by mobility. One major concern was just that possibility, that newcomers were innocent prey for charlatans and confidence men ready to present a false front of sincerity to bilk them of their money. Or if not theft and sham, the newcomer was *impression*able, subject to direct manipulation or control from shady characters.[33] We think of this today in caricatured terms as the country rube in the city or the slack-jawed tourist staring upward at skyscrapers, an easy mark for those who might dupe her or him. It's caricatured the other way, too: the urbanite out in the country, the city slicker unfamiliar with patterns of small-town life, equally dupable.

In either case, the challenge is one of unfamiliarity. Opportunists like Barnum exploited that scenario, mixing the genuine and the fake.[34] In 1849, a *Herald* reporter in New York coined the term *Confidence Man* to identify what had become a visible public problem.[35] It was confusing, observers thought, because social mobility and geographical expansion challenged the very notion of "a public" that could be defined by common attributes and collective experiences. In turn, it was unsettling because the ways people understood things in a fully agrarian world were coming undone. People feared experience alone wasn't enough to deal with the new world.

By midcentury, Melville was trying to come to terms with a disrupted world in *The Confidence-Man*. His novel spoke to concerns regarding purity, trust, and authenticity. As Bruce Franklin wrote in the introduction to my edition of the book, "all varieties of fraud and swindling" were on board the main ship in the story, the *Fidele* (faithful), including "herb-doctors, land agents, counterfeiters, impostors, charity agents, card sharps, divines, transcendental philosophers, [and] con men of all kinds." A sign on one of the boat shop doors said "No Trust"; an exchange soon into the voyage has one character arguing to another—that other having asked for charity—that "looks are one thing, and facts are another." Two characters fight over the purity of wine later in the book, an argument about expressions of authenticity and protestations of health and truth. Quoting *Hamlet*, they bring Polonius into their debate to align their view with a call to sincerity, to thine own self be true.[36]

Melville had intimated those contrasts in *Moby-Dick* six years earlier: confidence, trust, faith, sincerity, and truth on one side; skepticism, disbelief, suspicion, distrust, and confusion on the other.[37] He identified the anxiety and suspicion that grew from uncertainty spawned by mobility and distrust. Even the geography was fluid in *The Confidence-Man*, set on a flowing river that was moving and unsteady. The steamboat never had a secure physical connection to the land, no chance to provide anchor for stable, tethered community or political relationships. There was no environment or local ground against which the traveling boat could identify itself. The *Fidele* was in a liminal state, always caught between two places; characters on deck or below, at one port and another, between the East Coast of the continent and the West.

Melville's younger contemporary, Mark Twain, similarly took pretensions to authenticity as the subject of many of his works, responding to mobility's challenges. His critique called into question prevailing norms about self-presentations in the public sphere. His semiautobiographical *Life on the Mississippi* (1883) also used the fluid and unmoored riverboat as its setting, taking aim at the perceived falsehood of manufacturing. He got right to the point, as one chapter discussed edible oils (the subject of chap. 5 in this book). Twain's characters wrestled with their awareness that new deceptions abounded with new manufactured foods. An oil salesman boasted about masquerading cottonseed oil for olive oil, "and there ain't anybody that can detect the true from the false."[38]

The confusion exacerbated by mobility, the intermixing of genuine and fake, the loss of unmediated sensory evidence, and the coveted object of character led to a culture of credulity on which writers and philosophers focused their attention and from which arguments for pure food took shape. Just as I came to ask who this guy in the café was, Twain's characters were asking what Melville had asked: Who are you as a public agent? Are you the genuine article?

It may be one of Twain's more enduring appellations, the coauthored *The Gilded Age* (1873), that spoke most overtly to the overarching theme of appearance and truth. The book was replete with grifters and shady characters, giving its name to the postbellum era from the 1870s to the turn of the century. In title and theme, it emphasized the glinting golden surfaces beneath which festered the realities of a culture of incredulity, asking readers what it means to trade gold for gold plating.[39] Things were not what they seemed on the surface. Knowing what lay beneath, the authentic thing, that was the difficult part. Making matters worse, the historically assured if imperfect methods of understanding what lay beneath were eroding, leaving a gap for new ways to construct knowledge of the sincerity of things.

Detecting Frightful Truths

Adulteration thus didn't just come from disreputable sellers, cheats, and swindlers, nor only from accidents or unexpected contamination. It also arose from the conditions of unfamiliarity and distrust amid the luxuries of the age Jefferson warned against. I noted two broad historical questions at the top of the chapter, one about how people expected to trust their food—what they asked—and the other about how they answered it. Here a return to that second question points to the new ways advocates for purity and against contamination sought to respond. In the face of losing the confidence of community norms, advocates for pure and sincere food worked to build new methods.

Analysis could help. With instruments and equipment, chemical analysts could look beneath the surface to expose reality. It's a premise of most modern scientific inquiry, using tools and techniques to uncover the presumed reality of a world beyond our direct, unmediated senses. It might not have helped detect the true identity of, say, the shape-shifting con man on Melville's *Fidele* or the true character of my associate at the café, but it could help with material items. Like food. Part 3 of this book speaks more fully to the growth of antiadulteration chemistry in its organization, occupation, and methodology during the pure food crusades. Beyond its technical merits, though, chemical inquiry was fundamentally a method to get beyond the naked eye. Chemical analysis provided new ways for people to respond to the problems of adulteration in an age of manufacturing and artifice, adding another layer to the elements of the nineteenth century that differentiated it from prior eras.[40]

Scientific publications investigating chemical evidence of purity grew into a definable genre in the later nineteenth century. They had their origins the century before in a pharmacy culture aimed at exposing poison. As early as 1757, an English physician published *Poison Detected, or Frightful Truths*, pointing his barbs at bakers and bread adulteration. A 1781 "essay on culinary poisons" anticipated the wave of antiadulteration arguments that cast the food problems as larger crises in public health. John Farley's *The London Art of Cookery*, first published in the 1780s, included an appendix with "considerations on culinary poisons." The poisons could come from cookware if tin, copper, or lead leached into the food; they could come from dangerous ingredients cooks should watch out for, such as mushrooms or berries or various herbs; and they could come from devious sellers. "Mealmen and bakers have been known to use bean-meal, chalk, whiting, flacked lime, alum, and even ashes of bones," Farley explained.[41]

The London-based but German-born chemist Fredric Accum, whom I referenced in chapter 1, moved to the front of antiadulteration work in 1820. He advanced the poison theme with *A Treatise on Adulterations of Food, and Culinary Poisons* (fig. 2.1). Accum didn't shy away from heavy-handed rhetoric. His exposé of "fraudulent sophistications" was in keeping with eighteenth-century predecessors. He treated the words *genuine* and *pure* synonymously, referring to "genuine" practices and products no less than two dozen times in the work. He also framed the problem in starkly moralistic terms: counterfeit, adulterated, nefarious, mercenary, criminal, unprincipled, fraudulent, and evil. And that was just the preface. He drew his biblical epigraph, "There is death in the pot," from Kings 4:40. The frightening cover graphic showed a hollow skull and intertwined snakes. Accum gave his audience sensationalist chemistry drawn on the assumption of a clear and detectable distinction between proper and improper foods.[42]

Writing confidently a century later, the British chemist Frederick Filby categorized three eras of food contamination with reference to Accum: the one before 1820, the one between 1820 and 1934 (when Filby wrote), and the new one from 1934 on. The first phase spanned all of human history

FIGURE 2.1. "'Chemical Lectures' Given by Friedrich Accum at the Surrey Institution, London," by Thomas Rowlandson, 1819. The gentleman in the lower left corner carries "Accum's Lectures" in his pocket. Courtesy Inv. 22885 © Museum of the History of Science, University of Oxford.

until the 1800s. The second, century-long era was a period of increased attention by scientists to the problems of contaminated food and its identification and eradication. During this time, analytical chemistry "caught up and finally outstripped the crime," Filby wrote. The new, third phase was one where "the grosser forms of adulteration have been completely abolished." Filby was wrong about that last phase. Adulteration has not ceased to be a public problem. But he deserves credit for identifying the Accum inflection.[43]

Filby's review stated that Accum's work "finally brought the storm over adulteration," marking a "turning point in the story."[44] As recently as 2009, the British food writer Bee Wilson began her study of the history of food fraud by noting that "the concerted fight against poisonous or superfluous adulterations" followed Accum. Contemporaries gave Accum credit as well, alluding to his pioneering work in antiadulteration texts throughout the century. Dickens modeled a character in *Our Mutual Friend* (1865) on him. In the novel, which touches on the strains between filth and cleanliness, he deployed "An Analytical Chemist" to illuminate the difference.[45]

Accum's crisp origin story doesn't hold up to historical evidence given the lengthy back catalog of efforts to denounce poisons and respond to the perceptions of insincere character in the food marketplace. Yet his moment and his text rightly served to mark a new phase in the relationship between modern chemistry and the political management of food. This was especially true as it spoke to the fears of lost trust and unfamiliarity that only grew in the century after his writing. The same features of culture and character charted above shaped Accum's place in history.[46]

An entire genre of writing devoted to chemistry followed during the nineteenth century on both sides of the Atlantic that resulted in a remarkable bibliography of new texts dedicated to classifying the adulteration problem. Soon after Accum, the West Point professor of chemistry James Cutbush brought the debate to the United States with his collected *Lectures on the Adulteration of Food and Culinary Poisons* (1823). An anonymous writer co-opted the phrase "there is death in the pot" for an 1830 title, *Deadly Adulteration and Slow Poisoning Unmasked, or Disease and Death in the Pot and the Bottle.*[47] English and American presses didn't monopolize the category. A fair portion of the work came from France and Germany. A French chemist published an adulteration detection volume, *Dictionnaire de substances alimentaires indigenes et exotique et de leur properties* (Dictionary of native and exotic food substances and their properties) in 1830. An anonymous author published the guidebook *Der Chemiker fur's Haus* the same year in Leipzig. The English press translated it as *Practical Advice for the Detection of Adulterations as Well as Toxic Substances in Food.*[48]

This attention to chemistry appealed to consumers in a food market plagued by fears and distrust. The texts above were the tip of the iceberg—I made the mistake of thinking I could catalog and chart all the adulteration texts of the century early on in the research for this book before a wise librarian dissuaded me from the potentially never-ending nature of the quest. A body of literature that began as public pamphlets protesting things like poisoned bread or tea grew to define an enormous library shelf by the end of the century. Whereas Accum's 1820 publication had surveyed about two dozen items, by 1831 another chemist had expanded the roster to assess two hundred separate contaminated foods, drinks, and drugs.[49] It was worthwhile to list specific adulterants in early century compendiums, but fifty years later there were simply too many. By then it made more sense to count full publications *about* specific adulterants or classes of them. When, for instance, the National Academy of Sciences took stock of antiadulteration literature in 1884, there were so many specific items that they instead compiled a thirty-eight-page bibliography from German and French journals to British and American texts on adulteration and the chemical detection thereof. The academy cataloged over seven hundred publications on adulteration in total, many with hundreds of foods, and that was only up to 1883.

One of the more relevant parts of that catalog is its focus on food rather than agriculture. Accum's work on poison and purity was influential for more than his sensationalism *because* he focused on the chemistry of the kitchen and market, not the field. In the agrarian world around him, it was unusual to proffer a study divorced from the fields and attentive to foods. The famous English chemist Sir Humphry Davy—Accum's onetime boss and a salient example of a chemist of prowess—had initiated a catalog of studies in agricultural and soil chemistry in the 1810s. Much of the related chemistry of the next decades attended to fertility and what became agricultural chemistry.[50]

In the next generation, German Justus von Liebig would become "the most celebrated chemist of his era" by building a remarkable record in an assortment of chemical fields. In retrospect, his career marked a pivot from chemistry's earlier attention to agriculture, fields, fertility, and the like to stores, kitchens, meals, and food. His *Researches on the Chemistry of Food* (1847) was the start of a longer series of publications in the second half of the century that tackled animal chemistry, nutrition, and physiology. They all provided chemical underpinnings for use by antiadulteration analysts to come.[51]

Of the many contemporaries in the second half of the century who wrote those antiadulteration volumes, a few gained staying power, influential in their own time and standouts still in the annals of history. One of particular note was Britain's Arthur Hassall. He was the preeminent public health official in Britain and, after Accum, the most influential contributor to the new library of adulteration analysis. He led a years-long effort to address the contamination problem in Britain. His efforts played a crucial role in establishing the analytical policing of health, purity, and food identity, appealing specifically to the differences between outer appearance and inner behavior.[52]

Readers understood the implications of Hassall's analytical work because they interpreted it with an eye to creating insights into seeing and believing. In a widely circulated part, Hassall included studies of water in which he revealed the theretofore unknown contents of the River Thames. Within a seven-hundred-page report, more than a hundred plates showed magnified views of various foods and organic products. London's satirical weekly *Punch* mocked the meaning of the exposures. As figure 2.2 shows, *Punch* understood that a lack of moral character accompanied contamination. A pestilent society was reflected in the pestilence hiding within the water. You might not want to find what lies there.[53]

188 PUNCH, OR THE LONDON CHARIVARI.

THE WONDERS OF A LONDON WATER DROP.

The freshest fruits of microscopical research are the wonders which have been revealed in a drop of London water through the Molecular Magnifier, illuminated by the Intellectual Electric Light. For the ability to behold these astounding marvels, a certain preparation is necessary, bearing, superficially considered, some resemblance to Mesmerism. The person intended to be the Seer is placed on a seat. Any competent individual then takes him in hand, and explains to him the composition of water, showing him how the pure fluid differs from the liquid constituting the Thames, and from that which exists in the metropolitan wells, when the former has received the contents of the sewers, and the latter the oozings of intramural graveyards. Some delicate subjects, even of the male sex, cannot endure this process, it affecting them with faintness and nausea.

Having been subjected to the above preliminaries, most people are in a sufficient state of enlightenment to discover, by the aid of the Molecular Magnifier, the curiosities contained in

A DROP OF LONDON WATER.

The drop to be magnified is taken from a mixture of the common well-water of London with that supplied by the various Companies. Mr. Hassell, it is already known, has enabled philosophers to discriminate between these waters, by the verminous and other peculiarities which he has demonstrated in each particular form of beverage.

The Molecular Magnifier differs from all other microscopes, in displaying the ultimate constitution of objects; a spectacle not only defying the naked eye, but all vision which is not in a measure psychical.

And wondrous indeed is the scene disclosed within the sphere of a little drop of water—of that water which Londoners drink, swallowing daily, myriads and myriads of worlds, whole universes instinct with life, or life in death! It transcends all that has hitherto been deemed astonishing. America herself will confess that it stumps the revelations of Andrew Jackson Davis.

Creatures—who shall name them? things in human shape—in all appearance London citizens—aldermen, deputies, common councilmen, —are seen disporting in the liquid dirt as in their native element. Behold them, fiercely hustling each other in competition for atomic garbage. What pushing, poking, fighting, kicking, scrambling! There goes an unfortunate wretch fast as if for dear life, with a hook-nosed homunculus—evidently a genuine water-bailiff—darting after him. Here a cheap slop-seller has caught a smaller individual of the same species by the head, and is trying to bolt him. There again, as plainly as possible, you see a funeral procession with an undertaker at the head

FIGURE 2.2. Microscopy could alleviate or agitate fears, as with this satirical interpretation of a drop of London water from *Punch*, May 11, 1850, 188.

Before a biological revolution in the second half of the century—when Pasteur, Koch, and others first opened up the scientific view of microbes, bacteria, germs, and viruses—the contents lurking invisibly inside everyday organic products posed a threat for their very unknown-ness. *Punch*'s take on the water of London poked fun at the prevailing view that a theretofore unknown world might expose mirror images of the world above.

Looking inside might not simplify things. In an age riddled with problems of character, people thought the truth was buried safely inside, waiting to be revealed. On the one hand, then, analysts sought to get to the bottom of the matter with their inspections, to find the truth within. On the other hand, the interior had a relationship with the surface such that a technical fix—instruments to go beyond the naked eye—might not resolve things. Prevailing social norms framed the ways people saw the interiors.[54]

As my failed bid to catalog the entirety of a century's texts devoted to the analysis of foods and drink would attest, a slew of similarly minded analysts and antiadulteration detectives soon joined Hassall. The four decades from 1860 to 1900 were key for such developments. In the United States, the largest of these jurisdictional efforts followed from work in the Division of Chemistry at the USDA, founded in 1862. In 1883, the USDA hired Purdue University and Indiana state chemist Harvey Wiley to lead efforts confirming the quality and purity of seeds, crops, and fertilizers at a national scale. He expanded his bureau and the agency as a whole to lead the fight against adulteration for decades at the federal level.

This kind of work proceeded at the state level as well. Ellen Swallow Richards, for one, worked as a chemist and pure foods advocate at MIT. The Massachusetts state board of health hired her in the 1880s, during which time she wrote the widely read and reprinted *Food Materials and Their Adulterations* (1886). This, too, sought to inform households about how to know what they were eating. It also took as a core premise that the task at hand was differentiating between surfaces and interiors.

Although I write at greater length about antiadulteration chemistry later, I note it here because all of these chemists operated under the assumption that Melville, Paraf, Barnum, and Twain observed and leveraged that seeing was no longer the reliable way of believing it once seemed to have been. Even under normal circumstances with well known foods—bread, milk, meat, and the like—the "misjudging eye" could be deceived or inexperienced. The new techniques of instrumental analysis responded to the conditions of adulteration by providing a different way to verify purity.

From the Bible to Plato to medieval butchers to Mississippi River con men, there has always been a fear that what we get from our interactions is not sincere. This is true in the moral sense, about what we expect from one another as community members. It is true in the material sense, about what people expect of the things we buy and trade. The two aspects have long been linked for food adulteration.

The modern phase of adulteration that the rest of this book addresses gained its meaning from cultural changes in the generations after Accum's work in the 1820s. It followed from new perceptions about where adulteration came from. This was especially true of the form that antiadulteration anxiety took in the United States. A broad cohort of observers worked to pronounce a shift in the middle nineteenth century, one animated by unfamiliarity and spawned by mobility, the distrust agitated by distance, the uncertainty about artifice brought by manufacturing. In short, a crisis of authenticity. All of this confused and upended norms that communities had understood in their measures to police the identity of food. When embodied experiences of sight, touch, taste, and smell could not be trusted—or when purveyors manipulated those characteristics to provide a false front—advocates thought new chemical techniques would cut through the surface and allow us to peer inside.

I sat in a café over a century later recognizing the same cultural dissonance, the kind familiar in a modern age of distrust. I wondered whether those modes of suspicion were playing out right in front of me. I didn't send money to the guy with the Stetson hat. I haven't seen him since. I'm not sure where he went or what his true motivations were. I didn't even need to doubt them. I could perhaps more gracefully have taken him at his word and given him the benefit of the doubt. But I questioned my judgment in our interaction. The fact that I lacked that confidence was the lasting takeaway from our conversation.

Those questions of confidence and trust felt like a historical string connecting me to the angst Melville and his peers characterized. It may be that we still ask why it's so hard to get an honest egg; we live in the shadow of the intermingling of enchantment and disenchantment they saw, a combination that set the nineteenth century apart from prior ones but that remains with us still. Fights over food have long been part and parcel of fights over credulity.

To those fights, I note Ellen Richards not just to highlight the surge in late-century antiadulteration efforts but because she placed her work into another distinct feature of the age: the setting of kitchens, grocers, and household consumers that distinguished the fight for purity and sincerity in the early age of food manufacturing from the eras that preceded it. The grocer's counter was perhaps the most dramatic space for community concerns over adulteration. It was the front line for the century-long shift from a producer-oriented culture of food and agriculture to a consumerist one. It provides the basis for the next chapter.

✷ 3 ✷
Household, Grocer & Trust

The prosperity of a nation depends upon the health and
the morals of its citizens; and the health and morals of the people
depend upon the food they eat and the homes they live in.

ELLEN SWALLOW RICHARDS, 1886

How is Ellen Henrietta Swallow Richards not a household name? In the pantheon of "people who should be much better known," Ellen Richards must be near the top. She was MIT's first female student and graduate. She would become the first female instructor at the school in the 1880s when she was forty and the school itself was less than twenty years old. She was a pioneer in chemistry, public health, and sanitary engineering, and she was a leader in the domestic-science movement and the official water analyst for the Boston Board of Health. She wrote fifteen books and coedited several more. She was the leader of the new American Home Economics Association and founded its journal, the *Journal of Home Economics*. She was the first woman elected to the American Institute of Mining Engineers, and she was a cofounder of the American Association of University Women. While her contemporaries Jane Addams worked in Chicago and Jacob Riis documented New York City, Richards labored tirelessly in Boston for the cause of social reform. She helped open the New England Kitchen in 1890, famous for its dedication to nutrition, working class diet, and respite for the poor. One 1911 obituary referred to her as "a powerful leader, a wise teacher, a tireless worker, [with a character of] sane and kindly judgment." She's sometimes celebrated by engineers, environmental scientists, chemists, and ecologists. She even appeared in *Wonder Woman* 1, no. 50 (1951) (fig. 3.1). Food historians, especially those studying domestic science, have done best to place her at the center of their histories. They have explained Richards's prominent though complicated role in domestic science, not to mention her complicated legacy in the

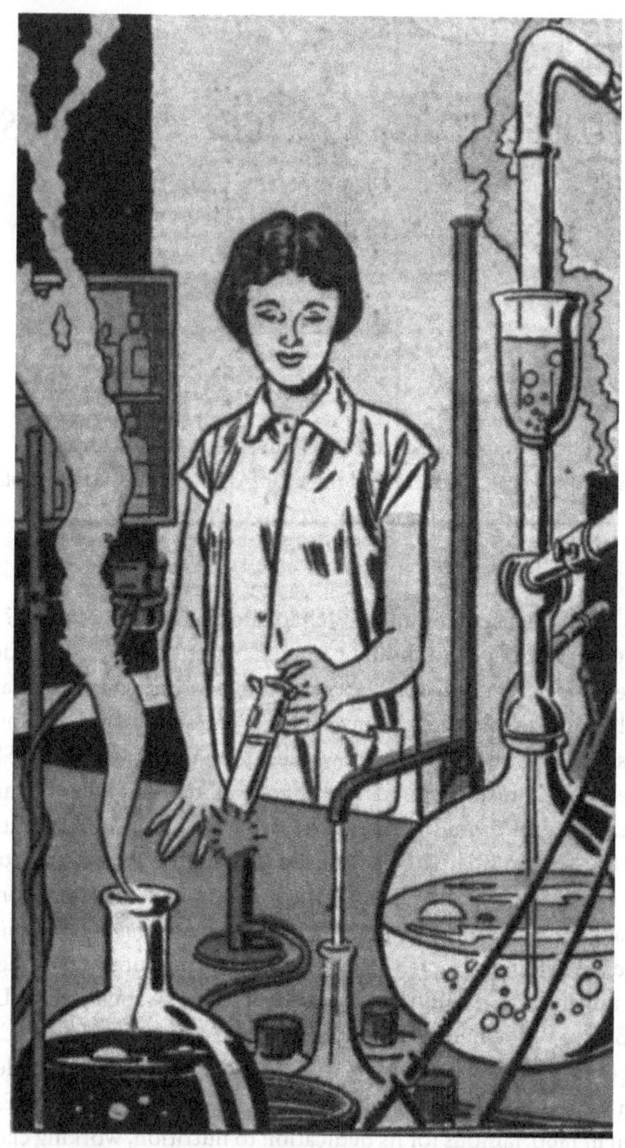

FIGURE 3.1. Ellen Swallow Richards was profiled as one of the "Wonder Women of Science" in *Wonder Woman* 1, no. 50 (November 1951). Courtesy of Michigan State University Special Collections.

quantification of food. Yet beyond the academic sphere of historians and perhaps experts on domesticity, she's not a household name.[1]

There's something to be said about the place of women in the history of science that accounts for this lack of public prominence. There's also something to be said about the muted historical prominence of grocers' markets and household management in the way people talk about food health and safety today. I mention a complicated legacy because that, too, might help explain Richards's lack of A-list historical fame. The home economics movement that defined a large part of her work "aimed to rationalize and homogenize the American diet," as historian Kristin Hoganson put it, in ways that eventually framed food control in narrower rather than broader cultural terms. That aim added visibility to the importance of food in the ecology of the home—as Richards framed it—but also laid the groundwork for racialized, ethnically narrow ideas about what counted as proper eating and what did not.[2]

Regardless of how we account for her current notoriety, Richards opened a window onto two dominant questions of her day: What was it like to buy food in the late nineteenth century? And how did that experience help give shape to the pure food debates? Answers to both required attention to the newly confusing food consumer experience that came from the rise of the modern grocer's market. The grocers' empire changed the infrastructure of food sourcing and distribution in the Gilded Age and in the process changed modes of trust and knowledge about food that we wrestle with still.

Consider it this way: here in the twenty-first century, our interactions in a complicated food market may still be confusing, but they aren't new. One of the methods of our generation's local food movement has been to reduce the space between farms and forks (producers and consumers), or at least make the parts of those spaces more apparent. In between the land and the kitchen is a complex infrastructure of processors, distributors, marketers, and retailers that make it hard to see or know food as an environmental item. There's too much clutter; the view is muddied; kitchen literacy suffers. Richards and pure food debaters were confronting the burgeoning modern face of that confusion. They understood that grocers solved one problem while exacerbating another: they provided access to the new range of processed, imported, and manufactured foods, but they also required yet more trust from the customer that these were honest goods, that the grocer was providing the genuine article. The tension between inside and outside from chapter 2 was an existential component of nineteenth-century life in general. It played out metaphorically, with respect to character, through fictional narratives, and in terms of face-to-face interaction, it always confronted the differences between what we see and how we know. The tension played out in the grocers' empire, too, though with more specificity and as directed at food identity. It was hard to trust the grocer on the inside; it was tricky to know what your market-bought food was as an outsider.

Richards pitched her 1886 text *Food Materials and Their Adulterations* to that interlaced problem of trust and knowledge. Her antiadulteration text brought the fight for pure foods to the households and kitchens of a new urban demographic, a demographic that was neither necessarily farmer nor agrarian. The cultural legacy of sincerity and authenticity helps explain what led to arguments over pure food. New spaces of exchange in grocers' markets and urban households add a layer. The grocer's counter contributed a critical feature giving shape to the era of adulteration.

I'd like Ellen Richards to be better known, and I return to her soon. I have more limited ambitions for Frazier Paige. Sixty-odd miles to the west of Richards's work, Frazier and his wife Wealthy had been shopping at the general store in Hardwick, Massachusetts, for nearly fifty years. Their involvement stood in relief to that of Richards's target readers. The Paige's experience charts a shift from rural general store to urban grocer's market, offering a backdrop against which to see the changes in food access and distribution to which Richards responded.

Back in Hardwick, Frazier's account name was on the ledger. Wealthy was a daughter in the general store owner's extended family, the Knights. She had as much responsibility managing purchases as Frazier. The two conjure the image of a rural household where food sources were close at hand, either brought from the cornfields, cow pastures, and butcher shops behind the frontage buildings of the small town or, soon, secured by new railroads from urban centers. Just east of Worcester, Hardwick was (and remains) a small farming town in the middle of the state that was home to an abundance of Paiges.[3] The store kept accounts of over four hundred families of customers in the 1840s, when the locally prominent Mixter family owned it before the Knights bought it in the 1850s. It held steady over the next half century with about six hundred families as customers, or nearly the entire town.

Frazier was born in 1822. He was a farmer, a dairyman, and a member of the extended Paige clan. He was town treasurer in the later 1860s. He was manager of the Hardwick Center Cheese Factory in the 1870s, the state's largest and most prosperous. Figures show that in 1875 alone the factory made about 124,000 pounds of cheese and 35,000 pounds of butter.[4] His father, also a farmer, was a customer at the store before him; his grandfather, as his great-grandfather before him, had been a successful farmer in the years before Mixter opened the store. Six generations earlier, Frazier's grandfather four times over was an original settler in the town. He farmed the land in central Massachusetts after moving with his wife and three children from England in 1685. In Frazier's day, barns outnumbered houses. By the 1870s the town was home to about two hundred farms greater than ten acres. Granted, most of the people in the country were farmers. Around Hardwick, those who weren't went into the clergy, practiced law, or plied a trade such as baking, butchering, or, as it were, running a general store. The mid-nineteenth century was not much different in this regard from the seventeenth century when the Paige family first settled.

Hardwick's store name was apt; it was general. A typical day in the typical year 1845 saw sales of salt, silk, beans, coffee, hats, salmon, raisins, nails, pins, sugar, cotton, gingham, "1 firkin" of butter, "2 bushel oats," "1 bbl" of pork, oil, candles, soap, and tallow. Customers, in other words, bought fabric, food, fuel, and household supplies. To set this into a wider context, a contemporary of Paige's just five years his elder was that same year about fifty miles east outside Concord tabulating what it took to grow and produce those supplies himself. In a self-built cabin at Walden Pond that cost twenty-eight dollars in construction supplies, Henry David Thoreau calculated a precise expense of $8.74 for food during an eight-month span there. Like Paige, Thoreau bought pork, molasses, lard, and salt as staples. He spent two dollars on oil for his two-year stay, which, if it was the same as Frazier's, meant a few gallons. It appears that they paid about the same amount for food, considering Thoreau's partial barrel purchase of pork and ample (several gallon) supply of molasses.[5]

The ledger for the store listed customer accounts by number and purchases paid. Frequent customers settled their accounts several times a year. Itinerant shoppers instead paid on the spot. The regulars got better deals. Frazier's cousin Stephen Paige bought two bushels of oats for sixty-seven cents to put on his account. Later that day Levi Chamberlain, an infrequent customer, paid eighty-three cents for the same thing. The next day, another Paige cousin, Luther, got the sixty-seven-cents deal. On another day, the itinerant Nathaniel Simpson paid thirty-five cents for butter; Mr. Berry, a frequent customer (#163 that year), paid twenty cents. Today, paying by cash might garner a lower checkout price for the customer; in Frazier's town, credit was a privilege of trust and community.

If the mid-nineteenth century was not dramatically different from the seventeenth century, it was Frazier's generation that saw changes. This was especially true by the 1870s and 1880s. From that period on, the store saw a shift in clientele, a decreased pace of purchases, and an increase in the range of processed and distant goods brought into the store. Those later decades of the century found more traffic in goods from wholesalers coming to the country storekeepers rather than rural general stores that stocked their shelves from local producers, like Frazier Paige himself.[6]

For one thing, changes in diet and product availability slowly crept into the accounts in Hardwick. Flaxseed meal was on the shelves, for example. This wasn't a particularly fancy item, but it was what people today call a "value-added" product, processed away from its harvested origins into a form more suitable for cooking and baking. "Pillsbury flour" from the upper Midwest was sold by name and brand by the mid-1880s. Customers were also buying cans of "extract of beef" before the turn of the century, when Count Rumsford and German chemist Justis von Liebig lent their names to such new products. As Frazier aged, he also took an increasing interest in "Graham." These were the so-called bars pioneered by Sylvester Graham's followers as an effort to provide a one-stop shop for all dietary needs. Calling it a graham cracker, as we do now, would be misleading. Graham's version was instead an early form of a health bar, a bulky, heavy loaf for which Frazier had a taste. He bought them frequently at four cents a pound in the 1870s and 1880s.

By the end of the century, the Knight General Store had slowed its sales pace. Neither Frazier nor Wealthy stopped by anymore. He passed away from dysentery at age seventy-four in 1896; there's no record of Wealthy's passing. Their son Timothy, still farming family land, remained a frequent customer. The store stood as a placid and connected witness to a transformative half century in food provisioning and agricultural livelihoods.

The store's accounts also offer a barometer of food sourcing and historical trends that paint a picture of community networks of trust and exchange. Questions of authenticity and purity were vetted within those networks. These general and dry goods stores were the small-town rural forms of what would later become urban grocers' markets and then, in the twentieth century, the grocery store. They were a kind of stereotypical classic Americana, with images from *Little House on the Prairie* and other pop culture sources filling in the picture for most people today.

Yet the store wasn't necessarily an idyll, and I don't mean to discuss quaint Hardwick townspeople as a gambit to frame an agrarian ideal of the past. These were sites of consistent and well-developed exchange. They were also fraught with the residue of the paths the products had taken to get to the store, thick with social and environmental entanglements. Supply lines were inconsistent. Unexpected weather patterns disrupted stability in the store's stock. And while the Hardwick General Store's ledgers accounted for customer interactions, they also shined a light onto the uncomfortable transformation of food distribution and access.[7] In the store's early years, a horse-drawn wagon could travel maybe thirty miles in a day, meaning Knight got imported products at best two or so days after shipment into Boston Harbor—Ellen Richards's backyard.[8] In 1870, "railroad fever raged violently in Hardwick," a fever that connected the town to "commercial citizens of Boston" with the newly chartered Massachusetts Central Railroad.[9] Thereafter, the railroad could deliver the same supplies in a few hours. Cheerleaders called this convenience and speed; critics were guarded, calling it confusing and unsettling.

The tea, sugar, spices, and coffee on Knight's shelves were visible traces of trade networks that were common across North America and followed patterns from European traders already established the century before. Those networks brought goods from colonial holdings in the Global South through port cities in England and the United States to, in Knight's case, central Massachusetts. They arrived by way of storeowner trips to port city wholesalers and from hawkers, peddlers, traveling salesmen, and via the new freight rail. In the middle of a market system between farm and fork, the general store functioned as a mediator managing the flow of goods from neighboring towns and foreign suppliers to farmers and townspeople in small town after small town. The lifeblood of the market was found in sacks of flour, bricks of tea, and casks of sugar. When expanding distribution networks brought more unknown goods to the shelves, customers might suspect their true identity. The challenge was in trust, not just source.[10]

Frazier and Wealthy Paige and Hardwick's case was a small one, but like other rural towns of the time, it exemplified the patterns of traffic raised in chapter 2. In this, Paige and neighbors replicated a well-told story about the new flow of goods and people from coasts to the interiors of the continent. The other end of that line and the metropolitan counterpart to Hardwick's general store was even more illustrative of the confusion over food identity in an expanding distribution system. At that end, in the city, was the urban grocer.

Most people are comfortable with the self-service grocery store today, so much so that the term "self-service" rarely factors into the description. Why the modifier? Shoppers grab a cart, stroll the aisles, stock up with goods, and cruise to the checkout lane. With fewer items, the grocery chain has us check out the goods ourselves, a thrill for scanning and beeping that wears off almost before the first experience of it is over. Before

FIGURE 3.2. Typical grocer's market of the later 1800s, like the one in Hardwick. Courtesy of the American Antiquarian Society.

innovations at the Piggly Wiggly and A&P early in the twentieth century, though, grocery stores were grocers' markets, and goods were delivered across a counter by the grocer himself. The storekeeper stocked wares behind him and out of reach of the rabble so that groceries came to customers like prescription drugs from a drugstore counter today (fig. 3.2). Grocers managed the flow of goods from wholesaler to customer. The transactions exacerbated questions of food identity and veracity because they asked the customer to trust the grocer as a representative of the supplier and the source. That wasn't always easy to do.[11]

We know that the managed form of interaction was not without precedent, of course. Frazier Paige had much the same experience buying food and products from Albert Knight in Hardwick. Focused in rural settings, questions of food safety and identity were common problems with drummers and peddlers and potential hucksters, real and suspected. The world of swindlers and dubious characters framed the moral background from which rural anxiety was borne. In such cases, though, townspeople were on their own turf to meet suspected swindlers to vet their sincerity. This was an agrarian problem in an agrarian world. In the city, the interactions came between urban traders and residents with less experience in a community. The immigrant or second generation resident was part of a world less carved out than Hardwick's seventh generation of farming neighbors and all those Paiges.

While a "grocers' movement" flourished in the later decades of the century in the United States, the eighteenth century was the real moment for grocery development.[12] The word itself, *grocer*, descended from medieval storekeepers who sold products in gross bulk. During medieval centuries, they were known as spicers or pepperers before their dealings in gross weights brought them the name grosser. (In the nineteenth century, French grocers were still called *épicers*, or spicers.)[13] The English spelling evolved by the time grocers were part of a food system that included not just the general or dry goods stores of rural livelihood but public markets, butcheries, bakeries, and other specialty shops of the urban streets. The public market was, as the name would suggest, a public space. Sellers rented booths and participated in a kind of exchange ruled by the conventions of the city and market overseers. In their American form, they evolved from the public storefronts of trading companies to market squares and sheltered stalls for hawking wares in a secure and reliable location much like farmers' markets today.[14] Grocers' markets more often operated as private shops subject to their own self-decided operations. As urban spaces organized and municipalities governed their local businesses, grocers professionalized to develop standards of practice and decor. One of their biggest challenges was to garner the faith of their customers. One main reason their customers didn't trust them was the often-leveled accusation of adulteration.[15]

By the later 1800s, the grocer's counter was a front line in the battle against adulteration. It was a place of interaction between producers and consumers, a node in the network of food access, and a space of exchange between the trusted and the suspicious. How did customers know how to navigate the grocer's empire? And why was it such a big deal?

Enter Ellen Swallow Richards.

Richards operated within a world of urban storekeepers. Her own biography underscored the transition from rural transactions in Frazier Paige's world to urban storekeepers. She thought that customers could navigate the grocers' empire if they had more tools to inspect their food. She offered a means of circumvention: customers didn't have to trust the grocer because they could come to know more about the food than he did. It was a big deal, Richards maintained, because ensuring the purity of food went beyond challenges of nutrition or cost alone. The question of individual health was no less than a subset of broader moral questions about "right living," to use the phrase she adopted in the 1870s. "The prosperity of a nation depends upon the health and the morals of its citizens," she wrote, "and the health and morals of the people depend upon the food they eat and the homes they live in."

She presented this as Russian nesting dolls of bodily health: the health of the individual body was tucked inside the health of the household body; the health of the household body was tucked inside a healthy nation, that of the body politic. Thus, if the microscale image of rural communities like Hardwick added a color to the picture of the changing foodways and trust mechanisms producing the pure food crusades, Richards's work added another. It wasn't just new forms of mobility and disconnections that raised concerns over food fraud. The health of the household and the health of the nation also motivated the fight over adulteration.

The Boston-based chemist was a self-proclaimed scientist of "oekology"—the "household of nature," as she translated it, before twentieth-century scientists called it *ecology*.[16] Born in 1842, she was raised in a small farming town, Dunstable, northwest of Boston, just south of New Hampshire, and only fifty miles northeast of Hardwick. As a teenager, she worked in the general store her father bought and operated in Westford, where she also enrolled in the coeducational Westford Academy. A "small, compactly built woman" known in her family for her energy and curiosity, her parents taught Puritan principles of moral living at home, where she read William Alcott's *Young Woman's Guide to Excellence* before enrolling at Westford. At the academy, she excelled in her course of study in math, science, and "principles of morality."

As a teenager, her ambitions and plans were shaped in part by working at the general store with her father, where she kept the books and counseled customers on cooking and housekeeping practices. Her ambitions also came from exposure to periodicals, *Godey's Lady's Book* in particular. The feminist and literary light Sarah Josepha Hale edited *Godey's*. Hale also lent support to a new school in upstate New York, Vassar. Richards enrolled at Vassar in 1868, immersing herself in the literature of household economy and excelling at science. She then frustrated the admissions officers at MIT, who saw her application, realized there was no explicit claim that women could not attend, and reluctantly admitted her. She graduated from MIT in 1873 and began a prolific life of applied science.[17]

Throughout her published life Richards had a sympathetic understanding of grocers garnered from a childhood working in a general store. She wrote the first of three editions of *Food Materials and Their Adulterations* in 1886 to aim for the coordinated health of the individual and household body anchored in the grocers' world. The book grew out of a paper she had published six years earlier on "The Adulterations of Groceries." As part of the Massachusetts state board of health's first annual report, the paper reviewed a survey of 141 dealers in forty towns across the state. While she found that the adulteration of flour, baking powder, and sugar was widespread, in large part she defended grocers, writing that "retail dealers, as a rule, sell what they buy, without change, and whatever adulteration exists is to be found among the manufacturers and wholesale dealers."[18] Two years after the 1880 report, she published *The Chemistry of Cooking and Cleaning* as a text on applied science for housekeepers. It sought to write up the defense plan for housekeepers. "It is time they should bestir themselves for their own protection," she wrote of the middle-class cooks and housewives who bought the book, from the "number of patent compounds thrown upon the market under fanciful and taking names."[19]

Her 1886 text expanded earlier arguments with a motivation to "arouse women providers to the need of a study of the materials they purchased, both from a sanitary and economic point of view." It was widely reviewed. One called its premise that the health and morals of the people depended on the food they ate and the homes they lived in "a self-evident one."[20] *Science*, then in its seventh volume, noted that it was "a little work intended for the intelligent housewife." Magazines and papers from Boston, New York, Cleveland, and Chicago to Germany and England all lauded the work for the clarity it provided the housewife and its easy to understand analytical directions.[21]

Richards was familiar with the long history of adulteration. Food fears were timeless, she said, a consistent part of community anxiety for centuries. But she recognized the changes afoot and placed her research squarely into a current debate about a food system whose central axis was the grocer's counter. Unlike earlier generations, she wrote, "Now, the food products of the whole world are accessible to the people of the United States through the use of improved methods of transportation—the refrigerator car and steamship compartment—and through improved methods of preservation—by cold storage and by the canning process."[22] Unlike previous eras or the experiences of Wealthy and Frazier Paige, "the conditions of life have changed . . . so rapidly and completely that the methods of their mothers and grandmothers will no longer answer." The problems of a new age of manufactured and artificial goods challenged familial understandings. "*They* had no trouble with their soap, for they superintended its making and knew is properties. *They* knew how colored fabrics should be washed, for they had the coloring done under their own eyes. *We* buy everything, and have no idea of the processes by which the articles are produced, and have no means of knowing beforehand what the quality may be."[23]

Those were rose-colored glasses suggesting an idyllic past, but they were the glasses Richards and her cohort wore. She certainly wasn't alone when speaking to fears about a loss of "olfactory vigilance." In congressional testimony the same year, 1886, a representative from New York had complained that "the fours senses which God has given us ... are completely baffled."[24] Another lamented that "the family dining table [did not] come equipped with either microscope or reagents for chemical analysis." Food producers were onto this change, too, proactively seizing the confusion to their advantage. The dairy firm Horton and Company, in a "word for our customers," knew well enough about the privilege of organoleptic evidence. Their ad boasted that "Taste, touch, sight, smell ... fail to note the difference" between their product and the real thing.[25]

Richards fought that loss of experiential knowledge by instructing the housewife on the chemistry to go beyond embodied sensory evidence. She didn't seek to privilege the experience of unmediated senses. Household chemistry and domestic science would fill the role, because housewives and housekeepers had a crucial role to play in policing the home's boundaries against suspicious products. "To the housewife and mother we say: For the sake of your children keep yourselves informed of the true state of the food manufacturer."[26]

In presenting an image of nested health—individual, household, national—she was part of the new generation of authors of texts on domestic science in their evolution beyond earlier and more broad-based treatises on household economy. Her use of the term *oekology* was but one indication of that evolution, changing "household of nature" into a scientific term. *Food Materials and Their Adulterations* was a more notable one. It offered a fuller enunciation of the ways one could trust oneself instead of the grocers.

Before that, Catharine Beecher's (1841) *A Treatise on Domestic Economy* was perhaps the model for home guides. It had been a publishing hit, reprinted for decades to bring a specific moral vision about the virtue of domestic labor. In 1869, when Richards would've been an early reader of it, Beecher cowrote another key text on household management with her sister Harriet, *American Women's Home*. The treatise similarly placed worries over adulteration into the context of fears of household degradation. In a nod to the changing nomenclature, publishers frequently reprinted it in the coming years as *Principles of Domestic Science*. Contributions in the new genre of domestic periodicals similarly crafted coordinated views of food and home. *Ladies' Home Journal* (1883), *Good Housekeeping* (1885), and other upstarts joined Richards's old favorite *Godey's Lady's Book* (founded in 1830) as new forums for sustained dialogue about eating right and living right.[27]

Richards focused on the urban household because she understood that "the intelligent housewife" had a pivotal role to play in the protection of nothing less than the nation and civilization. This was where the nested dolls stacked together, of household and nation. As historian Charlotte Biltekoff captured it, Richards and her cohort would "articulate the moral dimension of nutrition in relation to emerging ideals of good citizenship." Their aim had been to create "better homes and cities" by protecting the household from the ravages of adulteration.[28] Here was another way to fashion the consistent inside-outside dynamic of purity and adulteration. In other examples, such as that of character and swindlers, advocates framed the problem in abstract terms: it was a difference in truth between outside appearance and inside reality. In Richards's case, the problem was that good foods came from inside the proper culture. Outsiders represented a threat to health and integrity.

In that vein, and in a feat of well-trod historical consistency, pure food advocates of her time had no trouble blaming adulteration on foreign sources. If it wasn't the grocer cutting sugar or adding rice to wheat, it was imports that sneaked past him. Two century's worth of anxiety from imports had animated antismuggling and customs import laws, so the foreign fears didn't come out of the blue. By the later nineteenth century, however, the political character of the problem and the new mediator of the grocer's counter ratcheted them up.

Adulteration was an affront comparable to a foreign invasion into the grocer's shelves and, by extension, the household's stomachs. A congressional committee in 1899 took it as given that "large amounts of imported goods are sold in this country the sale of which goods would be prohibited in the country from which they come." Representatives used the term *foreign* interchangeably. There were "foreign substances" such as flour, coffee, and meat. "Foreign countries" shipped them and "foreign" people promoted them.[29] Wisconsin representative Robert La Follette argued on the floor of Congress about the household food's importance "for providing a few home comforts, but even for the security of the home itself." Michigan senator Thomas Palmer played up the fear of foreigners angle. He asked colleagues why it was okay to beat back foreign armies but not adulterated "foreign" foods. "If a 'foreign assailant' should attack such a 'vital part' of the nation," he said, "'no time would be lost before beating to arms for its defense.'"[30] If they were to protect the borders from military invasion, they asked, why wouldn't they protect the borders from insidious impure imports just the same?[31]

With barely concealed racial and ethnic connotations, others blamed immigrants—the German shopkeeper, the Irish merchant, the Jewish peddler, or an uncivilized native. Racism was at the root of a notable strain of adulteration fears. And the "other" didn't have to be foreign. He could be a so-called enemy within.

Two staples of the grocers' market, soap and bread, made this point. True, soaps were not edible, but they derived from the same animal fats and slaughterhouses as other processed foods, and their place on the grocer's shelf landed them squarely in the same debate over purity. Soap ads consistently blurred the difference between racial and product purity. They were racially coded in ways that should be stunning to a twenty-first century reader in their attempts to parse the difference between purity and contamination. Dirt was by definition "matter out of place," to quote a popular Sapolio Soap advertisement at the time. Civility was the absence of dirt. Excising dirt was a matter of preserving purity.[32] Another ad blared, "barbarity and dirt are closely allied." Ivory Soap reverted to the metric of appearance. They claimed that "the whiteness of the 'Ivory' is prima facie evidence of its purity." They famously advertised their product as $99^{44}/_{100}$ percent pure. Their competitor Dreydopple Soap marketed "An American Soap for American People!" One of their trading cards—an early form of viral marketing, like bubble gum and baseball cards—included a cartoon of a young black child who, when he used the soap, could "change from black to white."[33] Not only did the soap make claims for purity, but using it would make *you* pure.

As for actual foods, bread followed similar cultural logic. It was valued not just for its lack of adulterations but its culturally appropriate whiteness. Harriet Beecher Stowe wrote of the difference between proper civilized breads and improper ones that "more or less attention in all civilized modes of breadmaking is given to producing lightness." And although Plato had argued millennia earlier that "dark, hearty rural loaves" were the basis of a strong civilization, by the Gilded Age and Progressive Era in the United States, "eating white bread was said to 'Americanize' undesirable immigrants." In that, the United States was following a broader Western phenomenon. Brown bread was so denigrated in the United Kingdom that prison codes dictated that prisoners were not allowed white bread. In one case, a British guard was fired for giving a dying prisoner white bread, thinking it would sustain and save him. The logic and practice of food identity and value—about the actual physical food—followed from perceptions of "pure" and "better" anchored in cultural judgments about better or worse people.[34]

Those were specific product examples that spoke to the larger point about defending the boundaries of proper and pure grocers' items. Politicians were not shy on the same matter. They argued over the larger points of national health and identity in lofty rhetoric. To them, this was about "putting one's house in order" and defending "the home front." Bad food could be coming into the country with possibly bad or unknown people. Even if they were being lofty, this was about defending the actual borderline of the nation.[35]

No matter; if that were too grand a scale to win the day, advocates for purity could focus more reasonably on the scale of the household. This even followed by seeing the boundary of the physical home as a concrete site for seeking household health. Specific house designs and architecture could promote a well-ordered life where cleanliness (next to Godliness) required clean foods. Where the soap makers wanted you to think dirt was matter out of place, in that schema adulteration was food out of place. The household guides of Richards's time were necessary to define the appropriate order of the home so that they could represent proper living. The Beecher sisters included floor plans in *Principles of Domestic Science* to suggest the proper flow of traffic inside the home and the appropriate arrangement of tasks by room. Andrew Jackson Downing, in the widely acclaimed *Series of Designs for Rural Cottages and Cottage Villas* (1842), prefaced his argument by claiming that houses are an "unfailing barrier against vice, immorality, and bad habits." One should seek the "grace and loveliness" of life in the moral constitution of the home. Put more directly, "A good house is a powerful means of civilization."[36]

We take food seriously today for so many reasons—taste, aesthetics, culture, economy, environment, hunger. They did then, too, if not more so. Pure food was a righteous matter, one of moral sanctity at the level of civilization. As Richards put it, "Good-tempered, temperate, highly moral men cannot be expected from a race which eats badly cooked food, irritating to the digestive organs and unsatisfying to the appetite."[37] Pure food was a matter of individual health. It mattered for household integrity. It was necessary for a healthy citizenry.

When Richards spoke of badly cooked food and ill-tempered men, she was recasting a common line by the French gastronome Jean Anthelme Brillat-Savarin. "The fate of nations depends upon how they are fed," he wrote, a line already widely quoted at the time.[38] Brillat-Savarin is probably more famous for another aphorism: "Tell me what you eat and I will tell you who you are." His suggestion was that food preferences and eating choices were signatures of cultural and personal identity. Antiadulteration writers noted the same thing in the opposite direction. Tell me who you want to be and I'll tell you what to eat. Eating the right, *pure* food would create the right culture. It shouldn't go without saying that this proffered a specific racially and ethnically coded vision of purity and what "right food" meant, a vision shared by Richards. (When I said at the top of this chapter that the legacy of this specific idea of a proper diet might explain Richards's lack of top-tier public fame a century later, I was thinking of culturally coded views like these.)

Yet there they were. Pure foods, advocates claimed, were at once a virtue of the moral cook and a weapon against foreign contamination. Rather than defining what purity meant—a more difficult task, to this day—they argued over what it stood for. The grocer's counter was a front line in these battles, a world of urban denizens where advocates posed and challenged the big questions. Richards knew this. She may not have blamed grocers for adulteration, but she knew their markets were a central axis for the problem. The fight for pure foods would have to go through the grocers' market if it were to succeed.

For those questioning honest food, adding a new node like the grocer was like playing the old childhood telephone game with an extra person to confuse the story when passed along. Richards and her colleagues tried to undo the confusion. Grocers tried to avoid it.

There were a lot of them. As the urban balance to Hardwick's general store, urban "grocerdom" was an increasingly prominent feature of the neighborhood landscape. In the postbellum United States, grocers were remarkable for their consistent per capita presence. In 1869, New York's three thousand grocers meant there was one market for every 314 people; Philadelphia's 2150 shops that same year placed them with a grocer for every 313 people; Chicago was only slightly sparser with one for every four hundred people. Baltimore, the sixth largest city in the nation, was home to one grocer for every 353 people by the mid-1880s. The figures were largely similar at the end of the century.[39]

They were quite visible, in other words. Because of that visibility, and even if Richards was cautious about blaming the grocer, others had no problem doing so. Theirs was a disparaged profession. They were maligned in papers and exposés as cheats; they were charged with trying to swindle housewives; their character was questioned (fig. 3.3, 3.4). They also didn't sit on their hands in the face of such accusations. Their perspective on the problems adds another color to the bigger picture of purity in the grocers' empire.

In the face of accusations about their integrity and honesty, grocers organized and professionalized to defend their interests and define their role in bringing healthy foods to customers. The emphasis here was on *customer*—someone who interacted with a trader through the customs of the market. Before the con*sum*er revolution of the twentieth century—

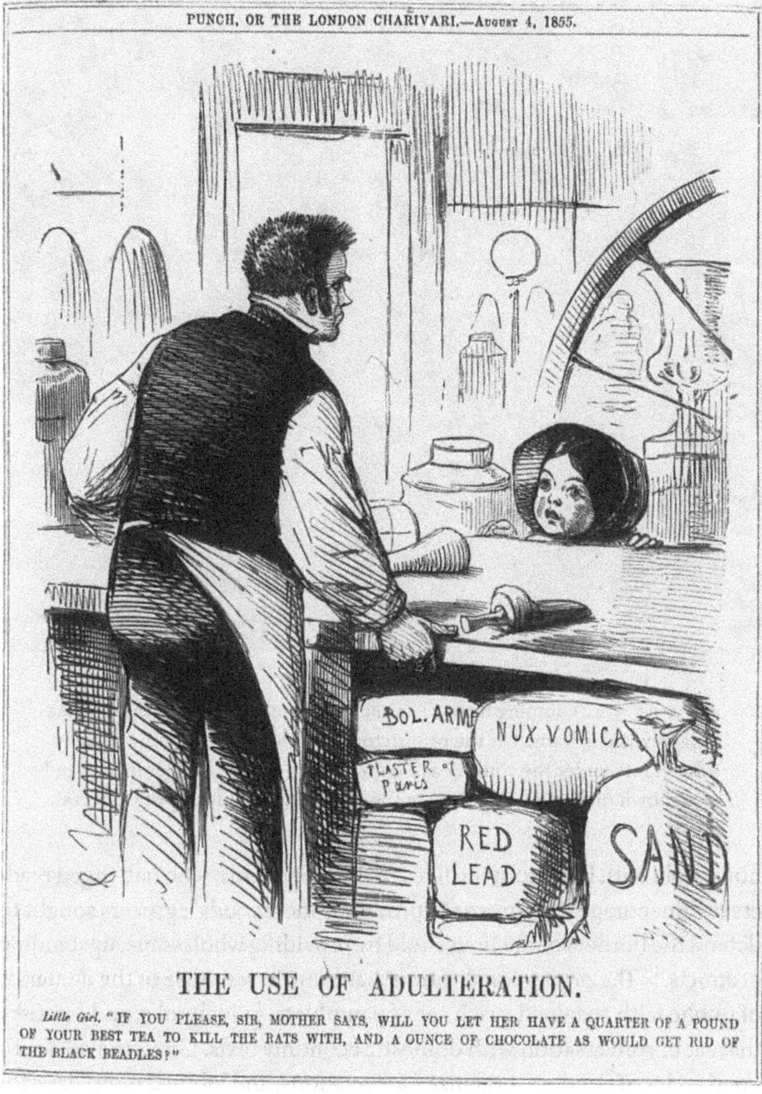

FIGURE 3.3. Those worried over adulteration often suspected the grocer was hiding adulterants behind the counter, as with this take from *Punch*. Source: "The Use of Adulteration," *Punch*, August 4, 1855, 47.

defining us by what we buy and consume—the reigning market paradigm was the *custom*er, defined by how people conduct themselves in the marketplace. By professionalizing, grocers were seeking to define the customs of the marketplace and thus control them.

Customers and their advocates sought to defend the borders of the

FIGURE 3.4. Clipping from a late 1890s newspaper showing the USDA's Harvey Wiley framed as the proprietor of A. Devil & Co., with the same adulterants under the counter as *Punch* showed decades earlier in England (as shown in fig. 3.3). Courtesy of the Wiley Papers, Library of Congress.

household with help from the likes of Ellen Richards, who had urged readers to "Encourage your grocer to provide honest goods"; grocers sought to defend the borders of the household by providing wholesome, upstanding products.[40] The professionalization that came in response to the demands of public faith spawned a rich set of pamphlets, guidebooks, and treatises that sat in conversation with domestic economy texts. Grocers also developed codes of conduct, organized associations, and lobbied politicians for support through city governance.[41]

A scene in a cold February of 1893, the high Gilded Age in New York, brought this professionalizing phase together with a zeal for respectable public standing. That day, Manhattan grocers were celebrating the opening of a new hall for their union in midtown on East 57th Street. They had a café on the first floor. The new hall had bowling alleys in the basement. The union commissioned a tight-bound book to tell its history ahead of the opening. They distributed it to all its members and used it as advertising throughout the city. The epigraph to the book, *Grocerdom*, read, "Down all the changes of the years / Across earth's mingled joys and tears / The star of endless progress shines."

On the day of the union hall's opening, in a building with "modernized Italian Renaissance" styling, New York's Mayor Gilroy stood to mark the occasion. A brass band led a group "hundreds strong" marching down the street from the nearby Lenox Lyceum. The pomp and circumstance built up to the ivory-handled silver trowel the mayor used to coronate the new site. The men sang and drank. Grocers wanted to appear thriving.[42]

Their professionalizing efforts were a defense against accusations of adulteration, price gouging, and dishonest weights. To hear the grocers tell it, theirs was a trade of "honest dealing in every way, and the agreement to sell only pure goods with full weight are the precepts of the guild." This particular union was founded in 1882—there were similar ones in cities across the nation—because "Honest dealing always wins. Honest goods always command a sale." In their bravado, they took credit for launching "the formation of the National Pure Food movement" in 1887; it was their lobbying that would bring Congress to "pass the Paddock Pure Food Bill" (Congress did *not* pass it); and it was their profession who agitated at the local level, not just in Washington, to "cut out the cancer" of dishonest trade, to prevent "poisoning the blood of the business current of society" with fair deals. In a list of aims and goals, including those for self-preservation, patrolling the marketplace, gathering information, and engendering trust, their purpose was "protection against adulteration of goods, fictitious labels," and more. Beer-battered singing may have been their passion, but subtle rhetoric was not: images of cancer, blood, and poison painted a grim picture for those outside their group.

Italianate union halls were one thing. A steady stream of published commentary on the trade's rectitude was another. The genre of the grocers' manual provided such additional forums heralding the honesty of the trade. It defined and defended the norms of food sales while staking claims to purity. Artemus Ward's 1886 *Grocer's Companion*, out of Philadelphia, was one of the more widely circulated ones (fig. 3.5). He set the pattern, announcing from the top banner, "the grocer is never an advocate of adulteration." Another handbook complained that "laws of a specially stringent character exist in States of the Union; but, unfortunately in most cases, the Retail Grocer, who is usually innocent and ignorant of the impurity, suffers the loss both of money, custom, and character."[43]

To their credit, many of the grocers acknowledged complexities. The most common refrain was to follow the point Ellen Richards made that they were victims of circumstance. They would thus admit the fact of adulteration before explaining it away. One thing to do if you are in a defensive

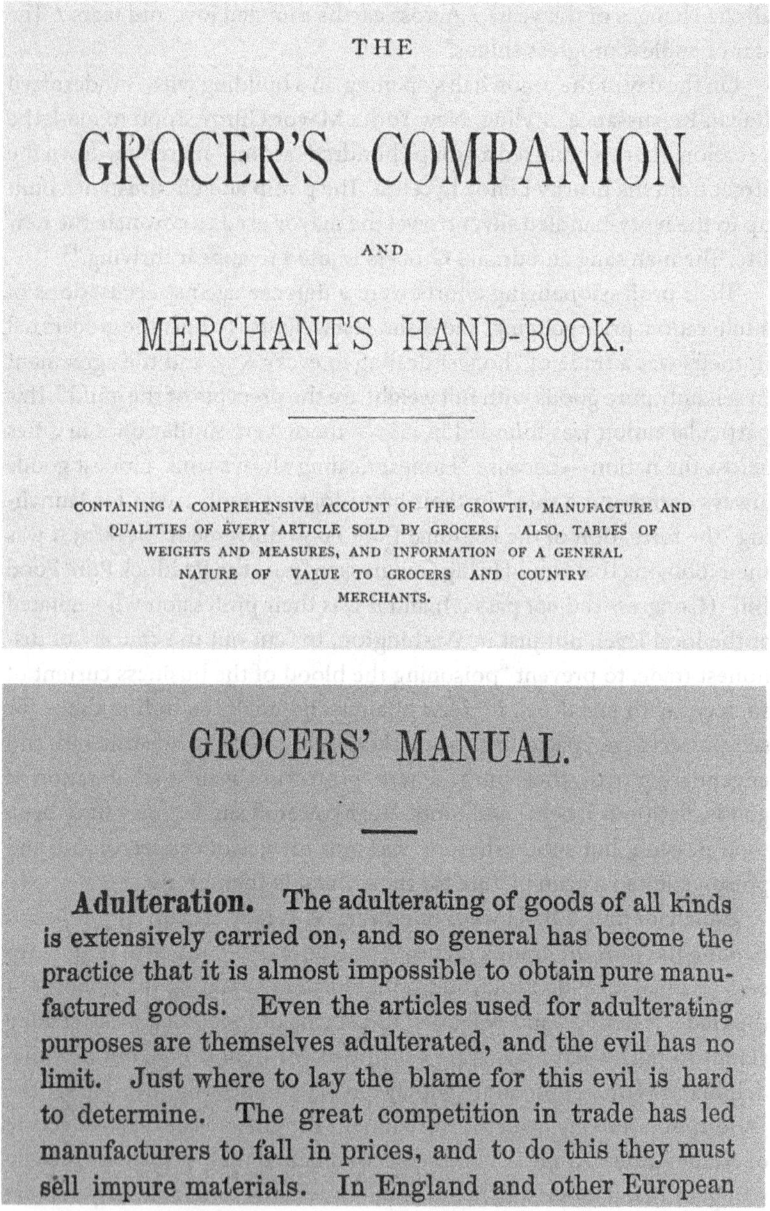

THE

GROCER'S COMPANION

AND

MERCHANT'S HAND-BOOK.

CONTAINING A COMPREHENSIVE ACCOUNT OF THE GROWTH, MANUFACTURE AND QUALITIES OF EVERY ARTICLE SOLD BY GROCERS. ALSO, TABLES OF WEIGHTS AND MEASURES, AND INFORMATION OF A GENERAL NATURE OF VALUE TO GROCERS AND COUNTRY MERCHANTS.

GROCERS' MANUAL.

Adulteration. The adulterating of goods of all kinds is extensively carried on, and so general has become the practice that it is almost impossible to obtain pure manufactured goods. Even the articles used for adulterating purposes are themselves adulterated, and the evil has no limit. Just where to lay the blame for this evil is hard to determine. The great competition in trade has led manufacturers to fall in prices, and to do this they must sell impure materials. In England and other European

FIGURE 3.5. Artemus Ward's *The Grocer's Companion. a,* title page. *b,* detail, with definition of *adulteration*. Courtesy of the Hagley Museum and Library.

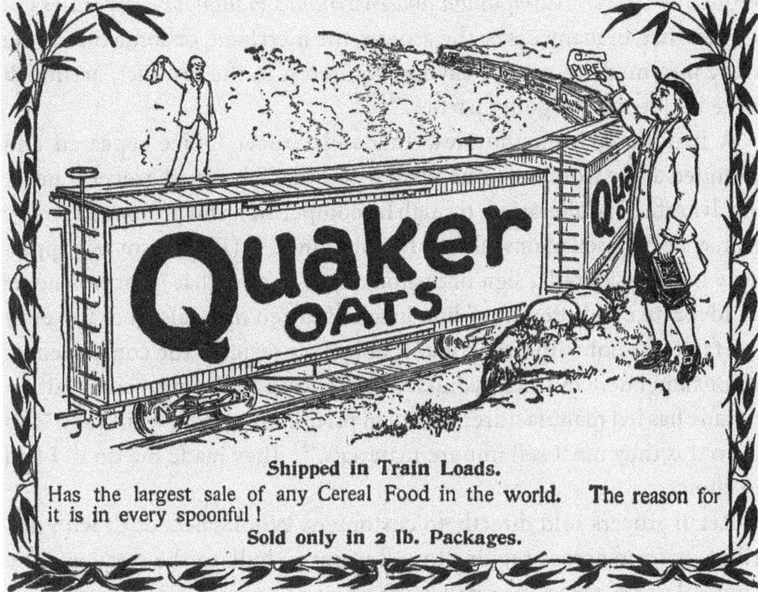

Quaker
OATS
PURE
Shipped in Train Loads.
Has the largest sale of any Cereal Food in the world. The reason for it is in every spoonful !
Sold only in 2 lb. Packages.

FIGURE 3.6. Quaker figure waving a certificate of purity as a trainload of product leaves the manufacturer, making the claim that it was pure when it left the factory. Courtesy of the Warshaw Collection of Business Americana-Cereal, Archives Center, National Museum of American History, Smithsonian Institution.

position, accused of aiding and abetting adulteration, is to go on the offensive: it was somebody else's fault. This was not just a grocers' tactic, of course. Farmers accused wholesalers, the ones they shipped their agrarian goods to in the city; wholesalers accused retailers, the ones who stocked wares in parcels on the shelves of urban markets; retailers looked both ways, blaming their suppliers and blaming the customers. Soon enough, processors would advertise the purity of their products right out of the factory door, as with Quaker Oats waving the flag of purity as its shipment leaves the station (fig. 3.6). At the end of the line, where Richards's readers lived, customers blamed everyone.

Grocers were in a particularly tight spot. They sought sympathy and protested innocence, acting as part of that telephone game of muddled communication between one side of a life cycle chain and another. The grocer had to assuage consumers' fears that his wares were unhealthy while keeping up viable connections with wholesale suppliers who were increasingly offering packaged and processed goods to the retailer.[44] It was "fashionable to blame the Retail Grocer for all the iniquities of adulteration," wrote *The Grocer's Companion and Merchant's Handbook*, "regardless of the fact that, in many cases, the grower, the merchant, or some one of the many intermediate parties having an interest in the product, are much more frequently the guilty parties."[45]

A handbook from the New England Grocer Office repeated and extended a line from Artemus Ward to say that "the retail grocer is never an advocate of adulteration, though by competition and prolonged rivalry he may be compelled unwillingly to sanction it."[46] (Plagiarism was apparently not considered a sign of dishonesty; I've seen that "grocer is never an advocate of adulteration" line in a half dozen manuals from the era.) The fault was not with the shopkeeper but the result of the consequences of consumer demand and manufacturing success. "The great competition in trade has led manufacturers to fall in prices," the grocer would say, "and to do this they must sell impure materials."[47] They made me do it. I had no choice.

Retail grocers sold directly to customers. Wholesalers dealt with distributors from across regions to sell wares in bulk to the retailers. The wholesalers co-opted the retailer's rhetoric of honesty in their own publicity. New York's *American Grocer* was probably the most prominent organ for those claims. Business magnate Francis B. Thurber published the paper, and stalwart retailer and industry spokesman Francis Barrett edited it.[48] Thurber and his paper supported the new grocers' union and their midtown Union Hall, endorsing their founding enthusiastically. He spoke at their meetings to say it was high time the grocers united. He donated furniture to their library. He advertised his "pure and wholesome" goods in their minutes. He had friends in other cities—publishers of the *New England Grocer*, *Chicago Grocer*, *St. Louis Grocer*, and more—share the enthusiasm.[49]

A grocer, be it wholesaler or retailer, could always claim honesty while accusing customers of mishandling food at home. There went the blame game again. This placed the culpability at the consumer end of the life cycle. Indeed, that's where Richards aimed her work. Grocers also looked back to the land to ally themselves with the farmer, suggesting that they were in league with the virtues of the agrarian community. The New York union sent delegates to an agricultural convention in Philadelphia in 1886, garnering praise from the producer class. Lauded for "doing all in their power to bring about a higher standard of food products and [educating] the consumer to reject adulterated goods," the conventioneers resolved "that the most grateful thanks of this convention are due to the Retail Grocers' Union of New York City."[50]

Granted, others in the food chain saw it the opposite way. Rather than noble farmers, the countryside was rife with growers and producers bent on duping gullible city folk. A look to the rural countryside perpetuated an age-old country-city dynamic. Figure 3.7, below, does just that, placing the blame on the rural farmer. Here a wise cosmopolitan calls out the dubious country bumpkin. The bespectacled city boarder, reading a very important book, barely peeks over his lenses to cast a sarcastic aside. With rain falling from the roof into milk jugs, the farmer has "increased [his] crop of milk to nearly double what it was before rain commenced." The image pays homage to the idea that farmers' milk was cut to dupe the urban consumer. Again, something like the cultural worry over character may have come across as an abstract social principle at times, but with specific fears over foods and milk, the question of character was direct and worrisome.

To say more about the milk case, by the mid-nineteenth century new

MILK
F.M. HOWARTH.
THOSE FERTILIZING RAINS.
FARMER—"I tell you, though, these ere rains do make the craps grow."
CITY BOARDER—"Yes; I just noticed that it has increased your crop of milk to nearly double what it was before the rain commenced."

FIGURE 3.7. City Boarder in the countryside drolly calling out his farmer host for watering down milk, making it clear he is dubious that country milk is necessarily pure. Courtesy of the Warshaw Collection of Business Americana-Cartoons, Archives Center, National Museum of American History, Smithsonian Institution.

PURE COUNTRY MILK.
UNDERHILL
"PURE COUNTRY MILK." FILLING THE CANS AT THE STABLES CONNECTED WITH HUSTED'S DISTILLERY, BROOKLYN.—FROM A SKETCH BY OUR OWN ARTIST.

FIGURE 3.8. Woodcut about swill milk controversies from *Frank Leslie's Illustrated Weekly*, May 8, 1858. The image promotes the view that pure country milk was tainted by corrupt city stable owners.

rail lines and quicker routes between town and country led to an ideal of "pure country milk." By the era of adulteration, urban dairymen knew to appeal to the gold standard of rural milk as healthy milk. Thus, while figure 3.7 suggested that honesty and integrity lay with the urban cosmopolitan, figure 3.8 proposed the opposite, that if you wanted the real thing you should escape the charlatans of urban streets.

Rampant metropolitan controversies over "swill milk" made this clear at the time. Swill was "a combination of processed corn, barley, and rye malt left over from the distilling of liquors such as whiskey." In what you might call the wise use of waste, distilleries fed the swill to cows. Which one was the real thing? Historian Catherine McNeur writes that "The vendors' carts and the groceries that advertised 'Orange County Milk,' 'Westchester Milk,' or 'Pure Country Milk' might actually have been selling milk from the distilleries on 16th Street or 39th Street, or from sister cities across the rivers."[51] Grocers watched from the middle of the spectrum, bemused that they could push the fault of adulteration away from their storefronts to almost any other place.[52]

Swill milk was just one case. For all of them—butter, oils, sugars, and more, as subsequent chapters explore—the response exacerbated preexisting modes of distrust between the city, the country, and the grocer in between. That's because buying food in the late nineteenth century was fraught with tension, distrust, and unfamiliarity. The struggles over adulteration were amplified in that era because the norms governing interactions between customers handing money over a counter were changing too rapidly for those customers to maintain. Frazier Paige's seven generations of interaction in rural Hardwick built a durable kind of transaction. It was a form of trade that Ellen Richards's readers would not have experienced as recent immigrants in Boston and other metropolitan areas.

The striking literary imagery of Melville, Twain, and other authors evoked moral and cultural dynamics of distrust and their corollary, angst for authenticity, as the backdrop to Richards's work and the rise of the grocers' empire. But while the literary diagnoses of the problem worked at a grand and perhaps conceptual level, the world of the everyday consumer found the same difficulties at the direct and concrete reference point of the grocer's counter. Ellen Richards operated in that forum as a trained chemist, a leader in domestic economy, and a voice fully aware that verifying the truth of one's food was part of a battle for verifying the health of the nation. The urban grocer sat as a counterpole to the rural general store. The fight for purity took hold in those urban centers precisely because of their tangled relationship with a fading rural landscape of idealized virtue and integrity.

The grocers' trade reflected a late-century growth in urban development that anchored an evolving consumer experience for housewives, cooks, and metropolitan eaters. In that manner it replicated a longer-standing forum for agricultural and food issues, the rural newspaper. Back in Hardwick, Frazier Paige read one of the oldest ones in the nation, the *New England Farmer and Horticultural Register*, a weekly published in Boston. Rural newspaper editors often pitched their pages to hearth and home, overlapping in content and tone with domestic-science texts and guidebooks on household management.

One of these was the *Rural New Yorker*. The paper's editors directed their commentary to the country home with reports on rural virtue and granger politics. Every few issues they ran a feature called "The Rural New Yorker Eye Opener." It included a header image of a small lamplight shining on a many-winged devil scurrying away. On the wings were written "frauds," "impostors," "humbugs," and "cheats." The paper sought to expose swindles to protect the farming class. If urban grocers would blame the customers after them or the suppliers before them, rural newspapers would blame the manufacturers and capitalists beyond the farm. At a wider level, these sentiments provided the seedbed for a Granger Movement and later populist politics in the decades after Reconstruction. At the more circumscribed level attentive to pure foods, farm advocates argued that "the monster" of "food frauds" was being "pushed on by those who have grown rich and powerful by handling and delivering" adulterants.[53]

The editors of the *Rural New Yorker* commissioned artist August Berghaus to run a series of illustrations to forward that point in 1887. He produced a twelve-week run. A panel in each issue spoke to contrasts between agrarian and urban classes. Some were paeans to the Granger Movement, offering the classic country versus city analysis of integrity (country) and duplicity (city) that came out in the swill milk debates. Some were more nuanced commentaries on bankers, farmers, and the corrupt politicians who irritated them both. All the illustrations were predicated on the view that rural virtue and honesty were under attack by manufactured sincerity at factories and in cities.

The eighth entry was titled "At Bay."[54] It showed the hydra-headed monster of adulteration (fig. 3.9). It played to the stereotype of the robber baron manufacturer bent on demolishing agrarian virtue. The scene shows a three-headed snake spurred on with spears by seven capitalists in stovepipe hats. The monster rears its heads against the farmer wielding a bayoneted rifle with "ballot" written on the blade. A cow and pig race behind him and away from the monster, driven out of the country by the principal scourges of the day: oleomargarine, cottonseed oil, and glucose (fake sugar).

Among extensive lists of accused adulterants and antiadulteration texts accounting for hundreds, the hydra-headed trinity came into the spotlight in the press. An 1884 summary of the Massachusetts board of health's recent report—this was the board to which Richards contributed—found that "of twenty-one [butter] samples, four were condemned," but the others were "more or less adulterated with oleomargarine"; honey analyses revealed that most of the "commerce consists almost entirely of glucose"; and "of 49 bottles of olive oil, nearly all of which bore foreign labels, genuine or counterfeit, 32 contained cottonseed oil."[55] Three years later, reporters came after Philip Armour, "the great pork packer" in Chicago, for his investment in a cottonseed oil manufacturer. It was "as great a swindle to deceive consumers by selling cottonseed oil for cheese or lard as it is to sell oleomargarine for butter." The reporter hoped that revealing the meatpacker's oil interest "may direct attention to other products that are manufactured in large quantities, but are not sold in the open market for what they are," such as fake sugar: "Like the manufacture of oleomargarine, the manufacture of glucose lives by fraud." The *New York Daily Tribune*

FIGURE 3.9. August Berghaus's hydra-headed monster of adulteration, from *Rural New Yorker*, May 1887.

rebuked grocery magnate Francis Thurber's antimonopoly politics by calling his new party "the Anti-Monopoly, Oleomargarine, Glucose and Hash party." Their reasoning was that Thurber's party "pretends to be one thing and is really another." Elsewhere, a quick satire, "The Grocery Order of the Future," mocked the trinity. A housewife asks the "grocer's boy" for "a can of glucose" among lists of "veritable poisons." When told they had no glucose, she concedes, "Well, cottonseed oil will do, and one pound of margarine."[56]

I take the three heads of the monster as case studies in the next three chapters. This and the previous chapters explored the legacy of trust, deception, and confidence to understand where pure food debates came from while examining changing food distribution, country and urban demographic changes, and the insistence of the grocer as a new mediator in the farm-to-fork chain. The next three explain where so-called adulterants came from in the first place.

II

The Geography of Adulteration

✷ 4 ✷
Margarine in a Dairy World

I trusted thee, and now, with pain,
I see how lost my trust has been;
I've been deceived—but never again,
My own, my Oleomargarine!

Puck, 1880

Why should not oleomargarine tell the truth? Why should it be allowed to lie itself into my stomach?

SENATOR BLAIR, 1886

I prefer butter to margarine, because I trust cows more than I trust chemists.

ATTRIBUTED TO NUTRITIONIST JOAN GUSSOW, 1975

Margarine is not butter. I learned that only accidentally as a child. I grew up in a Jewish household, although we didn't keep kosher in any systematic way. One time we had a guest over for a steak dinner. I was probably eleven. My father asked me to grab the margarine from the fridge, I grabbed the butter instead, and I was reprimanded for bringing the wrong dish. I protested, but it's the same thing? I was then told that it was, indeed, not. Margarine was not made from milk and so it was appropriate for steak dinners. Not only did I learn that margarine was *not*, as I had previously thought, a synonym for butter; I also learned that when we had guests for dinner apparently we were kosher.

Had I been quicker on my feet (thirty years quicker), I would have commented stoically that I had the original formulation of oleomargarine in mind. Producers first made it by substituting a portion of milk with melted animal fats. It was certainly not kosher. The earliest patents for oleomargarine, artificial butter, or butterine—people often used the terms interchangeably for the first few decades of the product's use—all emphasized the incremental change over common butter in their legal cases for innovation. If butter was natural, they argued, and it was a product made from manipulating cow's milk, so was margarine. This is quite like arguments today that genetically modified organisms (GMOs) offer differences in degree, not kind. For oleo advocates, these were "improvements" to the butter process; they were in the category of animal fats research. They invariably included a recipe that reduced milk use by mixing in boiled and refined suet (cow fat) or lard (pig fat). So actually, at its inception this was probably the least kosher thing imaginable.

Recall that our rogue from the introduction, Alfred Paraf, stole the patent for oleomargarine in the early 1870s from its inventor, the French chemist Hippolyte Mège-Mouriès. The *oleo* came from the Latin *oleum*, meaning oil. It was dropped for the most part in common parlance by the later twentieth century. The word *margarine* is derived from the Greek *margaratis* and referred to the "pearl-like" appearance of the oil expressed from the animal fat. (Chemists originally thought animal fats were made up of margarine, oleine, and stearine.) Mège-Mouriès intended the pearly reference to give it a kind of public luster.

Margarine was the most famous, most enduring, and most legislated adulterant of the era.[1] The first factories producing it came on line by 1873; within a decade there were margarine manufacturers in a dozen countries. The product was also quickly maligned, banned, or taxed in locales around the world. The British discussed it in the House of Commons in the early 1870s, as did the German agricultural council. The Dutch Department of Foreign Affairs debated the new product as early as 1875, when the Italians also debriefed on it. The French had a special relationship with it, to be sure, given that Napoleon III had earlier decreed the search for a butter substitute. After 1870 and the end of the Second Empire, the product was a topic of consideration not just for the Third Republic's dairymen but for pharmacists, agriculturalists, chemists, and politicians alike. The Turks considered inspection codes by the early 1880s. Canada, a major dairy and butter exporter, banned oleo by 1886. The Indian forestry service reported on microscopic studies of it that same year. In Russia, when new provincial councils called *zemstvos* struck some as not quite legitimate, they called them "margarine zemstvos." Even Barbados had laws on the books to ensure their citizens could distinguish between butter and the upstart artificial butter.[2]

In the United States, public debate began with Paraf's exploits in the mid-1870s and escalated to 1886, a date that looms large in the nineteenth-century fervor over the pure and the adulterated. Senate hearings on "Imitation Dairy Products" stewarded by New York's Warner Miller and parallel testimony in the House of Representatives led that year to the passage of the first federal antiadulteration law targeted at a specific commodity. It would be a launching pad for the next twenty years of agitation for addressing the confusing character of artifice, nature, and food.[3]

Attention to the new fake butter was intense. When the Miller hearings took place in the spring of 1886, President Cleveland was preparing his June wedding in the White House just down Pennsylvania Avenue, the Statue of Liberty sat in the harbor in Miller's home state on its way to opening that fall, and the Brooklyn Bridge stood proudly as a three-year-old. Benz patented an engine in Germany, the French were in Vietnam, the Belgians were in the Congo, the British were in India, and Geronimo surrendered in Arizona. As Senator Miller began the testimony, a monumental labor action, the Haymarket Affair, was brewing only to erupt in a week. The austere dairy hearings convened in Washington while in Chicago a largely immigrant population with leadership from unions and anarchists was about to demand an eight-hour workday and justice in labor practices across the country. Chicago's Haymarket Square, near the home of McCormick Harvesters and the Union Stock Yards—a source for much of the suet used in oleomargarine—hosted the violent protests that erupted on May 4. The strikes eventually led to the creation of a Labor Day holiday and changes in workplace practices, though if Upton Sinclair and Jane Addams and Jacob Riis were any witnesses, those practices were hardly more acceptable twenty years later. All of that and margarine took priority over other pressing national agenda items that spring.[4] A not insignificant part of the United States government was hunkered down on Capitol Hill arguing vociferously about cows, milk, and butter. Dairy was a big deal. Margarine was a big problem.

The debate over margarine's identity captured in the Miller hearings included the very question of whether or not it should even be called an adulterant. Was it not just an improvement over common practice? Was it not, as its advocates claimed, a product of innovation that fit snugly inside the Age of Edison? Allies considered it "one of the greatest boons that modern science has wrought for the benefit of the poorer classes." Some applauded it as "a blessing to the public."[5]

Others saw it differently. A typical view cast the story as a simple contrast between a dairy lobby back on its heels and an urban capitalist class charging into modernity. Berghaus's hydra-headed monster took that view. Notice that oleomargarine's neck is bandaged in the 1887 cartoon. The wound is a reference to the sharp slice of dairy-backed 1886 legislation. *The Rural New Yorker* may have been a bit on the nose, but they were no less subtle than comments elsewhere. The secretary of the Chicago

Produce Exchange, for example, called the issue at hand "a fight between 7,500,000 agriculturists and a score and a half of capitalists . . . who would see the world sink if they could get a golden canoe to float to heaven in."[6] Another dairyman laid bare the moral context, using his testimony to castigate margarine makers as the "disgrace of mankind . . . the iniquity of business [and] the train of evils."[7] Margarine was "like a villain with a smiling cheek," said Rep. Hopkins, quoting *The Merchant of Venice*, to swipe at the false face value of the artificial product.[8]

Was oleomargarine a con? Along with its monstrous brethren cottonseed oil and glucose, oleomargarine challenged norms of dairy production and offended codes of proper agrarian conduct. Cottonseed oil (chap. 5) offended customs of the food marketplace because producers surreptitiously slipped it into other common products. Glucose (chap. 6) offended pure food advocates for its outsized industrial scale and seemingly secret process. Margarine was offensive and controversial from the start because it competed with prevailing agrarian practices. It was born into a world of dairy prominence, a product of changing environmental relationships between cows, dairies, farms, slaughterhouses, grocers, kitchens, and a new industrial order. Making things worse, both sides of the presumed divide, butter and margarine, were arguing about different things because they were framed within different food geographies. Historian Kendra Smith-Howard writes astutely of commonalities, that "both butter and margarine were mixed products of environmental forces, technological interventions, and social mores." It was the differences between environmental forces, technological interventions, and social mores that led to the rancor. Exploring those differences helps show how my confusion between butter and margarine more than a century later was created at the font of our still-current age of manufactured food. The environmental geographies of dairy provided the proximate backdrop against which the debate over margarine's validity took place. I'll get back to the self-proclaimed war on oleomargarine later in the chapter. First, though, the question on the table was where did all that margarine come from?[9]

An Ecology of Butter

Talking about margarine means talking about butter, and talking about butter means talking about cows. The butter empire was huge because the dairy industry was huge. In the year after oleomargarine's invention, 1870, there were close to ten million milk cows and thirty-one million cows total in the United States. This was for a population of thirty-eight million people. Thirty-one million cows; thirty-eight million people. With about five people per family at the time, that means there were one and half milk cows and, overall, four cows for every family in the country. That ratio was about the same at the end of the century. In 1900, there were sixty million cows and seventy-six million people, also averaging out at four cows per family. Cows were ubiquitous, a visible part of daily life. They were in backyards, small towns, and rural pasturelands. They were even on college campuses—in the late 1800s faculty at my college had the privilege of grazing their cow on the quad. Like hogs, sheep, chickens, horses, and mules, cows were part of the landscape of everyday life.[10]

This wasn't only a rural picture. Livestock were common in cities throughout the nineteenth century too. The swine of Manhattan ran unfettered until later decades in the century; Atlanta was until the 1880s a commons for people and cows; residents in the metropolises of London, Paris, and Edinburgh all walked among urban livestock. Baltimore, Cincinnati, Chicago, Philadelphia, and San Francisco had their own statutes regulating cows within the city streets. Hogs and cows lived in urban spaces until housing development, public heath statutes, and zoning pushed them to the periphery or behind closed gates.[11]

Where there were cows there was milk, cheese, meat, and butter.[12] Butter was a staple. It was a key ingredient in kitchen recipes, common edible grease for cooking, a product of sustenance accompanying other foods (e.g., buttered bread), and a regular part of culinary life. When grade school primers needed math examples, they would use butter as an example—"If 2 pounds of butter cost 50 cents, what will 8 pounds cost?"[13] The United States produced over a half *billion* pounds of butter in 1870. In 1880, it was 800 million; by 1900 it was almost twice that at 1.5 billion.

Butter production had been concentrated in the Northeast for much of the nineteenth century. Upstate New York was a leader in production for decades. By the 1880s this was shifting under the Great Lakes and into the Midwest in a kind of dairy belt. A dairyman in New York defined it well: "confined principally within the latitudes of 41 and 44," the belt extends "from 125 to 150 miles in width and is about 1800 miles long."[14] Figure 4.1 shows the environmental migration across that swath in the second half of the century. It was as if someone moved an entire region like a checker on a checkerboard, sliding it under the Great Lakes and across the continent with Manifest Destiny.[15] In 1880, New York, Pennsylvania, Ohio, Illinois, and Iowa were the top five producers. The famed cheese- and butter-making prowess of the Midwest, especially Wisconsin, was yet to arrive when margarine made its first appearance. By 1900, midwestern states were overtaking the Northeast.[16]

So much of the nation's butter came from household production and

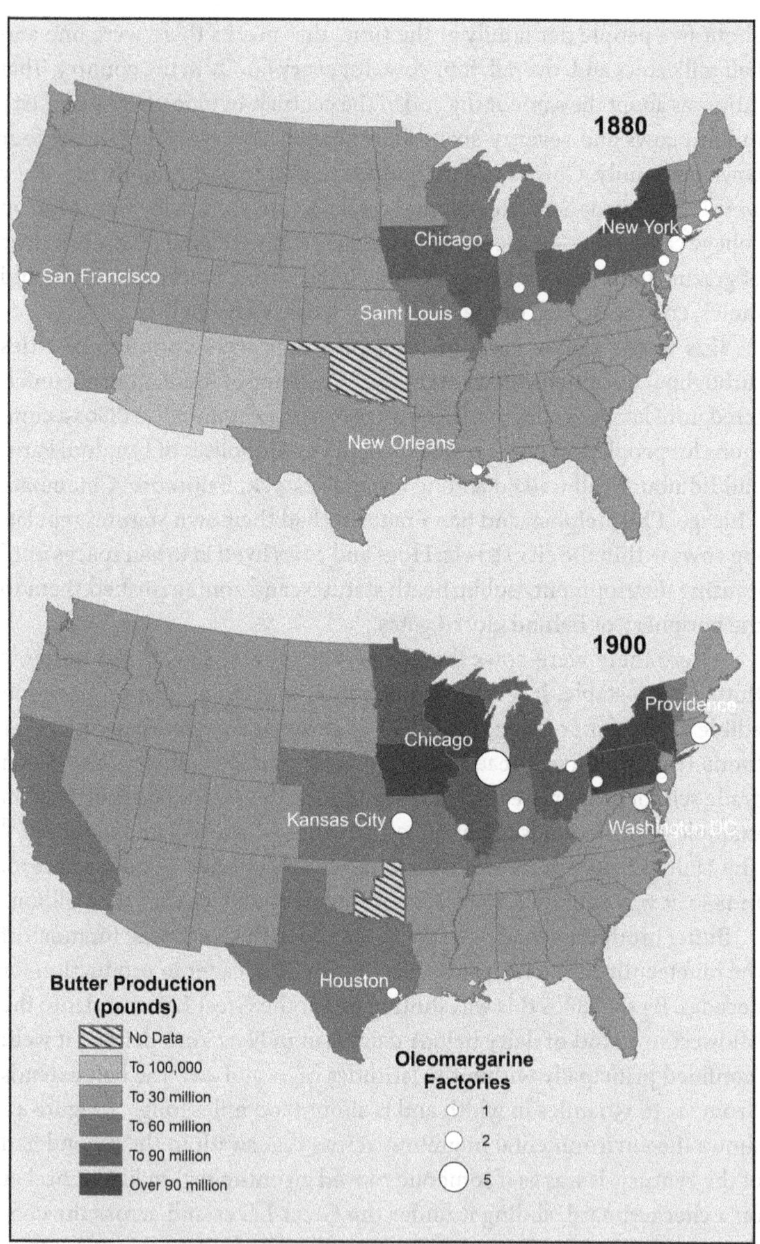

FIGURE 4.1. Butter production by states, 1880–1900. The darker shades in the upper Midwest reveal a changing intensity of production and the shift in concentration from the eastern side of the Great Lakes to the western side. As well, oleomargarine factories that dotted the east coast gain prominence in the Midwest near packinghouses by the end of the century. The online digital companion site at https://purefood.lafayette.edu includes a fuller set of maps of this transition, showing changes by decade from 1870 to 1900. Maps produced by John Clarke.

local creameries—like the one Frazier Paige knew in Hardwick—that we have to give the government credit for tabulating what it did. That got easier after the 1880s, when dairymen started to produce butter in "manufacturers." Like Midwestern elevators that collected grain, the butter manufacturers were often small factories on the edge of town aggregating the dairyman's supplies. Butter factories slowly developed in the final decades of the century, moving the locus of churning out of homes and into workshops. Iowans, for example, founded their first butter factories in the 1870s; Minnesota did, too, with its first one in 1877.[17]

There was a lot at stake for an industry that tripled its production in three decades to well over a billion pounds a year. It stood to reason that dairymen would jockey for market share. A fervor for butter improvements and patents followed in kind. There was even (and still is) an entire subclass of inventions in the US patent office called "Renovated Butter," Classification 426/530. *Renovate* is used here in its older sense: to refresh or reinvigorate, to revive.

The late 1800s were rife with butter improvement schemes. In 1869, a dairyman in Illinois patented a method for a "new and useful improvement in a process of purifying and preserving butter." A gentleman in Iowa invented a process for "the full recovery of pure butter-globules" from "old and 'tubby' butter." George Kirchhoffer patented a new technique on "Improvement in Refining Butter" in 1873, for butter is graded, like milk, like eggs, like an exam. Kirchhoffer's process could improve an "inferior grade" to a "good merchantable" one. By definition, this was a difference in degree, not in kind. Kirchhoffer framed the entire process as adding onto a normal activity "done in many households for cooking purposes."[18]

Butter color offered another way to gain competitive advantage (fig. 4.2). Colorants were an established element of the market. Mège was working as an animal researcher and dabbling with dyes before he invented margarine. Paraf, it's worth remembering, came from a dye-making family.[19] Like the butter improvement innovators, agrarians saw colorants as part of proper dairying, securing the view that the environmental ethic of dairying was one of intervention. They, too, had a patent class: 426/262.

The butter colorant industry responded to environmental variety in butter quality and character. As cookbook writer Amelia Simmons, dairymaids, and dairymen well knew, milk's color varied depending on the cow, the season, the grass, the farm, the feed, and the care. So, too, for the butter produced from that milk. A chemist in New York, Garret Cosine, thought those differences were based on variations in levels of oleine in cow fat. He earned a patent in 1875 that claimed to improve butter. "It [is] a fact," he wrote, with the then accepted view that "margarine" was one of

FIGURE 4.2. Readers will recognize the bucolic imagery used in early ads from common advertising iconography today, with images focusing on pastoral scenes, cows, and rural settings. Bean & Perry's is from the 1890s. Alderney is from 1902. Courtesy of the Warshaw Collection of Business Americana-Dairy, Archives Center, National Museum of American History, Smithsonian Institution.

ALDERNEY
Butter
Color
We are satisfied that
ALDERNEY
THE ACME OF PERFECTION
represents the highest standard of Butter Color.
We heartily recommend it to our patrons.
THE DAIRYMEN'S SUPPLY CO.,
1937 MARKET STREET,
PHILADELPHIA, PA.

FIGURE 4.2 continued. Although rendered in black and white here, both images were colorful and bright in their original form. Courtesy of the Warshaw Collection of Business Americana-Dairy, Archives Center, National Museum of American History, Smithsonian Institution.

the three principal fats in milk, "that butter made from cow's milk does not contain the same proportion of oleine and margarine in summer as in winter, it having a larger portion of oleine in the winter."[20] Nature, the dairymen well knew, had variation. Colorants were an acceptable intervention to iron out the wrinkles of the seasons.

There was a gold standard for color, "butter yellow," what many people today think of as a preordained definition. In the nineteenth century, that standard was more specifically "June" yellow. One guide, *First Lessons in Dairying*, took it as a commonplace to instruct young readers that "The general market requires that butter be as nearly the June color as possible throughout the year."[21] Organic plant extracts from annatto (seeds harvested from the South American achiote tree), marigold, safflower, saffron, and turmeric were most effective in building June yellow. "Twelve to fifteen drops of color for each gallon of cream that will churn out two and a half pounds of butter will be about right in the fall and winter," the guide said, "while less may be required during spring and early summer."[22] Or, "sometimes the dairymen pursue a little different plan by applying the color through the cow. They do it by feeding the cow on carrots or some other colored matter."[23] There were two ways to get the right color, in other words, either by adding an extract to the butter churner or by feeding the cow foods that would tint their milk. Color additives introduced a point that would become contested when tinted margarine presented a new form of competition: was adding colorant to butter a sign of adulteration? Before that question became controversial, the addition of color was part of a dairying practice people considered natural.

For butter churners, making a pure product followed from seeing the whole dairy as of a piece, from the season to the cow's diet to the proper organic plant extract. Dairymen understood well that superior butter was downstream from the milk of superior cows. They knew, too, that superior cows were part of healthy relationships with the land. Its process, not just the yellow-bricked product that resulted from it, defined butter and its maker's claims for purity.

Butter was about cows and milk. It was also about land use and farming patterns. Better cows were the result of better agrarian practices. Better grass made for better health, better milk, and better butter. It is important to examine dairy ecology because arguments for butter purity were connected to claims about the right way to make butter. Like the coloring method, this was as much about process as it was about product. We need to understand the geography of dairy lands to understand the vision of butter purity against which margarine as impurity was set.

That vision followed from competition in the butter industry that came from regions that were more effective at integrating the whole ecology of the dairy. Grasses and cows worked in tandem. As Rep. Scott of Pennsylvania reminded his colleagues during artificial butter hearings, "the cereal and dairy branches [are always] moving hand in hand." His colleague from Iowa, Rep. Frederick, amplified the view that "much of the wealth of this continent [is] in the growing of the grain [and] raising of cattle."[24] Expansion into the Midwest, upper Mississippi Valley, and plains states had opened up new growing patterns. Farmers put "cereal" lands under increased cultivation, especially in states such as Illinois, Iowa, Wisconsin, and Minnesota. Frederick admitted a worry that came from this ecological view that "if the farmer is stripped of his direct profits derived from the products of his dairy he must part with the cattle, and thus lose the incidental advantage that accrues from their natural contributions to the soil."

He was saying that farmers had cows because cows fertilized fields. The profit margins of wheat were too thin, as they remain today, so that milk, cheese, and butter—value-added products—kept the farm running. Without the sales of milk, cheese, and butter, farmers would not be able to keep planting the fields.

Farmers knew "our butter, our cheese, our milk, our beef, and our mutton" all formed a kind of mutual interdependence. A gentleman from New York stated that "an injury to either one of these products affects in some degree the whole." Grass and wheat strategies reflected this awareness. The best wheat states were becoming excellent dairy states because the excellent feed they provided cows produced excellent milk for people who produced excellent cheese and butter. To be sure, the New York farmer testified, "the production of butter is an adjunct to the crop of grass produced, one of the most important crops."[25] Historian James Young captured a congressional exchange about this: "A number of... congressman argued the crucial importance of manure from the dairy herd as fertilizer for the fields of grain, although the word 'manure' was not permitted to besmirch the Congressional record, euphemisms always being employed."[26] This was to say that although it was fine to castigate opponents with the foulest accusations of moral depravity—a "disgrace of mankind"—you couldn't say *shit* in Congress.

Pure butter at the store was a result of effective agricultural activities by those dealing with issues such as land-use policy and trade practices. Politicians and dairymen understood that land-use policies related to cereal and dairy lands were part of greater environmental networks. As noted, upstate New York's leadership in butter production in 1870 was ceding to states on the other side of the Great Lakes by the 1880s. This was at least partly because "the wheats [ran out] in the Genesee Valley some years ago." That change led to increased dairying at first and then increased motivation to develop new wheat and dairy lands in the upper Midwest.[27] One representative pointed to the "advantages that nature has provided" with the watering of wheat lands in Minnesota. He then emphasized and cautioned against the advantages of the watershed in Canada, fed by "three large lakes," Winnipeg, Manitoba, and Winnebago. The lake lands were fertile grasslands, leading to healthy, thriving cows and superior milk and butter. We see vestiges of those watershed connections on store shelves still—the "Land O'Lakes" brand of butter that I pulled from my refrigerator when I was a child was well named.

The dairy industry in the late nineteenth century benefited from a host of integrated practices. Patented innovations were one; a thriving colorant additive industry was another; and healthier cows eating richer feed from better-irrigated grasslands were another. Together they helped form the fabric of an agriculturally and economically important butter market. And then global competition and foreign pressure factored in.

Canadian agriculture was expanding in strength and adding competition to the global grain and butter markets. Manitoban lakes could take credit for such advances. In the least, they influenced the decisions of the American farming class and its lobbyists. With the mind-set that "an injury to either one of these products affects in some degree the whole," dairy supporters knew the equation: increased Canadian wheat farming meant increased dairying, which meant more butter and more competition. Statistics bore this out. Canadians increased butter exports by twenty-seven million pounds between 1871 and 1881. By 1882, the poundage of US butter exports had dropped from its peak of thirty-nine million down to about fifteen million, nearly the same amount added by production in Canada.[28] As one spokesman noted at the time, "a large supply [of butter] from Canada has doubtless interfered seriously with our exportations."[29]

Not even considering the margarine—coming soon—that would make all of these competitive problems worse, the ambitions of transatlantic exports also bound local dairy farmers in the upper Midwest of the United States with market forces abroad and financial structures of British colonialism. This happened in one way because changes in Canadian butter production were financed and encouraged by British banking houses intent on capitalizing on new wheat lands. In another way, the British also bought up American interests in the upper Mississippi Valley. Pillsbury is a case in point. John Sergeant Pillsbury moved from New Hampshire to Minnesota in the early 1870s anticipating the westerly shift of agricultural activity seen in figure 4.1. There he opened a flour mill that by the early 1880s was the largest in the world. His product was ubiquitous enough to make it to Hardwick's General Store. In 1889, British bankers bought him and his sons out, putting the vast flour processor into the orbit of British syndicates. For the benefit of the British Empire, the investors sought the farmlands in order to feed the British working class and bolster the capitalist linkages that would return profit to the banking class. This point was not lost on American politicians and lobbyists who saw that the Canadian threat to the US dairy was a subtle capital fight between the British Empire and US aspirations for export possibilities.[30]

All of a sudden, questions about butter propriety were connected to matters of global imperialism. Even the recently opened Suez Canal became a subject of discussion concerning dairy products, as if things were not cumbersome enough for policy makers. "The construction of the canal has revolutionized the maritime trade of the world," Rep. Scott told the House. The move by the British "brought the wheat fields and the cheap labor of the Indies into serious competition with the enterprise of our country."[31] A hit in the ability to sell wheat on global markets, Scott and colleagues understood, meant pressure on farmers to keep their lands profitably under cultivation. And that meant fewer cows and thus pressure on the butter market. A Memphis paper had reported before 1886 that because of the Suez Canal, "the imports of Indian wheat into great Britain increased from 291,200 bushels in 1872 . . . to 9,283,130 during the first nine months of 1877." Politicians took note. As the paper dryly put it, "There is matter here for American reflection." The Chicago Board of Trade, among others, did reflect. They kept their eyes on the canal's operations, hoping for closures when there was unrest amid the British Empire in the 1880s in Egypt, and speculating this would help their wheat and butter exports.[32]

Those were the circumstances in which margarine made its entrance onto the global stage. The elaborate geography of cows, milk, butter, and the dairy industry provided the background against which alternative butter arrived. Dairymaids, dairymen, creameries, and factories were already engaged in competitive alignments that were about more than butter alone. Butter and dairies were big business and successful sectors supporting thriving livelihoods. They had long been engaged in arguments over dairy quality and claims for purity within the patterns of common agrarian activity, patterns that defined the contours of trust and familiarity. Dairymen and their lobbyists objecting to the new upstart, margarine, were also dairymen aware of their participation in political agendas for global trade and continental expansion. Their eye to Canadian and Atlantic trade routes; their worry over water, grasses, and manure; and their attention to measures of improvement and the shades of seasonal variation were all nods to a fight both local and global. In this case, an argument over pure butter—properly produced butter—was part of arguments over fair competition in a global marketplace and a fight about maintaining the stable geography of dairy lands.

A Geography of Margarine

And then came margarine. What of this reviled upstart? In a competitive world of butter production full of purifying patents, renovation, colorants, and grass-fed cows from lush lake lands, margarine challenged the conventions of an agrarian world. If it competed with butter only as a kitchen staple, market mechanisms and consumer choice might have settled out the controversies. But margarine's presence was more problematic. It was a butter disruptor. It unsettled common means of agricultural production, upsetting the ways dairies, cows, pigs, grains, and landscapes interacted. Butter and margarine were analogs, though the era of margarine began not on farms but with slaughterhouse interests organizing in cities. Artificial butter was still an agrarian product but also a product of a new urban world that brought cows and milk together in different ways. The view from the oleo maker's side of things shows how and why we got it.

New York's Oleomargarine Manufacturing Company, founded 1873, was the first artificial butter factory in the United States. It was the result of Paraf's instigation and his investors' interest. Once shareholders detected Paraf's fraud, Henry Mott became the chief chemist of a new concern, the United States Dairy Company, which operated in Brooklyn. The French had already begun production at that point. By the mid-1870s, there were seven factories around Paris employing four hundred people to manufacture what they called "Mège's product."[33] Mège also sold his patent in 1871 to the Jurgens family of Oss, in The Netherlands. Jurgens would become the principal producer outside the United States, competing with New York and Chicago agents to export oleomargarine to the United Kingdom and other European nations.

Oleomargarine production increased rapidly in its first decade along with a confusing proliferation of synonyms. Mège at first considered oleomargarine the combination of two of the three oils that make up cow fat. Chemists soon understood from further analysis that margarine and olein were not distinct chemical compounds, but the term *oleomargarine* stuck in the United States as a general name. The English used the name *butterine* more frequently. Some US observers defined butterine as artificial butter that came from hogs' fat (lard), whereas oleomargarine was from cow fat (suet).[34] Both Swift and Armour promoted that distinction. Swift had a butterine building near his hog slaughterhouse and a seven-story oleomargarine building by the cattle slaughterhouse (fig. 4.4). Others, though, considered that "the two names are used in this country synonymously."[35] Sometimes the cow-derived version was "suine," named for suet. Others called it "Bosch" or "Bosh," a reference to Hertogenbosch, the capital of

North Brabant in the Netherlands, where low-grade butter had long been sourced. (This was adjacent to the Jurgens firm in Oss.) Because people are handy with nicknames, the product also came to be called "bull butter" as a reference to its meatpacking derivation and, for those not worried about concealing their distaste, "bogus butter" or "sham butter" or even "bastard butter."[36] We have, then, artificial butter, imitation butter, oleomargarine, butterine, suine, bosch, bosh, bull, bogus, and bastard butter. In any case, after the 1886 Oleo Act, the general term *oleomargarine* mostly took hold until the *oleo-* prefix eventually fell off some time in the twentieth century.

After 1873, the factories grew as prolifically as the diversity of names. There were fifteen in the United States across thirteen states by 1880 (fig. 4.1, above, shows these). A decade later there were thirty-six, ten of which were in Chicago and six of which were in New York and Brooklyn. Paris alone had ten by then. There were many more "at or near the abattoirs of... Vienna and Munich." In 1886, the New Jersey dairy commissioner reported that there were "hundreds of factories in Holland, a country about one-half the size of Ohio," most of them under the umbrella of Jurgens's patent rights.[37]

Artificial butter from these factories was the result of a more elaborate process than butter making. Mège had crafted a nine-step process for his invention. He did so while working alongside another animal chemist in France at the time, Louis Pasteur. Pasteur's early research had dealt with anthrax vaccines for sheep as a means to improve the economic core of the French Republic, its livestock, animal herds, and agricultural health. Mège's work was similarly dedicated to the health of French livestock and people. His oleo process recovered previously unused animal fats and reduced the draw on milk, anticipating the efficiency mantra of the Progressive Era soon honed by American meat-packers.

Mège's process included a series of simulated digestions of animal fat to achieve the same effect as the cow's natural activity. He cut away the fat from a dead cow, softened it, and hashed it with blades ("macerated" it) over specific heat ranges. Later, the Chicago houses specified that manufacturers use the "caul fat of beeves" for the best artificial butter. Caul fat is the thin layer of translucent fat surrounding stomachs and intestines from beeves; *beeves* is the now little-used plural term for beef. *Wallace's Monthly* reported in 1879 that the caul fat of beeves was the exclusive source of oleomargarine. (They too defined butterine in distinction to oleo.)

Some detractors took the complexity of the process as the very basis for their disdain. Chapter 1 showed a Macbeth-like cartoon mocking this with vats, cauldrons, barrels, stirrers, and a full chemical regime. *Puck* may have exaggerated the causticness, but the cartoon captured some of the truth of the constructed system. Artificial butter pulled together milk supplies from the countryside, oils and fats from cattle in slaughterhouses and abattoirs, and a thriving craft workshop business in firkins, tubs, tierces, and barrels to hold the oil and artificial butter. Maybe not Macbeth-like, but certainly intricate.[38]

If the geography of the richest butter was grounded in grasslands and lake-fed watersheds, margarine's stirred cauldrons were concentrated in league with slaughterhouse complexes. "As will be seen by the location of these factories" reported the New Jersey dairy commissioner, the margarine producers were "as a rule, adjacent to large stock-yard centers or slaughter-house." The *New York Tribune* saw the broader geographical consequences of this margarine economy, noting that "The Eastern manufacturers cannot compete with the Western in making this compound, because the Western men have the offal and refuse near at hand." The fat was only accessible soon after slaughter, leading it to become solely the product of an urban abattoir. As the product of an urban abattoir, advocates said, it was thus fresh and presumably safe.[39]

Chicago was the flagship of margarine controversy in the United States. Its meatpacking prominence more or less made it so. A handful of major margarine manufacturers operated in the city's stockyards, the "Yards," by the 1880s, either adjacent to packinghouses or part of them (fig. 4.3). Some of these names—such as Armour and Swift—are familiar still. The other

FIGURE 4.3. Overview of the entire Swift packing house complex at Chicago's Union Stock Yards. Building 13 (on the far right of the scene) and 14 (front right corner) are oleo oil and butterine facilities. Courtesy of the American Antiquarian Society.

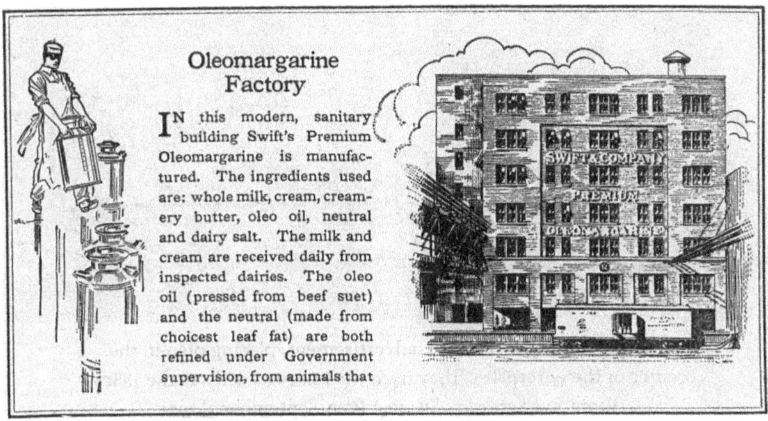

FIGURE 4.4. From a 1903 visitors guide to the Swift and Co. Packing Houses. This front elevation gives the factory a more approachable view than the birds-eye view from the entire facility of figure 4.3. Courtesy Duke Archives.

big three were the companies of George Hammond, N. K. Fairbank, and Samuel Allerton. By the end of the century they would be the Beef Trust in the public's eye. When a lawsuit challenged the meatpacking monopoly in the 1905 case *Swift & Co. v. United States,* the press had labeled them "the Big Six" or "the Big Four," depending on how you like your monopolies.[40]

The issue with meatpacking monopolies was one of scale. They were too big. They centralized too much. They had too much power. And the scale was one of networked complexity. As historian William Cronon summarized it, "Chicago livestock dealers and meat-packers established intricate new connections among grain farmers, stock raisers, and butchers, thereby creating a new corporate network that gradually seized responsibility for moving and processing animal flesh in all parts of North America."[41]

The scalar issue was important in the war on oleomargarine because, in addition to animal flesh, the dealers and packers collected, processed, and retailed the fats and bodies of those animals as by-products. Gustavus Swift's boast that he used "Everything but the squeal"—a line Sinclair would deploy in *The Jungle*—was apt. In the 1870s, the industry had developed over 120 by-products that helped facilitate its economies of scale. A survey in the early 1900s listed over 600 by-products. A new by-product industry was the eventual result of this capitalized management of scale and expanse. Margarine was one of the principal outcomes of that ethos (fig. 4.4, 4.5).[42]

Margarine wasn't just one by-product among many. It was a principal object of innovation in manufacturing. Like those for renovated butter and improved colorants, the first American patents for margarine were many.

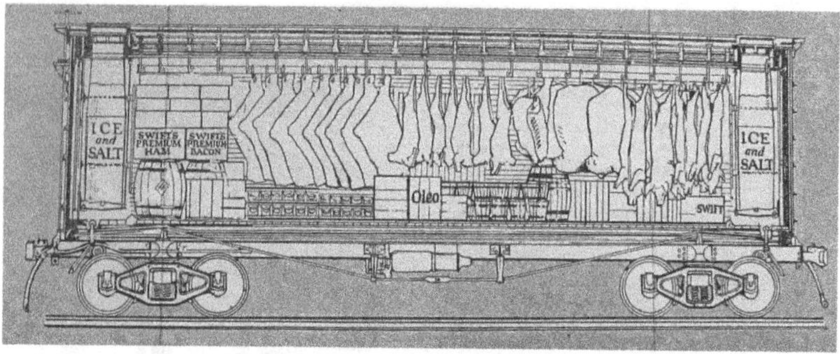

FIGURE 4.5. Swift & Co. advertisement, placing oleo in the center of the enterprise. They used the image as early as the 1880s. From the *Saturday Evening Post*, September 3, 1921.

The US patent office fielded 180 separate applications before 1886, thirty-four of which they granted. In an Edisonian age that may evoke images of electricity, factory ingenuity, and networks of railroads and petroleum industries, food innovation stood among the leaders of industrialization. This was still an agrarian world. Oleomargarine ranked behind only rubber, bicycles, photography, and glucose in the rankings of patent activity.[43]

Margarine advocates thought their innovations ran in parallel with the rationale of butter. The dairy industry considered butter improvement schemes—the patents, the colorants, the renovation—acceptable alterations to a normal, natural agricultural process. So did margarine advocates. Mège thought he was doing no more than "artificially producing the natural work which is performed by the cow." He had experimented on "milch cows" at Napoleon III's royal farm at Vincennes, leading him to describe a "natural process performed by the cow" whereby the animal "elaborat[ed] her own fat through her cellular and mammary tissues at the low temperature of the body."[44] He meant only to offer a quicker path to the process by extracting the fats before they were expressed in milk. Manufacturers like Armour took this to heart. They proudly advertised their view of a sensible intrusion into the supposed "natural" actions of the animal. "We proportion the oils instead of letting the cow do so" (fig. 4.6). He later wrote, "The animal oils are churned in milk, worked, salted, and handled precisely as butter and the product is butter made by chemical methods rather than by Nature."[45] Like the colorant industry, margarine advocates took an incrementalist view to claim that the new contrivance offered a difference in degree over butter, not in kind.

The cow wasted all that time digesting the fats before expressing—"elaborating"—milk, margarine makers thought. Why not cut to the chase, use the beef fat and get the butter faster and more efficiently. It would be artificial, they said, which they meant as a positive. It was an improvement on the inefficient pace of nature.

In fact, now that they were producing oleo oil with which to make the oleomargarine, they realized they could contribute to any number of markets. Rather than using the processed oil to make margarine, the packers could keep it in its unrefined state as barrels of raw oleo. Oils kept houses warm; they were the raw material for candles to keep homes lit; when made into soap they could keep people clean. To its advocates, it seemed reasonable and no less natural to use oleo like any other animal fat or oil. Armour sold "oleo oil" to thirty-eight separate creameries in the surrounding Chicago region in the mid-1880s for the explicit purpose of producing "butterine." Competitors like Swift, Hammond, Fairbank, and Allerton sold oleo to soap and candle makers too.[46]

Unfinished oleo flowed in impressive quantities from the American Midwest and eastern port cities to Europe and the world. Environmentally, that meant oleo and oleomargarine paired in interesting ways with the geography of butter sketched above. The butter industry was entangled in global trade practices that put American dairymen in competition with Canadian and British wheat and dairy syndicates; the margarine industry was entangled in global trade that combined two commodity flow streams: finished oleomargarine shipped in blocks and pounds and raw

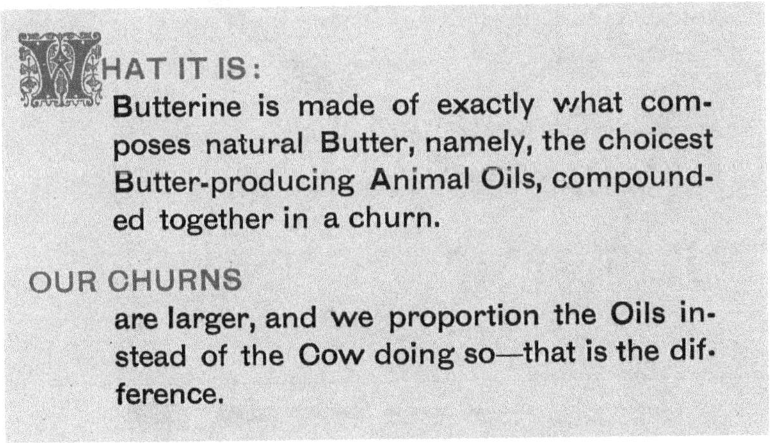

WHAT IT IS:
Butterine is made of exactly what composes natural Butter, namely, the choicest Butter-producing Animal Oils, compounded together in a churn.

OUR CHURNS
are larger, and we proportion the Oils instead of the Cow doing so—that is the difference.

FIGURE 4.6. Ad for Armour Butterine, 1900. The claim is for similarity in product through difference in process, with Armour intervening to mix animal oil instead of waiting for the cow to do it. Courtesy of the Warshaw Collection of Business Americana-Dairy, Archives Center, National Museum of American History, Smithsonian Institution.

oleo oil shipped in barrels and gallons. Those two export streams brought the processed fats of animals from the interior of North America to the import docks of the world.

As figures 4.7 and 4.8 show, this was no trifling matter. It involved dozens of states, port cities, foreign nations, and their markets. Chicago agents got into the pattern early, shipping one million pounds of margarine to England in 1877.[47] In the second half of the 1880s, U.S. manufacturers exported another six million pounds of margarine to foreign destinations. But raw oleo oil was the larger market as, during the same period, packers shipped 168 million pounds of oleo oil across the world. In the five years after Congress passed the 1906 Pure Foods Act, the United States exported twenty-six million pounds of finished margarine while nearly tripling raw oleo oil exports to 589 million pounds.[48] Figure 4.7 provides a snapshot of the oleomargarine export market in 1910; figure 4.8 shows the oleo oil export at the same time. By the early 1900s, exporters shipped oil from Chicago, Detroit, New York, New Orleans, and a half dozen other cities on the Eastern Seaboard to northern Europe, southern Europe, South America, Central America, the Caribbean, and East Asia. In total, eleven port cities were sending hundreds of millions of pounds of oleo to thirty-nine different nations and colonial holdings.

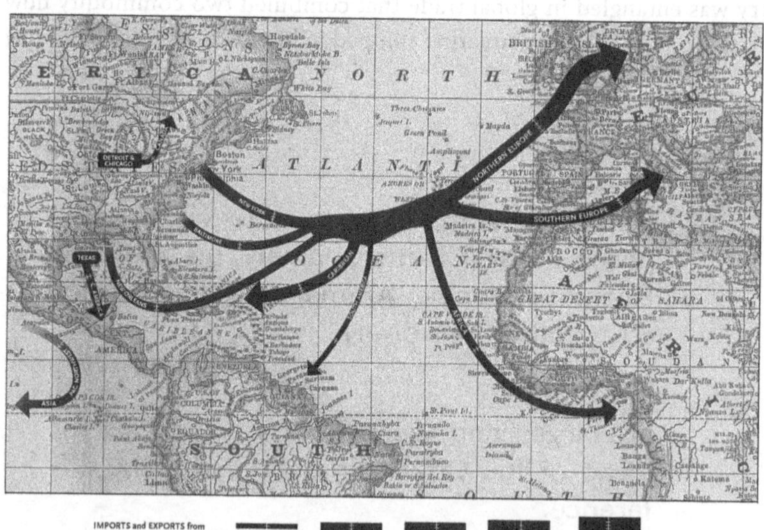

FIGURE 4.7. Map showing export streams of finished oleomargarine from US cities to global ports, 1906–1910. The maps aggregate foreign destinations into regions. The digital companion site at https://purefood.lafayette.edu shows changes across thirty years of export patterns. Map created by Kristen Lopez.

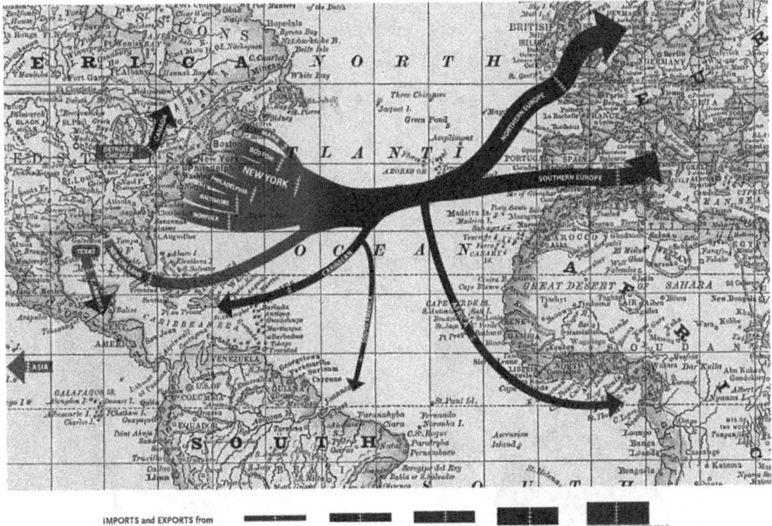

FIGURE 4.8. Map showing export streams of oleo oil from US cities to global ports, 1906–1910. The maps aggregate foreign destinations into regions. The digital companion site at https://purefood.lafayette.edu shows changes across thirty years of export patterns. Map created by Kristen Lopez.

The traces of these export patterns remain with us still. It is in the soaps we use and the margarine I should've grabbed in my childhood kitchen. More than a century ago, the British consul general was watching carefully as new trade routes grew, aware of the implications for colonial strength, market dominance, and imperial status. They saw that most shipments from New York's Commercial Manufacturing Company went to Bremen, Hamburg, and Rotterdam. From there, the oil found its way to Oss, where Jurgens and others processed it into the artificial butter (or butterine, or bosh, etc.) in establishments known as *Kunstboterfabriek* (fig. 4.9). As this oil supply chain gained consistency and with British encouragement, Jurgens ventured a key partnership by taking up trade practices with a similar shop in England, that of William Lever and his brothers. The Lever brothers, sons of an English grocer, had the same interests as Jurgens and the same motivation for capital investment in oil sources. The firms joined forces in 1930 to create Unilever, the third-largest consumer goods company in the world as I write and to this day the largest producer of margarine. Their product line includes Imperial Spread, Brummel & Brown, Promise, Country Crock, and the perfectly incredulous I Can't Believe It's Not Butter.[49]

FIGURE 4.9. Margarine-Boterfabriek, made by Jurgens in their Osch and Rotterdam facilities, 1893.

Like butter, margarine developed as part of intricate geographies. Because they were *different* geographies, the alternative butter created problems for those who considered margarine a sham product. Animals, oils, fats, stockyards, slaughterhouses, by-product ambitions, and trade patterns marked the features of those new environmental networks. The Lever brothers understood this when they acknowledged that "the growth of the soap and margarine industries in the thirty years before 1900 . . . depended on farming, especially American farming." The flow of millions of pounds of fat from the slaughtered cattle of Midwestern America brought artificial butter to nearly every continent on earth.[50]

If you were at the grocers' market or even the general store in Hardwick, Massachusetts, you just wanted to know whether the yellow block of "butter" behind the counter was the real thing. But in the fields and global marketplaces of a transformative agricultural world, a world that produced butter and margarine, the axis between pure and adulterated was harder to parse. Advocates and critics waged war on each other over oleomargarine within the prevailing culture of angst about deception and confidence and against the backdrop of such geographical complexities.

"The War on Oleomargarine"

Oleomargarine was controversial because its identity was unclear, its origins were murky, and it frustrated stable norms of agricultural activity. It also fit unsteadily between local landscapes of animal fats, on the one hand, and global commodity flows tying American slaughterhouses to foreign dinner tables, on the other. Take all of that into account and you get a better sense of why Congress could shutter all other issues to focus on it for days on end. Cutting through any one of those issues required the kind of trust and knowledge that was under more duress with each passing season of the later nineteenth century. The "War on Oleomargarine" spawned from those concerns justified extensive congressional hearings, untold numbers of governmental and private laboratory investigations, and the rancor of the nation for decades.[51]

The "war" had police-like regulators and government officials patrolling the border between acceptable and unacceptable butter-making practices. In terms echoed by Prohibition-era G-men and drug wars after that, regulators led raids, cultivated secret informants, and sought out illicit trade. Government inspectors found that "traffic is steadily on the increase." Authorities were "powerless to stop this illegal sale." A report by the USDA ended with a kind of dragnet-like announcement, telling readers "any information leading to the illegal sale of oleomargarine . . . will be thankfully received and accepted in strict confidence." District attorneys prosecuted violators; enforcers fined and jailed criminals. And yet, near the end of the century a trade paper wrote, "the war on oleomargarine shows no sign of abatement."[52]

Participants framed that war culturally, environmentally, and economically, but basically it centered on the question of poisoning versus cheating. The main cultural point followed from the argument over adulteration writ large. Was it safe? Or were consumers being tricked and defrauded? Senator Mason of Illinois aimed at answering that question when he asked to distinguish between "first. What food is sold that is deleterious to the public health; and, Second. What food is sold in fraud to the consumer."[53] Poisoned or cheated. The environmental element spoke to what counted as natural and proper—this was to ask, was oleo an offense not just against health and honesty but nature itself? And the economic element was about the ways answers to those questions informed public policy, specifically, the management of new immigrant populations and socioeconomic trends.

To the poisoned-or-cheated point, those supporting the new product were often manufacturers and chemists who supported the health of the alternative butter. In the 1886 oleo hearings, Professor Babcock of Boston testified that the new product was safe and healthy. He said that although he hadn't personally run experiments on the health of oleomargarine, the public "had been making them without knowing they were making them." The proof of its safety was that "not any of these people have died in ten years."[54] So long as there was "strict police surveillance" of the process, an agent from Armour and Company added, the public could be assured of oleo's health.[55] The *New York Times* reported that "Oleomargarine butter is the product of the oil from perfectly fresh beef suet, divested in the process of manufacture of the animal tissue and other possible impurity, and mingled in the operation of churning with pure, fresh milk, and is a wholesome and nutritious food product."[56]

Chemist Charles Chandler was an amiable witness in the 1886 hearings. He had a reputation for aiding the prosecution of fraudulent manufacturers, one that gained him credibility in the antiadulteration movement. He also supported the claims of oleo makers that it was not dangerous. With test results from a dozen chemists and references to a Berlin board of health report at his side, he assured senators in a testy back-and-forth that the imitation dairy product was "wholesome."[57] His colleague at Stevens Institute in New Jersey, Professor Morton, offered a similar line of argument to support oleo's wholesomeness. An exchange between Morton and Senator Jones (D-AR) captured the tone and frustration of the line of questioning[58]:

SENATOR JONES: You have stated that the pure fats of animals do not contain... germs or disease?

MORTON: They never have been in the history of science.

JONES: We have been told [that the industry uses] the fat of cats and dogs and animals which died of disease.

MORTON: [T]hese stories are simply absurd. It is utterly impossible to do any such thing.

This was a time when tainted milk and diseased meat scandals were common features of the daily press. Swill milk controversies were still common occurrences; bob veal—calves slaughtered under four weeks old—took the blame for health outbreaks in episodes across the country; imported American meat had been banned for trichinosis numerous times in Germany, Italy, France, and Belgium; and then this new artificial butter made from the fats of slaughtered animals jumped onto the public stage.[59] Critics saw margarine as a known adjunct of the meat industry and part of the same terrain.

The nascent recognition of germs, germ theory, and bacterial contamination entered the debate as part of an understanding of what led to sickness. Butter advocates—oleo opponents—considered margarine's suspected health problems the result of bad slaughterhouse practices. Morton's assertive claim that such things were "utterly impossible" belied publicly experienced realities. Babcock, Chandler, Morton, and others spoke in defense of oleo's health because they had to rebut the common view of its danger built on sustained public experiences leading to such an assessment. In other words, customers had legitimate reasons to be worried. The witnesses sought to bring in new testimony to assuage the fears.

Others added the environmental view that margarine was as natural as butter. It was a difference of degree, not kind. Professor Morton made this point, testifying to fellow congressmen that margarine was a variation of, not a product in distinction to, butter. He quoted nine chemists on its health effects to show that "all expert evidence upon this point strongly confirms the testimony of the manufacturers of this article, to the effect that it is a healthful food product." Mason, heading the hearings, then did well to represent his Chicago constituents, forcing the point about similarity. "The testimony shows that this product is the result of a combination of beef and pork fats, butter, cream, and milk with coloring matter," he said, "which is similar to that universally used by farmers and dairies engaged in the manufacture of butter for the coloring of that product."[60]

It couldn't be surprising that animal chemists and industry spokesmen were fine with a more affordable substitute for butter. They made an economic argument. They cared about the poor, they said. Margarine's purpose was to provide an inexpensive version of butter for the working class or, as with Napoleon III's decree, to help free up milk for military provisions. Mège wasn't subtle. His original name for the product was *beurre economique*—economic (affordable) butter. Those who would argue for the benefits of the substitute thus tended to emphasize lower costs, greater access, and more ready availability as its virtues. Supporters also folded the health and economic arguments together. Margarine was hardly a cheat at the marketplace. Quite the opposite, it was a healthy godsend. For its advocates, any regulation or taxation should be about ensuring that customers knew what they were getting. The product was safe. Those opposed were simply fighting progress. Such was the fight on the pro side.

Opponents were unconvinced. They questioned the advocates' metric of progress. The cheated-or-poisoned framing was too limited from their view. They thought of the disrupted agrarian environment. They thought of dairy livelihoods. They didn't trust the new experts. Was oleomargarine a con? Rather than an improvement, was it another low-quality metal covered in gold plating to trick the eye? Was it "natural," as in produced through proper agrarian methods? They embraced a wider array of cultural, environmental, and economic arguments while the proponents mostly left it centered on the cheated-or-poisoned axis.

Margarine antagonists saw a pitched battle between the trust found from appropriate ways to work in nature and the deception that came from a Gilded Age culture of anxiety. In a race of "Chemist v. Cow," the *Chicago Tribune* found "the Chemist ahead."[61] An article about "Western Bogus Butter in Eastern Markets" in the *New York Tribune* took the new knowledge of germs as a reason to dispute margarine's health. You couldn't trust the process or the product. "Spurious butters are dangerous on account of the disease germs which they contain. They look all right and smell all right, but when heated they give off foul odors."[62] In Massachusetts, the president of the Society for the Prevention of Cruelty to Animals (SPCA) and chief agitator for the state's antiadulteration efforts, George Angell, wrote prolifically about the germ profile of butterine. The microscopic investigations he had commissioned revealed living bacteria that also, he believed, caused disease.

The margarine antagonists didn't think it was a choice between being poisoned or cheated because they thought it was both. In just two pages of testimony, Joseph Reall, president of the American Agriculture and Dairy Association, managed to malign margarine industrialists as "enemies of honest industry, demoralizers of public morals, and robbers of the people." They practiced "dishonesty"; they "defrauded" the public; they were the "disgrace of mankind... the iniquity of business [and] the train of evils." Because they peddled "a noxious compound... it is a serious question whether or not public policy, public health, and public morals do not require the total extermination of imitation butter." Imitation butter was a "curse of God."

Dairymen hammered on the deception and trust theme. Victor Piollet spoke for Pennsylvania's State Grange. He asked, "Why should not

Congress institute laws to punish a man who counterfeits butter as effectively as you would punish a man who counterfeits money?" A wholesale merchant in Washington, DC, cried similarly that "the cool impudence and insolence of these men" busy selling "counterfeit goods" was a crime against the country. "If counterfeit food is to be sold in the Center Market," the merchant argued, "why not allow counterfeit money to be sold there?" Senator Blair (R-NH) was direct: "Why should not oleomargarine tell the truth? Why should it be allowed to lie itself into my stomach?"[63]

Taking a stand against oleo was also taking a stand for appropriate ways to work in nature. This often gets read as an argument for nature as a thing, that policy makers should defer to nature, as in *not human*. But butter doesn't make itself, nor does it get that perfect June color on its own. The question on the table was instead about the proper way to manipulate nature, not whether or not one intervenes. Butter advocates—margarine critics—protected an idea of purity that followed from culturally acceptable dairy practices, what was often summarized too simply as "natural."

In that sense, the war on oleomargarine was a war against unacceptable environmental practices. Speakers at a meeting of the Butter and Cheese Exchange in Manhattan described butter as being "made from milk, which was designed by the Creator as the primary article of food" and margarine as being made from "numerous mysterious fats of animals, hard to describe."[64] A Philadelphia Milk Exchange ad campaign took the same tack: "Nature, the Original Chemist, yields the secrets of its Laboratory grudgingly—without experiment or guess-work its products are perfect." There was a good deal of construction in this, the outright claims to "nature" as being God given and preordained. It took a lot of agricultural and political work to secure the idea that the creator designed milk and butter, but that was precisely the work the probutter crowd was doing.[65]

At the 1886 hearing, Piollet of Pennsylvania advanced these points as a farmer. He *could* have easily diluted butter with "imported grease from Italy" or "the deodorized grease of Chicago." But why would he? He was "scrupulous in having everything neat and clean and in subsisting our cows upon the best material." His farm bordered New York, where it was "a well known fact that the whole tier of counties on either side [were] capable of producing the very best butter." They have "water and grass that make extra butter," rendering the "counterfeit" entirely unnecessary. Senator Gibson of Maryland wanted the audience to know his local butter came from a county where "the grass is as good as New York." Witnesses spoke of "grass butter" throughout the hearings to indicate the tight connection between cow, grass, and butter. Their point was that margarine was a violation of long-standing agricultural principles.[66]

Others would go further to advertise the ecology of butter as the underpinning of its natural advantage in ways that are well familiar to consumers today. As one campaign put it, "The goodness of MERIDALE BUTTER begins with the very grass, the pure mountain air, and the pure mountain water of Meridale Farms. These and well kept cattle make that marvelous cream that makes MERIDALE BUTTER." Meridale then tied its virtues to the questions of trust and character, telling their customers, "once stamped Meridale, every print can truthfully be said to the best quality of butter than can be made." To their mind, the question of character and authenticity was part and parcel of the question of nature and "natural." Advocates considered "natural" to be that which was outside inappropriate human contrivance—the emphasis being on how to gauge propriety, not whether or not someone had intervened in the dairy environment. Here again it was an agrarian concept born of the daily experience of working the land, one of acceptable intervention rather than an ethic of strict separation.

One final economic aspect of the oleomargarine wars was about public policy. Margarine advocates defended fake butter precisely because it was cheap. Armour claimed that if unheeded by undue regulation, margarine would "be of great advantage to the poorer people of this country and to the cattle raisers."[67] Yet advocates for the poor, those *not* in the Beef Trust, questioned the meat-packers' sincerity, still suspecting that margarine was a con. The urban poor had their own ideas about what they were being fed, suspicious of cheap alternatives that were probably also dangerous ones. This dynamic set in place an aspect of food policy that ripples through our culture still: is low cost also low quality?

A different rogue new product, Woodeo-Sawdusterine, made the point well by bringing together all three elements—environmental, economic, and public health—to the war on oleomargarine. Woodeo-Sawdusterine was an innovation in bread for the modern age, in fact "the greatest invention of the age." It would help the working class move beyond "wheaten flour, [that] relic of the dark ages." Nor would the modern family remain bogged down in "the whole system of antiquated and cumbrous breadmaking, the wheat cultivation, the turning of the grain into flour, the kneading of the dough and the baking of the loaves." Such wasteful practices "ought to be relegated to the limbo of the past."

Launched in 1880, this was supposedly Thomas Edison's newest invention. Edison was then at an early peak of public prominence, fresh off the light bulb, the phonograph, and electric power distribution. Woodeo-Sawdusterine was made from electrified sawdust. It was perfectly pitched to a culture awed by invention and dazzling new technologies, none more than electricity. It was also aimed at a culture that sought agricultural innovation as a means to address social problems of poverty and labor unrest.

The new bread went by several names, "Breadarine, Lumberine, or Rot." It was "one twentieth cheaper than natural bread, but five times as wholesome and palatable." The benefits were outstanding: its "discovery is a boon to humanity at large. Starvation will be a thing almost unheard of." Anticipating critics, the report noted there was "no appreciable difference in appearance" between the upstart and traditional "wheaten flour" bread. Testing by an English Public Analyst confirmed it with printed results of a thorough analysis. Granted, "the growers of wheat are in despair," as their product would now be worthless. But was this a big loss anyway? What was lost in amber waves of grain was "likely to promote robust vigor in people of sedentary habits and . . . probably act as a strong moral tonic upon those whose religious education has been neglected early in life."

The report went on to explain that it was "the brilliant success of Oleomargarine [that] suggested to him [Edison] Woodeo-Sawdusterine." The inventor was "rejoiced at finding so pleasant a companion to artificial bread as an equally artificial butter." Win win.

This was not a con. It was a parody. Or, better put, it was a parody of a con. The editors of *Puck* magazine hit adulteration hard throughout the latter decades of the nineteenth century. Oleomargarine became a favorite target as, to *Puck*, it elicited the hallmark virtues of a scam. If animal chemists and butterine capitalists were pitching their new product to the working classes, then *Puck* was ready to speak on behalf of those workers. It pitched back a counterargument to prevailing stereotypes of technology and progress. They had an especially well-written take in the spring of 1880 with the satirical Woodeo-Sawdusterine.[68]

Less facetiously but just as emphatically, a chorus of witnesses in government hearings trumpeted the time-honored line that the poor and marginalized were hurt, not helped, by capitalist innovations in the dairy world. "Yes 'cheap food!' The stomachs of pigs, sheep, and calves reduced by acids, and then bromo-chloralum used to destroy the smell and prevent detection of the putrid mass. . . . Yes, 'cheap food' in the shape of putrefaction rendered odorless by a powerful disinfectant. Yes, 'cheap food' in the form of an apothecary's shop in the poor man's stomach!"[69] In these responses the legacy of distrust over food purveyors and manufacturers pervaded views about margarine. Long-held frustration over systemic problems could not be waved away by claims for novelty or sudden concern for the poor. The antimargarine contingent was unswayed by the social pretensions of the industry.[70]

And yet there were larger forces at work. The antimargarine dairy lobby was caught inside different arguments happening at the same time. In one way, they sought to protect a butter industry that was intimately tied to milk, cheese, cattle, and grain landscapes. In another way, they were pawns in a larger political fight for continental expansion, settlement patterns, and global commodity flows. To farmers who were trying to produce butter to make a living so that they could produce wheat to compete with Canada, margarine was more than a mere competitor for butter. It was a challenge to a political economy of land use and agrarian contours. To congressman trying to quell the rise of margarine production so they could build capitalist links between the Atlantic coast and export destinations across the globe, margarine was more than an industrial pest. It was also a new agent of commerce undermining treaties and trade regimes. One could say to the Midwestern farmer, grow more wheat, raise more cattle, churn more butter; one could also say watch out for the Suez Canal and lakes of Manitoba and British imperial networks bankrolled by syndicates in London. The geography of margarine shows that one could not say, though people did, that the dairy lobby was a mere reactionary movement. Examining that changing geography of butter and margarine undermines overly simplified claims. The situation was far more than an argument between progressive industrialists and traditional grangers set in their ways.

As we contemporary eaters know, oleomargarine eventually became part of the modern food pantry. It took a while, but it got there. By the mid-twentieth century it was the acceptable butter substitute its advocates claimed it should have been in the later 1800s. There was even a Purity brand oleomargarine by the dawn of the First World War. Their advertisements challenged customers to "See if they can tell the difference" between their product and butter. They promised customers they couldn't because there wasn't any. This was to say that by the early twentieth century there could be a purely fake product. Like "real" cubic zirconia diamonds, Purity brand margarine was authentically not butter.[71]

Basic industrial experience helped explain margarine's slow success as did accruing consumer familiarity. Margarine production had become a fixed and regulated industry by the early twentieth century. The World Wars helped. They reintroduced the argument Napoleon III had made in the 1860s about saving dairy for the troops. Purity margarine sold this idea. They showed images of Mège at the royal farm in their ads with text summarizing the French invention story. Chemical developments mattered too. By the 1930s chemists had mastered principles of hydrogenation so that the range of oils useful for margarine grew wider. Whereas (unkosher) animal fats fit the bill for the first few decades of production, manufacturers soon found that vegetable oils served the same purpose if properly processed in vats and boilers. Tropical oils like palm and Mediterranean oils like olive were some of the first to perform that function. Another new ingredient was the oil expressed from cottonseeds that I discuss in the next chapter. When chemists began hydrogenating oils in the twentieth century, they drew from basically anything available. As I write, the tub in my fridge is made of soybean, palm, olive, and canola oil.[72]

One final thread in the margarine debates cinched together its various strands. These were the oleo color laws. It took a while—public acceptance of margarine—but it got there in part because of color. Color laws epitomized the late nineteenth-century acrimony over artificial butter predicated on a view long entrenched in modes of understanding food that visible appearances matter. The problem was, if margarine looked like butter, it was being deceptive. As produced, it was white. So you couldn't make yellow margarine. Color laws lasted so long that some readers may have childhood memories of the last vestiges of this from the twentieth century.

Various proposals beginning in the early 1880s suggested that manufacturers color margarine red, blue, violet, strawberry, buttercup, pink, or nothing at all. In the 1886 hearings that monopolized Congress's time, New Hampshire's Senator Blair asked, "Is it not fair that I should know what I'm eating?" Color's deception served no purpose but to trick the consumer, he argued. Why would coloring oleo red or violet be bad, Blair asked chemist Henry Morton, since it would help consumers understand they were not eating butter? Morton thought that would create "a prejudice and a disgust in the minds of many men against it." But that was Blair's point—that there was something disgusting about it, something he thought unnatural. Morton thought he had a good comeback: "We have ice cream and candies the same color as butter, and candies also." Blair wasn't compelled: "But we never understand that we are eating butter when we are eating ice cream."[73]

Then there was the pink controversy, until the Supreme Court ruled it unconstitutional in 1898. *Puck* played a part in this one too. In a curious historical circumstance, what began as a satirical aside ended up being taken up by legislators for real. In a March 1881 issue, *Puck* poked fun of oleomargarine makers for their royal pretensions, suggesting they tint their product purple or perhaps pink to signify its regal status. Soon after, the *Sun* (New York) took up their proposal as their own and broadcast it as a policy recommendation. Seemingly unaware of or unconcerned about its facetious origins, Vermont took the bait in 1884. It was the first of many states where pink laws took root in actual state legislative chambers. Senator Blair was arguing for more of that in the 1886 hearings. Margarine makers tinted their product pink in Vermont so customers at the grocer's market could tell the difference based on sight alone.[74]

"Color laws" fell into one of four legal categories in the later 1800s. Figure 4.10 shows snapshots of the legislative maps in 1877, 1887 and 1897. New York passed the first state laws in 1877. Others developed the full slate of four levels of regulation: the lightest were label laws; more intense ones defined measures for color and taxation; the most severe were outright prohibition laws. Within five years of New York's example, ten states had labeling requirements. By the 1886 hearings, a few states had color laws and several more had prohibited margarine outright. By the end of the century, forty-four of the forty-eight states had some form of margarine regulations on the books, thirty-two of them color laws.

But by then, not pink. Government officials seized a shipment of Armour's oleo in Minnesota in 1895, which became the precipitating cause for a lawsuit meat-packers brought against the state. They knew that Minnesota—alongside New Hampshire, South Dakota, West Virginia, and Vermont—required pink coloring, the sarcastic origins of which had by then fallen out of view, but chose to ignore it to agitate the courts. The case did not go their way. Two years later Minnesota's State Supreme Court ruled in favor of "the pink law."[75]

But the industry persisted until New Hampshire became the next battleground to test the constitutionality of margarine legislation. This time it worked. 1898's *Collins v. New Hampshire* ruled that mandating the pink coloring "naturally excites a prejudice and strengthens a repugnance up to the point of an absolute refusal to purchase the product at any price." This time the meat-packers won. The Supreme Court of the United States deemed the mandating of the color pink for margarine unconstitutional.[76]

That was a small victory in a longer war. Color laws persisted well into the twentieth century. Wisconsin was the last holdout in the United States, overturning its margarine color law in 1967. Quebec still had laws on the books, largely ignored though they were, until 2008. I point to the century-plus legacy of color laws by way of conclusion because they do well to show how manufacturers of artificial versions competed with a prevailing view of natural goods.

Armour, Swift, and the promargarine faction sought to pull the debate out of its cultural and environmental context, instead framing it through market and consumer health terms. For the pro-oleo side of the margarine wars, this was about cheating or poisoning. They protested that they did

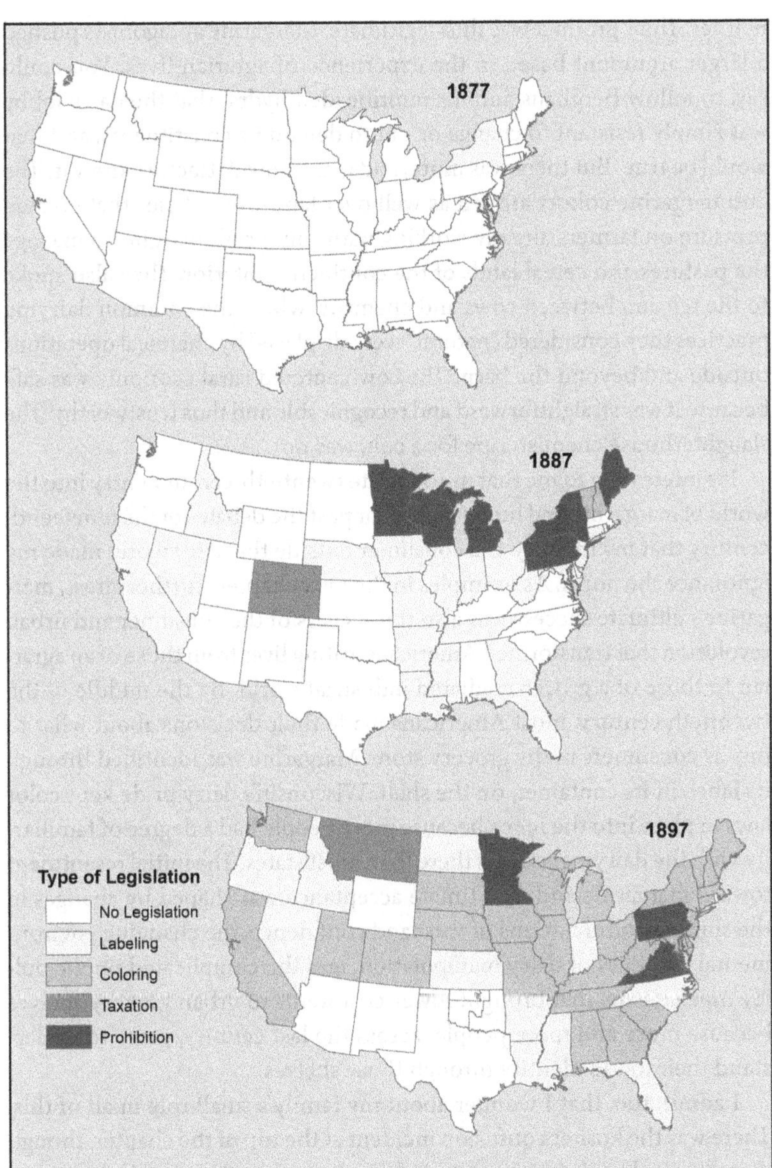

FIGURE 4.10. Maps across three decades showing the kinds and occurrences of state-level oleomargarine regulations, 1877–1897. The digital companion site at https://purefood.lafayette.edu provides a dynamic view of annually changing color laws, 1877–1899. Maps created by John Clarke and Matthew Plishka.

neither. Their product was thus legitimate. Margarine antagonists pushed a larger argument based in the experience of agrarian lives. You could say, to follow Berghaus and his multiheaded hydra, that the dairy lobby was simply resistant to change or out to defend its constituents, and that would be true. But there was more, and margarine detractors knew it. The antimargarine cohort argued as well over land-use policies that created pressure on farmers already working with slim profit margins to manage the pastures and cereal lands of the continent's interior. They also spoke to the tension between cows and chemists, where the common dairying practices they considered "natural" were displaced by chemical operations outside and beyond the barn. The cow-centered rural economy was safe because it was straightforward and recognizable and thus trustworthy. The slaughterhouse chemist, ripe for a con, was not.

It's interesting to me that my own late twentieth-century entry into the world of margarine and butter was so far past the debates of the nineteenth century that my identity as a consumer outside the dairy world made my ignorance the norm. As examples in the next chapters further attest, margarine's ultimate success was also the success of the consumer and urban revolution that transformed Americans' eating lives from those of an agrarian to those of a grocer-mediated industrial world. By the middle of the twentieth century, most Americans made their decisions about what to buy as consumers in the grocery store. Margarine was identified through its label, in its container, on the shelf. Wisconsin's dairy pride kept color laws in place into the 1960s because more people had a degree of familiarity with the dairying process there than most states. The initial resentment toward margarine and its ultimate acceptance was shaped by changes in the shifting cultural terms of trust and confidence, the changing environmental practices of dairy manipulation, and the complicated public policy mechanisms that brought lower-cost items to urban grocery shelves because more and more people, across the last century, came to understand their food's identity through those shelves.

I admit, too, that I wonder about my family's small role in all of this. There was the kosher confusion incident at the top of the chapter, though that must be largely inconsequential. That was the 1980s, exactly a century after the first federal oleo law. But there was also a dark family secret that my mother let loose one day, almost offhandedly. It happens, I learned only in adulthood, that my grandmother was a bootlegger. She was Canadian. Much of her extended family lived in Winnipeg, Manitoba. My mother grew up in Rochester, New York. Her mother would visit relatives on the trans-Canadian rail smuggling in banned oleo. She would hide illegal blocks of margarine in her suitcase, often in exchange for Canadian salmon or lox. Whether or not this was an ongoing illicit trade (doubtful) or a sometime activity that made for good stories (plausible), it now feels like a much more meaningful connection to the Manitoba watershed, its fertile wheat lands, the dairy cattle that helped create those fields of rich soil, and the butter that kept the cattlemen in business.

5

Oil without Olives & Lard without Hogs

> There now, smell them, taste them, examine the bottles, inspect the labels. One of 'm's from Europe, the other's never been out of this country. One's European olive-oil, the other's American cotton-seed olive-oil. Tell 'm apart? 'Course you can't. Nobody can.
>
> MARK TWAIN, *Life on the Mississippi*, 1883

Oleomargarine grew in relation to vegetable oils and cows. Cottonseed oil grew in relation to olive groves, pigs, and, of course, cotton. There's no stirring invention story for cottonseed oil, but in a postbellum Southern economy advocates dubbed the use of formerly discarded cottonseeds for the oil they expressed "the Cinderella of the New South." It was the neglected stepchild who began to shine for her economic potential. It seemed to come out of nowhere. "Mills sprung up as if by magic," wrote one observer in 1897. In 1874, growers crushed 5 percent of harvested cottonseeds for their oil. Twenty years later, it was 42 percent. By 1915, it was 84 percent. That translated to 1.5 *billion* pounds of cottonseed oil.[1]

Much like today's apparent passion for making "milk" out of any item that can be crushed and diluted (soymilk, almond milk, rice milk, flaxseed milk), late nineteenth-century manufacturers were enthusiastic about using any oil they could find to proliferate their product lines or, more problematically for pure food crusaders, to adulterate a range of consumer products both edible and inedible. In the history of adulteration, the first great potential for the Cinderella of the New South came as a substitute for olive oil. When crushed and refined, the oil from cottonseeds had reasonably similar consistency, density, and taste compared to the oil pressed from olives. This spawned a new sector of industry, the cottonseed crushing mill. It then had southern cottonseed crushers brokering deals with European olive growers. In a zigzag that went east across the Atlantic into the Mediterranean and back west to North America, the olive groves blended it with their product before sending so-called pure olive oil back through the ports of New York.

A second potential came as a substitute for lard. While brokering deals in southern Europe, cottonseed crushers also developed supply networks with meat-packers in Chicago to cheapen the cost of pig fat. As someone who grew up thinking lard was a generic synonym for "something you don't want to eat," I was initially surprised to learn there was a demand for "pure" lard, because why would it matter if something you didn't want to eat was pure? But I was as wrong here as I was with my margarine-for-butter mistake. Lard was a common dietary feature on nineteenth century plates. Part of the $13.34 Thoreau earned at Walden went to buy lard. Leaf lard was the highest premium type. It came from the soft fat surrounding the kidneys of the pig. It was a standard ingredient in nineteenth-century kitchens, serving as a spread (often substituting for butter), as grease for cooking, and as an ingredient in baking. Consumer trust in that kitchen staple was undermined when manufacturers added cottonseed oil as an adulterant.

Many critics considered cottonseed oil one of the principal adulterants in an age known for adulteration. A groundswell of people decried the new product, making the case that using cottonseed oil was a sign of chicanery. Cotton oilman Edward Atkinson (friend of Ellen Richards) admitted in the 1860s that he "dare not call it cotton-seed oil for fear it might prejudice the sale." Selling it outright was too risky, given the widespread public outcry over food fraud. Mark Twain's reference in *Life on the Mississippi* (1883) played into the feared deceit of confidence men. A riverboat traveler challenged a fellow passenger to detect distinctions: "One's European olive-oil, the other's American cotton-seed olive-oil. Tell 'm apart? 'Course you can't. Nobody can."[2]

Twain was echoing sentiments from Chicago and New York to Italy, France, and England. "The sophistications of olive oil are numerous," wrote Paris's *Correspondence Scientifique* in 1879. An 1883 report from the burgeoning olive oil industry in California worried over the competition from Europe that came from trade with southern crushers. "New Orleans annually exports cottonseed to Bordeaux, etc., to fill, 15,000,000 oil bottles." That was problematic. The rural agronomist and horticulturist Liberty Hyde Bailey wrote later to consider it yet another in the laundry list of questionable modern contaminants: "Hamburg steak often contains sodium sulphite; bologna sausage... contain[s] a large percentage of added cereal.... Lard nearly always contains added tallow [, and] Cottonseed-oil is sold for olive-oil."[3]

This being the era of adulteration (or this being humanity), there was also a booster for every critic. To that end, an industry observer gave the by-product a retrospectively positive spin of progress and triumph: "In 1860 cottonseed was a garbage; in 1870, a fertilizer; in 1880, a cattle feed; in 1890, a table food and useful by-product."[4] He was mostly on the mark, if not backdating the "useful" claim a decade or so earlier than it actually was. *Scientific American*, *Popular Science*, the *American Journal of Pharmacy*, publications by the USDA, and a range of rural newspapers echoed the notices about cottonseed oil as a viable newcomer in the oil and drug marketplace.[5]

At the front of the period in the 1860s, oilmen kept a mask on their product, aware people thought it was waste. At the end of the 1890s, "dealers," Harvey Wiley told Congress, were still "being careful not to put the name 'cotton'" on bottles of the oil. But as the new century began, they would dip their toes into promoting it as a virtuous product. To its advocates, rather than a secret to be hidden, cottonseed oil was a point of progress to be lauded. A sign of the rise of cotton oil as "table food" came from a new product in the early twentieth century. Its makers still sort of cloaked the product's identity, worried about being too forthright with a still-suspicious ingredient. If you thought you'd never heard of a cotton-oil product, you, too, have probably been deceived. Procter & Gamble developed a new consumer product, the shortening Crisco, in part to find a use for all the cottonseed oil it owned. "Crisco" was a contraction—a trade name derived from its original main ingredient, *cry*stallized *co*ttonseed oil.[6]

Berghaus fashioned "cotton-seed-oil-lard" as the second head of his monster. The center beast hissed from on high against the farmer's bayonetted rebuke. Berghaus placed cotton oil in an image of doubt and mistrust. Before the twentieth century, the oil was one of Berghaus's big three adulterants not because it looked odd at the store, like margarine, but because customers didn't know whether it was in other products on the shelf. In other words, while Berghaus and others lambasted the three big adulterants as of a kind, the three necks were offensive for different reasons. The process of oleomargarine production challenged the conventions and ethics of a complicated agricultural practice; anti-oleo activists in turn challenged its legitimacy. Cottonseed oil was in league with oleo for challenging food identity, but its primary offense brought a different angle to the fight against adulteration. Opponents objected to cotton oil because of its stealthy commingling and undetected mixture. At the root of disapproval of cottonseed oil was a concern that customers didn't know what they were getting in their bottle of olive oil or pound of lard. They couldn't tell just by looking.

Like the dairy background that stood before margarine, an environmental backdrop stood before cottonseed oil that explains how the oil became part of a new food marketplace. The geography of cotton throughout the US South, the bottled products of European olive groves, and the processed fats of midwestern packinghouses all gave shape to a new industry. When viewed by the customer, that might have been just another suspicious addition to the grocer's shelf. Public commentary was less emphatic about a full-out war on cottonseed oil compared with oleomargarine's ordeal, but the for-and-against debate in the skirmishes over cottonseed oil help paint the picture of cotton oil's role in the era of adulteration. Before looking to the end of the line where consumers waged that battle against cottonseed oil, exploring its derivation provides the fuller environmental foundation that gave meaning to the adulteration fears.

The Geography of Cottonseed Oil

Where did billions of pounds of cottonseed oil come from? An entirely new sector of the cotton economy was emphatically a big deal. Meatpackers had defended their view of oleomargarine by reasoning that byproducts created new revenue streams. Cotton growers had the same mind-set, viewing cottonseed oil as a smart and sensible development.

Their activity followed from a two-step process. First were the production-side cotton fields that created the supposed adulterant. Second were the agricultural products in which it was mixed, the oil and lard. I will get back to that second point soon. On the first point, the derivation was straightforward: cotton oil was made from cottonseeds. Farmers harvested the cotton, separated the seeds from the fiber in the ginning process, picked the remaining lint off the seeds (delinted it) in another process, and then crushed those seeds in a press to extract the oil. A staged refining process separated the crude version into higher-grade off-white oil. It sounds like modern petroleum refineries, but it was organic oil, this was the nineteenth century, and the processes for refining and barreling oil in fact preceded those for petroleum.[7]

The 1860s were the coming out decade for cottonseed oil, but advocates first took note of the process of crushing and pressing cottonseeds in the 1600s. At the time, they used it toward pharmaceutical ends. Colonial accounts show that the oil was expressed from cotton for medicine in the British West Indies as early as 1667.[8] Salves and vegetable oils were common in the apothecary's kit. A century later, the Moravian community in Bethlehem, Pennsylvania, gave exhibits of cottonseed oil to the American Philosophical Society in Philadelphia. They were the first to build pressing mills.[9] The Royal Society of London soon offered a prize for the first person to produce reputable quality oil in the later 1700s. In 1793, the US patent office granted its first patent for a process of extracting oil from seeds. By 1799, there was a patent for a mill to extract oil from hulls.[10]

The next half century found scattered notices of seed processing but nothing organized or widespread. In the antebellum South of the 1820s, a South Carolina paper reported, "millions of bushels of cottonseed are annually used for manure for corn, wheat, etc." But, while "the oleaginous [oil-like] quality of cotton-seed has long been known," most farmers were slow to consider managing that oil. The paper advised better ways to use discarded seeds than letting them lie in moldering masses by the side of the gins. "Whether it can be made to take the place of linseed oil, in painting, or of olives in manufactures, remains to be determined." The next decades marked off that determination.[11]

A quarter century later, Georgia planter and slave owner J. Hamilton Couper took on the challenge. He pursued his own investigations into the possible new market for a paint and fuel source. From his experiments on land he bought in Natchez, Mississippi, the *Southern Agriculturalist* wrote in 1846, "the oil from upland cotton-seed was found, when well refined, to burn as well as spermaceti; it made also an excellent paint oil." The refining process was still crude, but Couper expected that organic chemistry would soon help improve it. He compared long-staple Sea Island cotton with short-staple Upland cotton in experiments to measure how much oil he could extract from different seeds. Sea Island cotton yielded more oil than Upland because it was less fibrous. Couper concluded that "one bushel of seed" produced almost a gallon of crude oil. He also made seed cakes from the crushed hulls, compressing them like a ball of dough into feed for animals. He fed the cake to horses, cattle, and sheep to good effect. It "was found to be excellent," he wrote.[12]

Those were early efforts in cottonseed oil history. They were also still part of community- and home-based cooking networks and removed from larger public debate over adulteration. Regional recipes included a range of oils. They followed from well-established household practices. In the South, enslaved people's knowledge led to oil extraction techniques and uses in recipes. "The West African method of extracting oil was designed for a household level of production and consumption," writes historian David Shields, where "the seed was pounded by mortar and pestle." Most of its uses were for paints, fuels, varnishes, and animal feed. This led to concerns over contamination aimed more at assessments of quality than health. This was by and large the "before" phase of public health adulteration.[13]

The 1860s were formative years for cottonseed oil and its public standing for a host of reasons. One was that its promoters placed it in competition with other lubricating, burning, and lighting oils. A new petroleum market, for instance, was one such challenger. The *Southern Cultivator* put cotton oil and petroleum in conversation only two years after prospectors tapped the first fossil fuel gusher in 1859. The editors thought that the "recent working of oil wells in Pennsylvania may reduce the price of Cotton-seed Oil." They also thought it was "by no means certain that the Pennsylvania Oil will at all interfere with this article, being greatly its inferior in most of the properties for which oil is valuable."

Another reason for greater attention to cotton oil was the dramatic set of changes in the cotton industry advanced by the Civil War. For one thing, the racial shift in labor of the 1860s South spurred further attention to economically beneficial by-products from cotton. Without slave labor, the market structure of cotton plantations in the United States shifted to sharecropping and the violence of Reconstruction and then Jim Crow labor practices. Sharecroppers continued to use household oils in their recipes. But in the face of changes in field management that followed from the war's end, proliferating market opportunities were increasingly alluring for cotton growers. In that way, cotton oil's status as a suspected adulterant in the coming decades was related to its standing as the beneficiary of racial politics.[14]

As might be expected, US cotton production fell considerably across the decade, cut nearly in half from 4.3 million bales in 1860 to 2.4 million in 1870. Yet with international markets and capitalist investments, it took only a decade for those production levels to recover, reaching 4.5 million bales in 1880. By 1900, growers produced 9.4 million bales of cotton. With each bale weighing about five hundred pounds, this was an annual production before the Civil War of two billion pounds and by 1900 nearly five billion pounds.[15]

All that cotton led to tons, literally, of discarded seeds.[16] Rather than waste, it was a new resource. When margarine advocates argued for the value of their product, they were trying to *reduce* milk use and offer a cheaper alternative. They thought there wasn't enough milk to go around. That worked for margarine makers, but it threatened the dairy industry. When cottonseed oil advocates argued for the value of their product, they were trying to *add* an entirely new revenue stream. With the development of a new crushing industry, cotton oilmen saw the change from small-scale regional oil use to global marketing.

If the 1860s ushered in cottonseed oil's possibilities, the 1870s saw the rise of crushing mills across the southern landscape to help its market viability. And if the earlier community-based use of the oil kept it outside larger public debates over adulteration, this new phase brought the ingredient to global consumer networks in ways that agitated the debate. Geographically, the seed-crushing industry fit as an interesting southern counterpoint to the dairy map of the North. It was as if together the two pieces formed a puzzle of the entire nation. While the dairy belt expanded across the Great Lakes, the seed-crushing geography flourished across "the black belt" of the South.

"In 1860," one reporter noted in retrospect, "there were seven mills in the United States manufacturing cottonseed oil. In 1867 there were four mills in the South, twenty-six in 1870, forty-five in 1880, (119 in 1890 and 252 in 1894)."[17] A quarter century produced a tenfold increase in the number of major cottonseed crushing mills. What is more, as cottonseed oil's principal historian Lynette Wrenn has noted, "overseas demand for cottonseed oil increased tenfold during the 1870s, resulting in profits of as much as 50%." In a mere five-year span from 1880 to 1885, capital investment tripled in new mills, from the hundreds of thousands into millions of dollars. And that didn't take into account the smaller mills sprouting up throughout the postbellum South. Advocates at the time foresaw "that every gin or neighborhood should have its own small oil mill, just as most communities had their own grist and saw mills."[18]

The growth of oil mills encouraged an evolution in the ways agricultural communities organized themselves. As opposed to oleo's incursion into the northern dairy geography, the mill expansion enlarged rather than disrupted existing agricultural patterns. Crushing season usually began in mid- to late summer (earlier in the Deep South, later toward upland Georgia and South Carolina) and continued until early spring. Those in today's local food movement often talk about measures for "season extension," using greenhouses or hoop houses as ways to begin the growing season earlier or help it last longer. The crushing season for cotton farmers did similar work in the nineteenth century. The rapid expansion in the numbers of seed crushers and their new mills were visible signs of this new economy, enhancing the contours of cotton-growing practices that came with milling.

The sheer scale of geographical expansion in cottonseed crushing across nearly half a century from 1860 to 1900 was incredible. The seven mills in 1860 were in New York, Providence, Memphis, St. Louis, and New Orleans (which had three). Figure 5.1 shows cotton production at the time. The pattern indicates the layering of geological, soil, crop, and labor

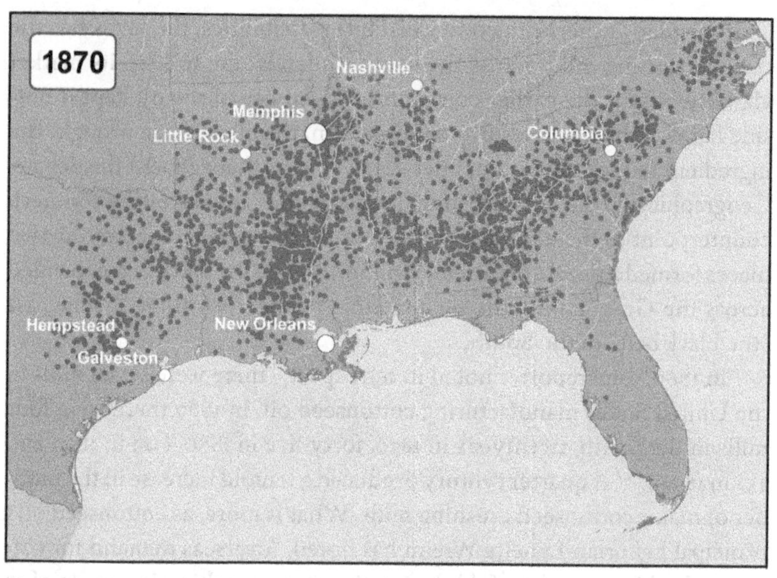

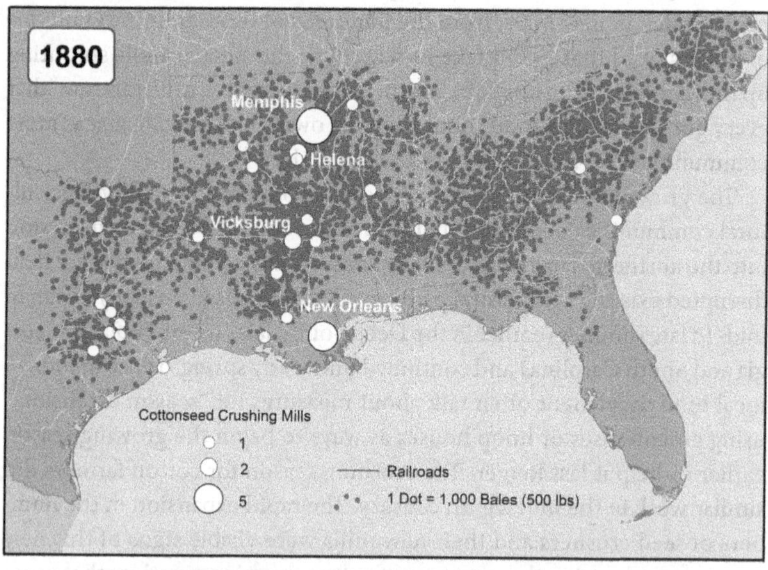

FIGURE 5.1. Cottonseed-crushing mills mapped against cotton production, 1870 and 1880. Each small dark dot represents five hundred bales of harvested cotton. Larger white dots show the crushing mills' locations. Data imported from the National Historical Geographic Information System database and data sets compiled from US census bureau statistics and annual reports published in the *Oil, Paint, and Drug Reporter*. The digital companion site provides further details about the maps at https://purefood.lafayette.edu. Maps created by John Clark and Matt Plishka.

patterns on top of one another, the so-called black belt. A band of rich soil resulted in part from the geological legacy of glacial retreat. It provided the agricultural conditions for productive growth. A policy of systematic racial oppression brought black bodies to that soil to do the work of cultivation as enslaved peoples. Perhaps unsurprisingly, then, the same belt was dotted with crushing mill after crushing mill. A crescent extending from South Carolina into northern Mississippi joined up with the deep zone of cultivation along the lower Mississippi. The map of 1870 shows scattered cotton production with only a few crushing mills. By 1880, Memphis and New Orleans hosted the majority of crushing mills because of their status as riverside trade depots. Figure 5.2 shows the continuing expansion and deepening of cotton production and seed-crushing capability.

By the turn of the century there were 715 mills across the South. Georgia, Mississippi, and South Carolina each had close to one hundred. Texas led the rankings. Its case offered a helpful view of the growth pattern. In one way, macropolitical motivations for continental settlement partly explained the east to west shift toward Texas. Expansion, Manifest Destiny, territorial politics. Environmental features helped explain it too. Texas cotton prospered in the specific Black Waxy Prairie region of the state, an area cutting a diagonal north to south nearly down its center. Note the deepening of cotton bales and crushing mills shown by figures 5.1 and 5.2 from 1870 to 1900. (Today that area defines the I-35 highway corridor between Dallas and San Antonio.) The area shared the same "clay loams from marly limestone" of the best fertile soil of the Mississippi delta, the *Oil, Paint, and Drug Reporter* (*OPD*) noted. It was there that the majority of growers fostered the state's cotton cultivation and defined Texas's first, prepetroleum reign as oil king.[19]

The mills also shifted inland in those decades, moving from coastal ports to the sides of cotton fields. That change was a consequence of expedience. This is visible as one difference between figures 5.1 and 5.2, showing the mills pulling away from the coast to the interior in the later 1800s. In the first generation of mills before 1880, farmers sent their discarded seeds packed in wagons to the mills. Mill operators—the "crushers"—processed the seeds, pressed the oil out of them, refined that oil, and barreled the liquid for shipping to markets across the Atlantic and to the US North. This was cumbersome. Crushers bemoaned how difficult it was to get to the coast "due to a lack of railroads."[20] By the end of the century, they had bankrolled new rail lines to aid local economic development and shuttle their oil, crushed on site, back to ports. The construction of rail lines connecting small inland towns to port cities facilitated a kind of pullback from the coastal crushing mills to inland, on-farm sites. Rather

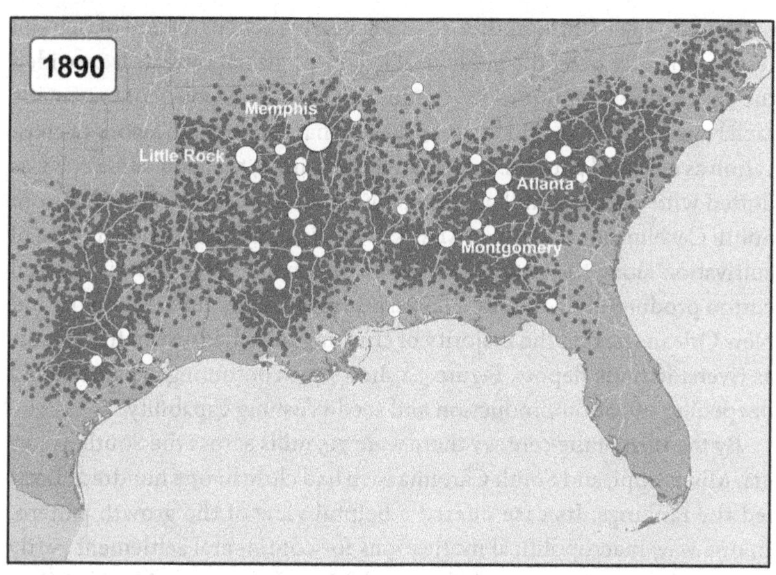

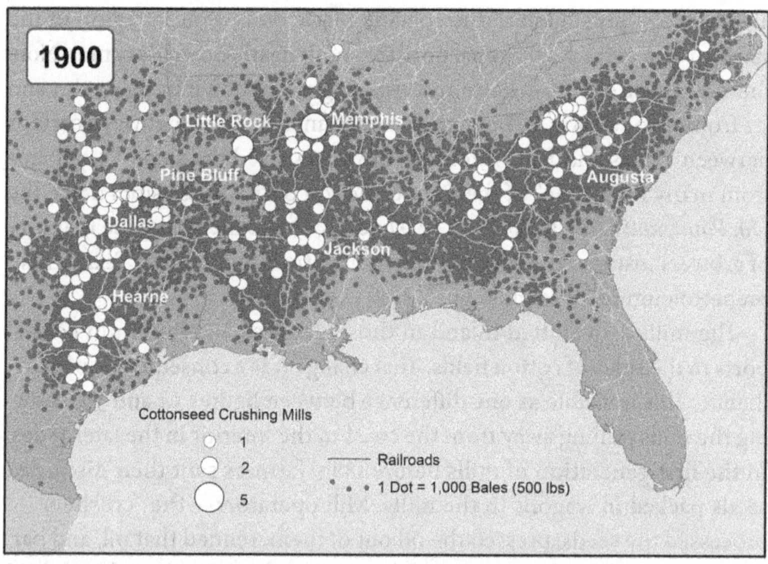

FIGURE 5.2. Cottonseed crushing mills mapped against cotton production, 1890 and 1900. Data imported from the National Historical Geographic Information System database and data sets compiled from US census bureau statistics and annual reports published in the *Oil, Paint, and Drug Reporter*. Maps created by John Clark and Matt Plishka.

than hauling seeds to a port city to be collected and crushed for oil, by the 1880s crushers handled the seeds on site and shipped the pressed oil in barrels to the ports by rail.

To cheerleaders and critics alike, the cottonseed oil industry seemed to have come out of nowhere, adding another suspected adulterant to grocery and kitchen counters alike. From seven mills to hundreds, from 5 percent to 84 percent of seeds crushed, from hundreds to millions of gallons, the industry expanded rapidly in just a few decades. The new oil was emblematic of the rise of processed and industrialized foods in the nineteenth century. Seasonal growing patterns, macropolitical factors, advantages of soil fertility, and new commerce-facilitating technologies—crushing mills and railroads—explain cottonseed oil's geographical growth and shift to the west. Postbellum labor politics and the new crushing industry explain how the pattern of oil extraction deepened across the entire South. Together, the extension and intensifying of oil extraction helped the southern cotton economy after the Civil War. Yet cottonseed oil's success was the consequence of more than southern politics and geography. It also had places to go. Those markets beyond the South made it easier for oilmen to increase their output.[21] They also made it harder for pure food crusaders to understand where the cotton oil ended up.

Where, then, did it go?

Olive oil and lard, for starters. Today's lively attention to the purity of olive oil offers a contemporary example of angst over adulteration. Tom Mueller, today's chief chronicler of olive oil adulteration, writes that as recently as the late 1990s "olive oil was the most adulterated agricultural product in the European Union." An investigator from the European Union's antifraud office thought that "Profits were comparable to cocaine trafficking, with none of the risks." Just a few years ago, Mueller could still write of oil adulteration as a tale of scams, dupes, fakes, crimes, and deception, casting the question of honest oil as being a quest for the "real" thing.[22] Cottonseed oil was a principal reason for this fear over a century ago. Olive oil, along with lard, was a thriving industry in the later nineteenth century. Both were suspected of being adulterated.

Europe was the primary olive oil producer then as now. Mediterranean nations were well suited in climate and geography for robust olive groves. At the time, Mediterraneans mostly used the oil pressed from olives domestically. They were not yet tied to the global market trade. Southern planters had experimented with olive groves in the antebellum period but without long-term success.[23] Californians began attempts to plant their own olive groves in 1860, the year a grower planted 503 trees. By 1900, there were 539,000 olive trees in the state. Competition from cheaper European imports, though, dogged the Californian efforts, which subsided by the early twentieth century. Exports from Italy, France, Spain, and England (via Gibraltar and other colonial holdings) into the United States led the rolls at the bureau of trade.[24] In the 1870s, the imports were on the order of one hundred thousand gallons per year; by the 1880s, this had risen to about four hundred thousand gallons per year; in the 1890s, customs was processing close to nine hundred thousand gallons of olive oil annually from Europe.[25]

Wholesalers sold the oil in two main categories, "salad oil" and "other." Salad oil was what we call salad dressing. "Other" uses included basic kitchen grease for cooking, as an ingredient in other foods, and as lubrication for kitchen equipment or domestic machinery. To assure its customers of the veracity of the oils, grocers sold the product in bottles and flasks labeled with certified claims of purity.[26]

When *Frank Leslie's Illustrated Family Almanack* reported on cottonseed oil manufacturing in 1882, it noted production levels of about "15,000,000 gallons in the United States," ten million of which "are yearly exported to Europe, where it is used to adulterate olive oil."[27] The state chemist of North Carolina lectured in 1883 that "Cottonseed Oil, when refined, is a pure, white, bland oil, a good deal like olive oil." He then reported that "the use of the oil is growing every day, and it is now largely used in adulterating olive oil."[28] Good old *Puck* took a swipe at olive oil adulteration too. They leveraged widely held beliefs that the imports were contaminated with cottonseed oil. They penned a parody that was a send-up of a speech by "Professor Riley." Riley—a thinly veiled stand-in for the USDA's Harvey Wiley—imagined that lettuce would object to having "rank cottonseed oil poured over me as a substitute for the best olive article." The lettuce, Riley tells his audience, would "feel disgraced and humiliated."[29] It was a matter of propriety.

The olive industry tied southern cotton plants to export markets in Europe and beyond. Meanwhile "the hog butcher for the world," as Carl Sandburg called Chicago, was already hosting a vibrant by-products industry that was ready to accommodate the cotton oil. It tied southern seed crushers to meatpacking facilities in the Midwest where they mixed cottonseed oil with lard.

Lard is the fat of pigs. Leaf lard has the highest value because it has little pork flavor. That makes it suitable to contribute to a wide range of recipes. Households used lard as shortening in baking in the 1800s, though it was also common grease, much like olive oil. When slaughtering pigs, meat-packers would separate the fat from the animal to render it with a wet or dry process. Wet rendering meant they boiled the fat and skimmed off the lard that separated at the top. Dry rendering was more like cooking it in a frying pan, where processors heated the fat and, as from a diner's grease pan, drew off the lard to collect and cool. Wet rendering led to "neutral" lard, a low odor product. Dry rendering had more flavor as a final product. By "more flavor" I'm trying to be subtle: that means the process produced an odor. The two biggest meat-packers in the world, Swift and Armour, were also the two biggest lard producers. These were the processes Upton Sinclair fictionalized in *The Jungle*, rendering vats so big and dangerous a man could fall in to be churned into the product.

Lard was big business. Finding a cheap substitute would be a boon for manufacturers. For consumers, unknown substitutes for lard were a threat to trust and possibly health. But there was just so much of it. Meat-packers processed three hundred to nine hundred million pounds a year from the 1870s to the turn of the century.[30]

Like margarine, it was a by-product of the meatpacking process. Margarine was tied to the practices that produced beef in the first place ("the caul fat of beeves"). Lard (and by extension its chief adulterant, cottonseed oil) was more closely tied to the processes that produced pigs. With the movement of livestock into penned stockyards and meatpacking facilities in the later nineteenth century, hogs, like cows, became a kind of machine in the eyes of the packers. They could be built in factories (the stockyards) and broken down into pieces in disassembly lines. Producers considered the hogs "corn . . . incarnate," little machines that took low-value corn and added value by making it into meat. Hogs were "but fifteen or twenty bushels of corn on four legs."[31]

Southern cotton growers interfered with the relationship between hogs and corn, between meat-packers and farmers. That was because cottonseed oil in meatpacking facilities disrupted the corn harvesting schemes of midwestern growers. The fortunes of one industry (meatpacking) depended on the fortunes of the other (cotton).

The relationship between northern corn and southern oil had a "greater influence on cottonseed-oil prices than any other factor," to quote historian Lynette Wrenn.[32] This was why: when corn harvests increased, so did the pounds of lard. More corn meant fatter pigs. Fatter pigs meant more lard. Thus, a higher corn yield led to more and cheaper lard. Meat-packers thus planned hog raising and slaughtering in league with corn planting patterns. This mattered for the upstart cotton-oil sellers. The patterns of hog raising and corn planting influenced the seed crushers' operations. "Knowing the size of the corn crop and corn prices aided crushers in estimating more precisely the number of hogs likely to be raised in a year and, consequently, the approximate amount and price of lard."[33]

It was basic market economics. When lard production was up, cottonseed oil production was down (fig. 5.3). With greater lard production, cheap costs meant you didn't need to adulterate. Cottonseed oil was therefore in less demand in such years. Lard supply goes up, cotton-oil demand goes down. When corn and hog production was down, so was lard, raising its price and encouraging the addition of a cheaper substitute (i.e., cottonseed oil). Lard supply goes down, potential for adulteration goes up.

At the risk of turning this into Econ 101, I can tell you that industry observers started to learn what seed crushers knew. The general interest magazine the *Forum* explained "Why the Price of Corn Is So Low." In the

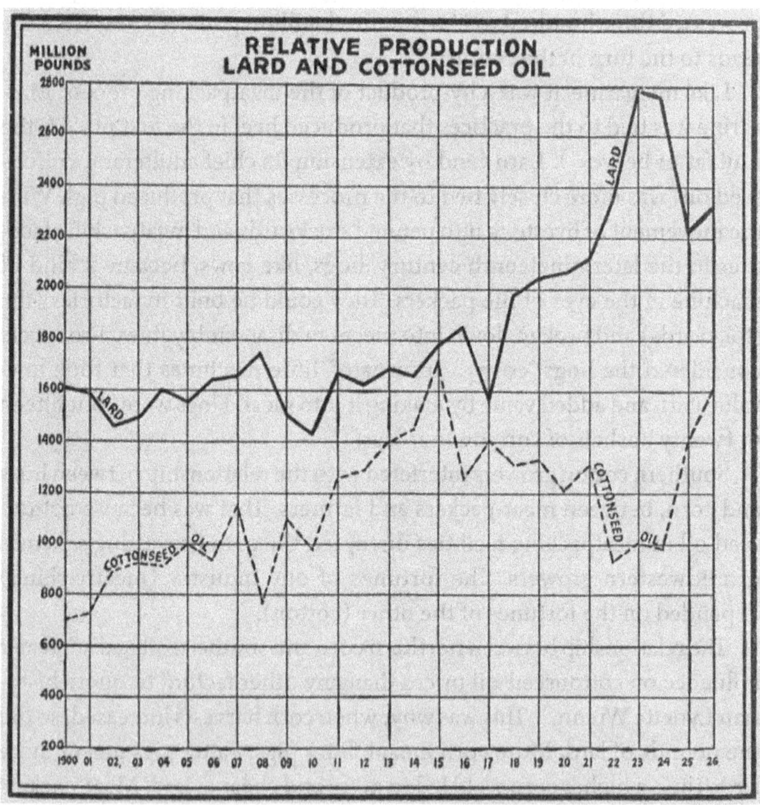

FIGURE 5.3. Relative production of lard and cottonseed oil. The x-axis shows yearly totals from 1900 to 1926. In general, when lard production was up, cottonseed shipments were down. Source: Clemen, *By-Products in the Packing Industry*, 121.

process, they identified the complex intermingling of the broader adulterant industries. "All forms of pork products, one half the beef, one-third the mutton, all the glucose, nearly all the spirits, much of the butter and cheese, as well as a large proportion of the oleo oil and margarine sent abroad are products of the corn field." The truth, they wrote, was "that the descent of the price-level has been directly due and readily traceable to the utilization of a by-product of the cotton-field"—cottonseed oil.[34]

What it meant for pure food crusaders was that it was getting increasingly difficult to trace back the singular source of foods. Pure food advocates had to decipher new integrated resource streams—cotton and crushed seeds, corn and lard, olives and foreign bottles—if they wanted to battle the final product at the grocers' counter.

The point is that cottonseed oil didn't come from nowhere, nor was it easy to disentangle once it arrived. It grew in tandem with southern cotton growers, the bottled output of Mediterranean olive groves, midwestern packinghouses—the hog butchers of the world—thousands of acres of corn grown to feed those animals, and the lard rendered and packed in containers once the hogs were slaughtered. Cottonseed oil's offense for purity advocates followed from its seemingly under-the-radar arrival.

So it was that critics were taken aback by the speed of the industry's rise—mills magically springing up, production increasing to millions of gallons in short succession. Billions of pounds of cottonseed oil came from the US South. Billions of pounds went to the Midwest and the Mediterranean coast. But the sudden growth of the industry wasn't itself a distinguishing feature of cottonseed oil. Oleomargarine was sudden too. From nowhere in 1869 it had gained global regulatory attention in just fifteen years. The seemingly sudden arrival of unfamiliar cottonseed oil was more notable to a cautious public because it was matched with a wide array of uses and end products.

The United States Tariff Commission conducted a survey of the American cottonseed oil industry in 1920 to find that "cottonseed oil is the most important vegetable oil used in food products." They calculated that the annual average of 1.5 billion pounds of domestic cottonseed oil produced between 1912 and 1919 was "greater than the combined production of all other vegetable oils, almost equal to the farm, small shop, and factory production of lard, and not so far short of the total production of butter." Although there was still a "popular prejudice against cottonseed oil," often leading its producers to conceal their wares under names like "table oil," "salad oil," "sweet nut oil," and "butter oil," the world war and a turning consumer tide had made the product more favorable. By the 1920s, the cottonseed oil industry was a dominant agent in domestic food production.[35]

A combination of foreign exports and domestic trade built that industry. The maps of figures 5.1 and 5.2 show the surge in production sites across the US South. Figures 5.4 and 5.5 chart an element of industry expansion beyond the fields of the southern United States and the packinghouses of the Midwest to the world. As an annual report on cottonseed oil explained in 1879, "a largely increased export is shown to England, Scotland, Italy, Spain, the West Indies and South American countries." That year alone, African nations, which "had not hitherto appeared among [the] export markets," took a considerable amount of oil.[36] The maps show a trend of increased scale and scope from 1880 to 1900. The number of export locations in the United States increased from two to eleven; the number of import nations and colonies across the globe doubled from twenty-five to about fifty. There was a nearly twentyfold increase in exports from about

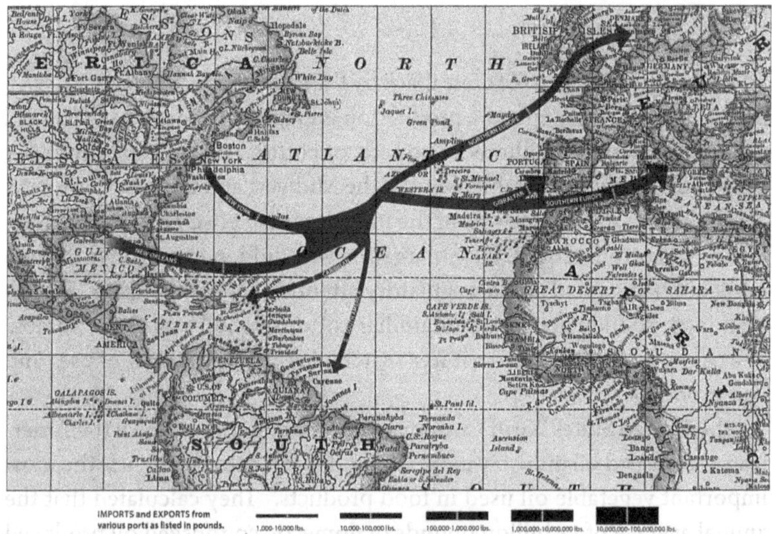

FIGURE 5.4. Exports of cottonseed oil from two US ports (New York and New Orleans) in 1880 going to six separate global regions. The largest export stream went to southern Europe (principally Italy and southern France). Britain brought most of its cottonseed oil in first through Gibraltar before shipment from there to the British Isles. The digital companion site https://purefood.lafayette.edu shows changes across twenty years of export patterns. Map created by Kristen Lopez.

three million gallons to fifty-seven million gallons in just twenty years. The industry spread its tentacles as if by magic.

The export streams show the scale and expanse of cotton oil's new global entanglements. By 1900 the bulk of the output, 86 percent of the fifty-seven million gallons, went to Europe. Oil sent to southern Europe went primarily to Italy and southern France. It could end up in olive oil that was sent back to the United States. Most of the oil sent to northern Europe—primarily Belgium, The Netherlands, Germany, and Austria-Hungary (44 percent of the total)—ended up in soap and candles. For purity advocates back in the United States, this meant that fighting adulteration at the grocers' counter was late in the game. Customers could suspect the grocer of adulterating flour with sand or sugar with grains, but they didn't usually think the grocer was mixing cotton oil under the counter with olive oil. That admixture happened far away and much earlier.

As early as 1880, Italian representatives acknowledged fear within the Italian government that the United States was becoming the chief supplier

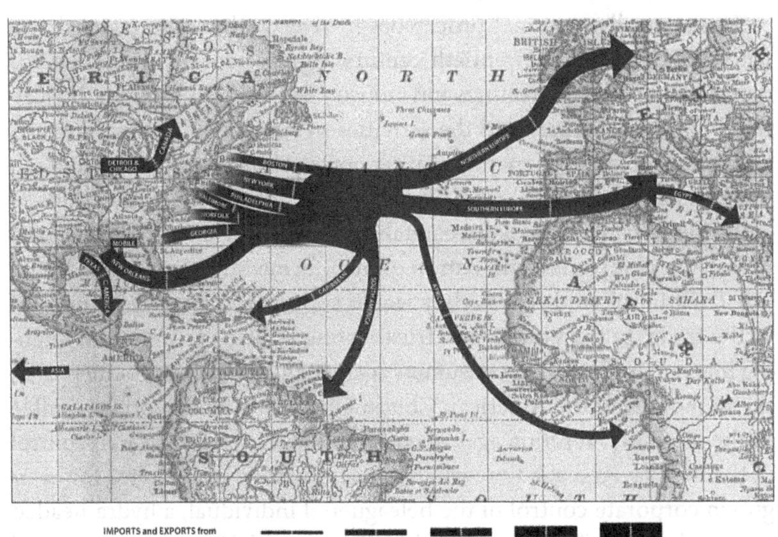

FIGURE 5.5. Exports of cottonseed oil from nine US ports in 1900 going to nine separate global regions. By this time, the largest export stream was still southern Europe (Italy and France), though northern European nations (The Netherlands, Belgium, and the United Kingdom) played a bigger part in the trade. By 1900, trade routes were restored to Egypt, having been closed off two decades earlier because of imperial British conflicts in the region. Map created by Kristen Lopez.

for the country's oil makers.[37] The Italian consulate wrote that "cottonseed oil threatens not only to make dangerous competition, as substituting the olive oil for various uses, but also in bringing the olive oil into disrepute as an article of food on account of its adulteration with the former."[38] It wasn't a one-sided worry; they saw this in America too. Mark Twain made the point in *Life on the Mississippi*, where his characters talked about how "the trade grew to be so formidable that Italy was obliged to put a prohibitory impost upon it to keep it from working serious injury to her oil industry."[39]

The export market from the US South to the world benefitted from new trade and lobbying groups and a vibrant consumer goods industry that expanded beyond lard and olive oil. A lively circulation in trade publications accounted for it all and hosted adulteration debates. New York's *OPD* was most prominent among them, a weekly paper edited and published in lower Manhattan. It broadcast commodity prices, advertised businesses in the oil trades, and hosted debates between members of the by-products networks. As figure 5.6 shows, oilmen of the prepetroleum age trafficked in dozens of types of oil for dozens of purposes, another sign that cotton oil ended up in a variety of products.[40]

Domestically, regional processors in river and coastal cities dominated the trade before twentieth-century aggregation (fig. 5.7). The *OPD* reported that regular meetings and industry conventions brought the scattered oilmen together until the activities of larger syndicates and trusts began to aggregate the manufacturing power of smaller crushers and oilmen into competitive blocks. That started to take shape in the 1880s. A collective of bankers, brokers, capitalists, Wall Street agents, officers from petroleum companies, and cotton oilmen formed the American Cotton Oil Trust in the mid-1880s. They would soon control nearly 90 percent of the trade. In the years when trustbusting politicians were first feigning interest in protecting consumers against unchecked corporate power, the monopoly of the American Cotton Oil Trust added another wrinkle to the consumer reaction against cotton oil because of its overall control of the market. And at a time when the octopus was a metaphor for overgrown corporate control of the beleaguered individual, a hydra-headed monster's neck of "cotton-seed-oil-lard" furthered the perception that oil was to be distrusted.

Part of the aggregation that led to the American Cotton Oil Trust followed from the needs of foreign demand evident in figures 5.4 and 5.5.[41] Part of it was the opportunity to expand oil uses for almost anything (again, the nineteenth-century version of soymilk, almond milk, rice milk,

MINERS' OIL CO.,

P. W. MEAD, Proprietor,

MAIDEN LANE, NEW YORK.

SPECIAL NOTICE TO THE DRUG TRADE.

We are prepared to put up PURE OILS in 5, 10, 15 and 20 gallon cans to suit druggists. Will sell at the lowest market price, and charge only the additional cost of packages.

OILS.	OILS	OILS.
LINSEED,	WHALE,	LARD,
RAPE SEED,	SPERM,	TALLOW,
MUSTARD SEED,	MENHADEN,	NEATSFOOT
COTTON SEED,	BANK,	PARAFFINE
OLIVE,	STRAITS,	ROSIN,
CASTOR,	COD,	WEST VIRGINIA
MACHINERY and	ENGINE,	ELEPHANT,
SALAD,	FISH,	COCOANUT,
RED,	SPINDLE,	SEWING MACHINE
	AXLE GREASE,	

FIGURE 5.6. Advertisement showing the extended array of oils available from Miners' Oil Co. Source: *OPD* 10 (August 9, 1876): 20.

etc.). These were efforts at edible and inedible markets. One member of the American Cotton Oil Trust, N. K. Fairbank, offered an example of how intertwined the cottonseed oil market had become with the noncotton range of products.

Fairbank pioneered oils, fats, and soaps in Chicago and became one of the more powerful members of the American Cotton Oil Trust. His lard was a popular kitchen shortening in the 1880s. (Shortening refers to short dough, which crumbles more readily and is thus useful for pastry, pies, or any crusted baked good. Long dough is stretchier.) Lard and soap, Fairbank's other primary product, were derivatives—by-products—of animal slaughtering that relied on animal fat (fig. 5.8). Not coincidentally, Fairbank housed his operations in the Packingtown neighborhood of Swift's meatpacking operations and Armour's vast animal slaughtering empire.

COTTONSEED OIL,
For Export.
J. F. O'SHAUGHNESSY,
P. O. Box 1,206. 142 Pearl St., N. Y.

A. A. Maginnis' Sons,
NEW ORLEANS, LA.
Refined, Crude Bleached,
COTTON SEED OIL,
Oil Cake, Meal, Reginnings
AND
SOAPS.
Liberal cash advances made on Consignments of Cotton Seed Oil and Cake;
AND OTHER MERCHANDISE.

Planters' Oil Company,
NEW ORLEANS,
Manufacturers of
Cotton Seed Oil, Oil Cake, Lint and Soap Stock,
ALSO COTTON SEED MEAL,
FOR CATTLE FOOD AND FERTILIZING.
AMES WINSHIP, Sec'y. JOHN W. TOBIN, Pres't

J. V. LEWIS & CO.,
Manufacturers of
Cotton Seed Oils,
Winter Yellow, Winter White, Summer Yellow, Summer White Oils
YELLOW AND WHITE
COTTON SEED STEARINE, SOAP STOCK, &C.
Works: 118 to 133 East Sixth Street, Cincinnati.
Orders from Jobbers and Manufacturers Solicited.

CRESCENT CITY OIL CO.,
NEW ORLEANS,
Manufacturers of
Cotton Seed Oil, Oil Cake, Lint, Soap Stock,
AND COTTON SEED MEAL.
FOR CATTLE FEED AND FERTILIZING
J. H. DUGGAN, Secty. D. F. KENNER, Prest.

DIXIE OIL COMPANY, Nashville, Tenn.,
And AUGUSTA OIL COMPANY, Augusta, Ga.,
MANUFACTURERS OF
Crude & Refined Cotton Seed Oil, Oil Cake,
LINT, AND OIL MEAL
For Cattle Feed and Fertilizing, and Superior Laundry Soaps.
HENRY SPERRY, Sect'y. ROB'T THOMPSON Pres

FUTURE CITY OIL WORKS, St. Louis, Mo.
and VICKSBURG OIL WORKS, Vicksburg, Miss.,
MANUFACTURERS OF
Crude, Refined and Bleached
COTTONSEED OILS,
Winter Yellow Winter, White, Summer Yellow, Summer White, Cottonseed Stearine, Soap Stock, Reginnings, Etc. Also Cottonseed Oil Cake and Meal for cattle food and fertilizing purposes.
J. J. POWERS, Proprietor,
605 South Levee Street, St. Louis, Mo.

UNION OIL COMPANY,
PROVIDENCE, RHODE ISLAND,
Incorporated A. D. 1856, Manufacturers of
White (Odorless) Winter Cottonseed Oil,
Yellow Winter Cottonseed Oil--Extra Quality.
Cottonseed Meal and Cake,
Fulling and Scouring Soaps,
of superior quality, and warranted of uniform strength. The only manufacturers and exclusive proprietors of the well-known brands—
PURE SALAD OIL ... OLIVE FLAVORED OIL
SALAD OIL IN BOTTLES—Quarts, Pints and Half [illegible]
CRUDE OIL DEPARTMENT—Gretna Mills, New Orleans, La.
REFINERY—Providence, R. I.
For quotations, circulars and further information, address
LYMAN KLAPP, Agent,
No. 203 & 205 INDIA ST., PROVIDENCE, R.I.

FIGURE 5.7. Front cover of the *Oil, Paint and Drug Reporter* showing the range of cottonseed oil and other prepetroleum oil companies. Source: *OPD* 15 (January 15, 1879): 1.

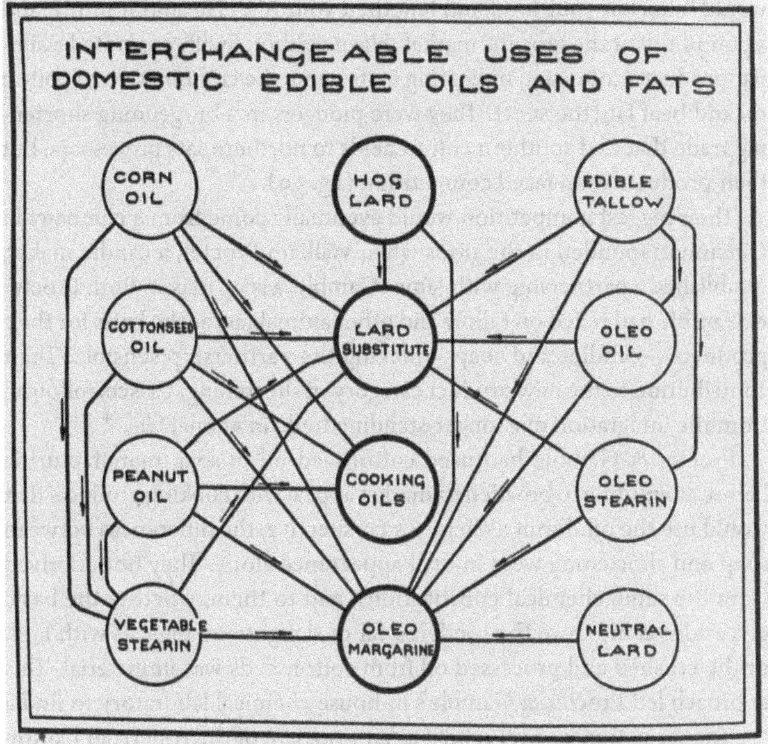

FIGURE 5.8. Interchangeable Uses of Domestic Edible Oils and Fats. Source: Clemen, *By-Products in the Packing Industry*, 150.

Fairbank thought shortening and soap could be made with cotton oil instead of, or at least with reduced amounts of, animal fats.[42]

He thus built an affiliation with the Cottonseed Crushers Association and developed a cozy relationship with the southern Cotton Trust to help his entrance into this new market. One of his managers was on the trust's board. The Cotton Trust incorporated as the American Cotton Oil Company in the late 1880s. (It was one of the twelve original companies listed on the Dow Jones Industrial Average.) As a holding company, it tied southern crushers to northern animal fats processors. Historian Susan Strasser explained the relationship well: whereas "The American Cottonseed Oil Company and the Southern Cottonseed Oil Company dominated the selling, Armour and Company and the N.K. Fairbank Company, both of which manufactured cooking fats compounded with lard, dominated the buying."[43]

Fairbank was an industry leader as he worked with southern crushers to provide an alternative to lard. The product line he developed, Cottolene, would "shorten your food and lengthen your life." He sold it in 1, 3, and 5 pound tins at the grocers' market. His neighbor, Swift, promoted a similar product, Cottosuet, indicating in its name the combination of cotton oil and beef fat (the suet). They were pioneers in a burgeoning shortening trade that tied southern cotton fields to northern fats processors, but their products soon faced competition (fig. 5.9).

Their biggest competition would eventually come from a company in Cincinnati founded in the 1830s when William Procter, a candle maker, established a partnership with James Gamble, a soap maker. Both Procter & Gamble had relied on tallow and other animal fats as the basis for their products—candles and soap—making the partnership sensible. Their contribution to the new product category of shortening, Crisco, followed from the integration of a longer-standing trade in animal fats.[44]

Procter & Gamble had used cottonseed oil in soap manufacturing before attempting to broaden its market appeal with cooking products that would use the oil. From a chemist's perspective, the differences between soap and shortening were in final appearance alone. They both derived from the same chemical constitutions, and to them, whether the basic glycerides came from the rendered fat of slaughtered pigs, as with lard, or the crushed and processed oil from cottonseeds was immaterial. This approach led Procter & Gamble's in-house chemical laboratory to find a use for the cottonseed oil it held as an associate of the American Cotton Oil Trust. As advertised, there's was "An Absolutely New Product, A Scientific Discovery Which Will Affect Every Kitchen in America." Neither Fairbank's Cottolene nor Swift's Cottosuet endured in the marketplace far into the twentieth century. Poor market plans and macroeconomic factors like the Panic of 1893 help explain the failure. The two products also stumbled because they arrived too early in the history of cotton oil for public acceptance to lock in. They were too open about the product's source.

When Procter & Gamble introduced their crystallized cottonseed oil in the early twentieth century with its less overt trade name, Crisco, they crafted a new market that lasted. Its success was instructive as a marker highlighting the fading era of adulteration. Two decades earlier, Fairbank boasted of the natural advantage of the sunny South's raw material, but that was too uncloaked. Procter & Gamble benefitted from the fuller, more widespread distribution of the ingredient into so many products. It could use the oil without having to tip its hand too openly. Although the Crisco on today's shelves is no longer made from cottonseed oil—the company shifted to cheaper vegetable oils in the twentieth century—it remains a legacy of the by-products industry of the nineteenth century.[45]

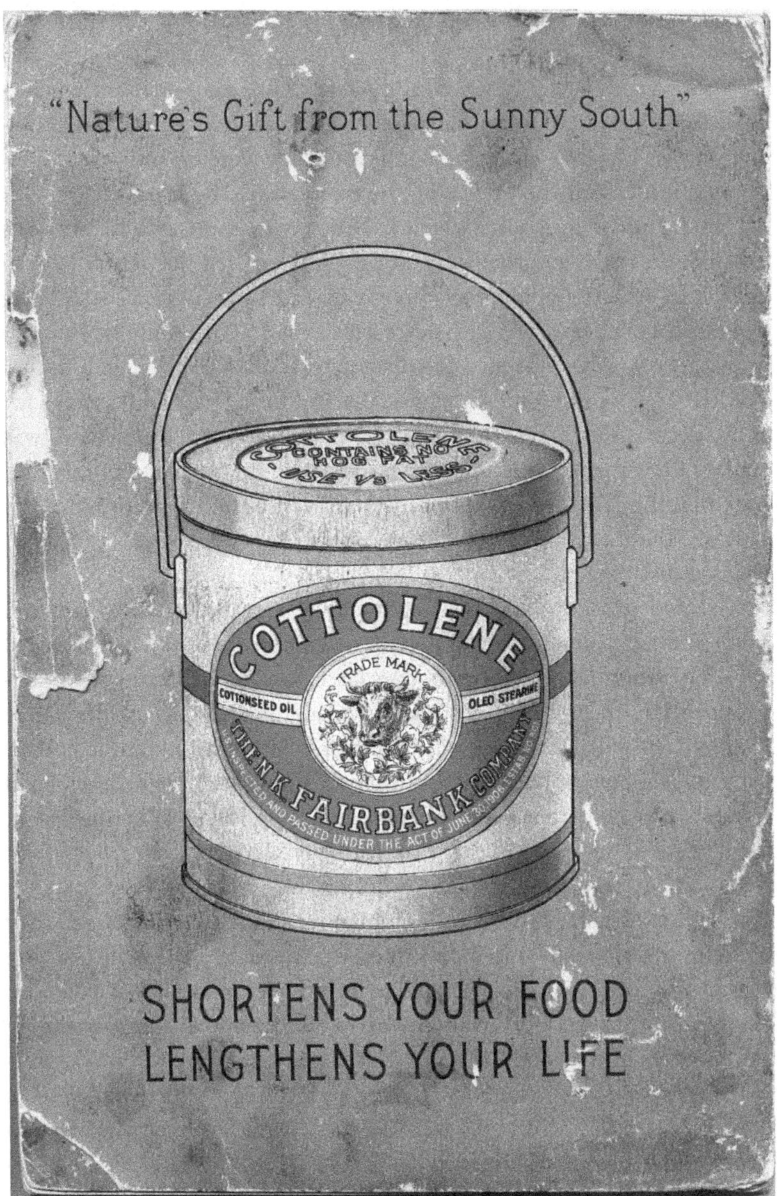

FIGURE 5.9. Advertisement for Cottolene, early 1900s, drawing connections to nature. Courtesy of Duke University Archives.

The Fight over Oil

The war on cottonseed oil was controversial. It was also less sharply drawn than that for margarine, perhaps less hostile, and not as protracted. Oilmen sat on one side. They were in favor. They were on the agricultural production end of the life cycle, the one canvassed above. Customers sat on the other. They were worried. Their contact was with the markets and storefronts of an urbanizing world. They entered the debate later in the game, once the oilmen had already made a range of choices to bring the product to the market. Grocers sat in the middle while working to appease both sides. Grocers, the distributors, often equivocated. In that way they were in the middle physically and morally. It might make for more dramatic reading if the story of cottonseed oil adulteration was strictly binary, good or bad, reviled or loved. The environmental complexities, though, were matched by complexities in the response to the new oil. The story of cottonseed oil adulteration is one with more nuance than an easy binary.

Granted, grocers could have legitimate reservations about cotton oil. Under new consumer-protection laws, some of them were being arrested for adulteration, leading them to more circumspection for the sake of self-preservation. In an 1897 court case in Philadelphia, to take an illustrative example, the judge found a grocer guilty of substituting cottonseed oil for olive oil in violation of the state's new pure food law. The grocer claimed ignorance as his defense, but the prosecution prevailed. Whether or not the ignorance defense had merit, the case revealed that even grocers, not just their customers, claimed to not know what was in their products.[46]

On the whole, the grocers were ambiguous about oil adulteration. They sometimes advertised impure products openly, defending their use. The Chicago *Grocers' Manual* explained to its industry readers, "olive oil pure is almost impossible to buy at any price." The guide even unabashedly told grocers how to deceive their customers. "Put on a foreign sounding label.... Pasting an inch and half of tin foil around the cork and neck completes the job."[47] Yet they were in a tricky spot. At once they tried to promote a professional outward character of honesty to consumers worried about impure products while working behind the scenes—under the surface—with wholesalers and meat-packers to secure contracts and shipments of new products. Their fidelity, in the most generous interpretation, was to garner trust so that they could find lower prices on two sides, both from their wholesale contact and for their customers.

The *Spice Mill* raised the issue of surreptitious oils too. Coffee and spice grocer Jabez Burns published the confusingly titled New York–based antiadulteration paper. He wrote repeatedly about the legal ramifications of unknowingly trading in illicit goods. (I say more about him in chap. 7.) In his experience, the problems of cottonseed oil adulteration were problems of an export and import market. The *Spice Mill* told readers foreign foods were a route to contamination and, to Burns, cottonseed oil was an especially egregious example. He saw that shipping oil from the mills of Texas to Italian and French oil pressers, who in turn shipped their by-product–mixed compound back to US customs houses as "pure," added to the oilmen's profits at the expense of consumers' trust. Burns argued that the benefit of economic growth that followed from global trade came at the cost of increased traffic in adulteration. What was an olive oil–selling grocer to do?[48]

For Burns and the *Spice Mill*, it wasn't even clear that cotton oil was illegitimate so long as he could tell customers what it was. To him and many others, the mere fact that it was a substitute did not automatically cast it as inappropriate. Burns saw that cultural specificities played a role in the assessments of cotton oil's propriety. The kosher questions that got me in trouble in my poorly litigated margarine versus butter case were relevant here because lard's presence in the grocers' empire had specific cultural and ethnic connotations. For some reason Burns, who was not Jewish, had a particular interest in "Hebrews" in the *Spice Mill*, so he was keen to address the kosher issue. "The bulk of olive oil was formerly used by the Jewish population," he wrote, "for purposes for which the uncircumcised would have substituted lard."[49] Thus, while some worried that cotton oil could confuse the identity of cooking grease, others thought that it could be a boon for the households of immigrant Jews. It would preclude the need for pig fat in baking.

Burns's contemporary in the antiadulteration press, the *American Analyst*, raised the same issue. In "Lard and Leviticus," the paper told of "the Hebrew housewife exulting in a cotton seed product." Quoting the *Hebrew Journal*, it shared in astonishment the possibilities of kosher lard: "The other day I received a circular extolling the merits of kosher lard—that is, lard made from cotton oil. And now the only thing left for the ingenuity of the American inventor is to discover some process by which bacon and ham can be rendered kosher enough to meet with the approval of Rabbi Joseph himself."[50]

I can't report whether or not the Hebrew housewife ever found kosher ham. Or I can: she never did. But I can say the point about cultural commitments was brought home through other ethnic considerations too. Bakers used lard as shortening, for instance. They used shortening more commonly when baking leavening breads such as rye and wheat. Those were dominant bread varieties for Europeans who immigrated to America. Those cultures were also the ones historically familiar with dairying and swine husbandry. In that way they already had butter and lard experience. Cooking aids—shortening for bread—were in this sense an ethnically defined need, the demand culturally situated rather than inevitable. Amid the grocers' equivocating replies to adulteration, cotton oil showed that for customers the argument over the purity and propriety of new foods was neither preordained nor obviously defined. It was instead wrapped in particular cultural commitments.

Other forums found speakers accepting the premise that oils and lard were adulterated with cotton oil. Rather than deny it, they leaned into it and argued it wasn't a big deal. That view pervaded a pure food convention in 1887. New York grocery magnate Francis Thurber (chap. 3) spoke to the forum, acknowledging "It is a well known fact that almost the entire product of lard is adulterated to some extent with cottonseed oil, tallow, and other ingredients." The National Board of Trade, a meat-packers' lobbying group, drew distinctions between different kinds of adulteration as fervor for federal legislation grew. Thurber took his turn to rank by-products on a scale of lesser or greater degrees of impropriety. He argued that while it was true "spurious lard" was problematic for consumer trust, cotton oil wasn't necessarily a bad thing. I don't know that it could be as easy to distinguish the spurious from the pure as Thurber suggested—chemists certainly tried—but Thurber thought it was, offering a booster's gloss on adulteration and passing off concerns, as he had in other forums, as the price of progress.[51]

Whether or not boosters or critics were right, in all of these views consumers and their advocates had a more nuanced approach than a for or against binary. Where margarine had been a purportedly insidious attack on dairy livelihoods and consumer health, cottonseed oil was suspect more because customers sometimes didn't know whether it was there. Where margarine came under attack for fears that it would make eaters sick, critics challenged cottonseed oil less because it was unhealthy and more because it was surreptitious. More cheating, less poisoning.

In his role as a consumer protection agent before such a label existed, the USDA's Harvey Wiley also offered a nuanced view of cotton oil. Wiley knew of corruption and fraud. "Dealers have left the word 'olive' off of their bottles and are selling such oil as table oil," he said, "being careful not to put the name 'cotton' on it. This is a fraud of great magnitude, and affects our consumers rather than our producers." But he clarified that using cottonseed oil was a matter of taste, not health. So long as it was properly labeled, customers could decide for their own sake whether they wanted it. "Personally, I prefer the flavor of the olive oil," Wiley told the chairman of the 1899 adulteration hearings. "I know people who prefer the flavor of the cotton-seed oil, but cotton oil sells for only about one fifth the price of olive oil, and hence if it is sold as olive oil, as is done to a very large extent, the profit is enormous and the fraud is correspondingly great." So long as a customer knew what it was, whether or not one bought cottonseed oil was "a matter of taste."[52]

Others made the same point from the kitchen side. A housewife in St. Louis thought cottonseed was superior to lard for cooking purposes. She reported that its taste was imperceptible except when it helped, as in flaky biscuits.[53] Domestic economists and cookbook authors defended its use in a similar fashion. Sarah Rorer's (1883) *How to Use Olive Butter: A Collection of Valuable Cooking Recipes* was one such case. Olive butter was a spread made from olive oil. Writing from her Philadelphia cooking school, she told readers how to use cottonseed in their recipes instead of olive. A series of local papers in Rorer's hometown noted the wholesomeness of cotton oil while acknowledging the fear of impurity. "Frenchman have for a long time resold the oil to us as pure olive oil," the *Evening Telegraph* said as a matter of well-understood fact. Unlike the Chicago grocers who knew the label was itself a questionable marker of honesty, Rorer told cooks that so long as the product was "warranted as pure" they need not worry.[54]

Given this open secret of cotton-oil mixture, manufacturers grew bolder as boosters for their new product even as broader public acceptance still awaited them. They went a step farther to brag about it rather than hide it. They imbued the oil with the unfettered past I noted at the top of the chapter. Giving a mystical, almost supernatural explanation of its astonishing growth helped make it seem like a marvel. Wrapping it in a story of progress also blunted the accusation that the unknown was unscrupulous. At the 1905 Interstate Cotton Seed Crushers Association convention, oilman John Allison thought that the "despised Cinderella of the Kitchen had become the radiant belle of the ball." A few years later, the group's spokesman liked the line so much he reworked it to say "that poor little Cinderella has been lifted from the dust heap and enthroned the royal consort of King Cotton."[55] Speaking for his company, Herman Armour happily praised "the lard industry . . . as a large consumer of prime cotton oil, not for the purpose of cheapening the product, but to enhance its quality and value."[56] Rather than a troubling cheat on society, its boosters thought it was a blessing.

The Cotton Oil Product Company of New York fit the larger trend

FIGURE 5.10. Label for the Cotton Oil Product Company, incorporated 1888 in New York for exports to Latin America and Spain. This "Butter of Cotton Oil" is "Made from the highest degree of Cottonseed Oil, yellow, refined." As the pig icon has it, the cotton butter is expressly advertised as an alternative to pork ("contra puerco"). Courtesy of the Library of Congress.

of leaning in to the novelty rather than hiding it (fig. 5.10). They were a Spanish firm incorporated as La Compania de Productos de Aceite de Algodon in 1888. They operated as an exporter to colonial Latin America, arguing in promotional brochures that their cotton oil was "pure, healthy, clean, and safe." This stood in contrast, they wrote, to the rampant adulteration of lard by the "heads, feet, and guts" from sick animals that contaminated lard as it was.[57] They would have shared in Thurber's easy distinction between spurious and pure lard.[58]

While boosters minimized the problems to foster the "pro" side, critics less ambiguous than grocers kept questioning the oil's legitimacy.[59] The view from the other side of the Atlantic offered another example of the "con" side. Trade barriers were one measure of this, a chance to look back on the United States from afar instead of from the United States outward. French, German, and English traders sought protection from what many of them considered deceptively presented American goods. They, too, worried that the oil might be unhealthy because of the way it was processed, that the new oil was unsafe and inappropriate, but mostly they just didn't want to be tricked at the customs house. The *Commercial Bulletin* reported that France, Germany, and England (along with Canada and Mexico) were unanimously upset with imports of fraudulent American lard. If they ordered lard, they wanted lard. The pro-US *Bulletin* sought to assuage their concerns by noting that proper labeling and an understanding of the healthful and cost-saving advantages would prevail.[60]

The same concerns worked in the other direction. Olive oil lobbyists spoke to the import trade, representing foreign houses at the New York docks as they sought to block the adulterant's entrance. They were just as worried about their industry gaining a reputation for fake olive oil as the consumer was worried about buying it. Antonio Zucca was an agent for a number of Italian producers. He was worried that his boss's olive oil was being unfairly maligned because of the actions of morally compromised adulterators. He spoke to the 1899 Mason Commission with "the honor of being president of the Italian Chamber of Commerce of New York" to defend his homeland's industry. It was too expensive to ship cottonseed oil to Italy and then back to the United States, he said. That's why "people who adulterate this oil import olive oil from Europe and adulterate it here in New York." In other words, it wasn't Italians, it was dishonest Americans. If pure food legislation could ferret out those port-side adulterators, Zucca and his Italian supervisors would be happy to support it.

The argument against cottonseed oil was in many ways structured like the war on oleomargarine. It, too, had a strong he said–she said dynamic—with cottonseed oil, it was meat packers, it was seed crushers, it was foreigners, it was customs agents. But the vitriol was more contained than that with oleomargarine. Oleo disrupted the agrarian complexities of dairy livelihoods and presented a confusing face to consumers unsure of its health status. But there weren't preexisting environmental connections to cotton fields that could be disrupted in the same way. The oil from crushed seeds was a new agricultural sector of the economy rather than one that displaced others. The angst over cottonseed oil, rather, was aimed at the products into which it was mixed, the lard and olive oil. The argument over adulteration was more about misleading mixtures and unclear processes, more about cheating than health. For all the rancor of the 1860s and 1870s, when the upstart oil first made its name at a broad scale, by the end of the century the supply chain, by-products offerings, and agricultural process were established firmly enough to overcome the resistance Berghaus still thought was in place in 1887.[61]

The journalist Tom Mueller has been cataloging corrupt olive oil practices and advocating for best practices for over a decade. His current project, "Truth in Olive Oil," seeks to advise "consumers about buying and enjoying oil, connecting them with skilled oil-makers, celebrating the culture of this storied substance, and calling out fraudulent oils which cheat consumers and undercut honest producers."[62] That Mueller is doing this work is admirable. That he needs to would be a disappointment to those from a century past. The world of fraudulent oils, cheated consumers, and honest producers is the world of modern food systems. The origins of cottonseed oil are not so distant a cry from the troubles brought by Mafia-connected Italian olive groves today. Those origins explain the modern infrastructure of an industry that makes it difficult still to maintain clean and clear lines from farm to fork for olive oil consumers. I don't know what role cotton oil plays in adulterated olive oil today, but I can speak to the prevalence and endurance of the metrics of honesty, integrity, authenticity, and purity that were built into Western culture in the late nineteenth century. That means the modern infrastructure of note is not just physical—the crushers, processors, shippers, and bottlers—but cultural and environmental as well.

The southern United States to southern Europe traffic evident in figures 5.4 and 5.5 was one route of oil distribution making the cottonseed debate complex. The other main avenue of traffic, from the crushers of the US South to the meat-packers of the Midwest, only added to the view that the grocer's new customers of the late nineteenth century were not buying what they thought. Most of the things that we now understand as "by-products" of the meatpacking industry were cut with cottonseed oil in the late nineteenth century. They were adulterated, as it were. Cottonseed oil was an impurity. Not just one of three heads on a monster of adulteration, though, it was an outcome of earnest development and market growth, postbellum Southern politics and agrarian transformation, midwestern agricultural industrialization, eastern urbanization, and expanded global trade networks. It was a member of a value proposition that sought to reduce wasted components and maximize efficiency, in this case the efficient use of resources in the cotton fields of the South, the new facilities of the big meat processors like Swift and Armour and even the olive groves of southern Europe.

To escape the debate over adulteration, cottonseed oil became normalized. The market success of Crisco, launched in 1911, provides a benchmark to recognize that transition to normalization. Cottolene came on the market in the 1880s, Cottosuet in the 1890s, both arrived in the middle of a crescendoing debate, and neither survived far into the twentieth century.[63] Crisco, by contrast, was the product of a well-developed network of suppliers and processors by that point, the consequence of an industry that fostered by-products as a point of repute, not denigration. Procter & Gamble, the soap and candle makers, had effectively grown their business beyond the limited range of animal fat uses that got them started.

Consumers didn't think margarine was being shipped from Europe to corrupt their plates. That battle was more domestic and more about disruptions to the dairy world. But cotton oil stoked fears of the foreign in ways that had it swept up in general paranoia about foreign contamination. The industry lobbyist who thought cottonseed oil was garbage in 1860, fertilizer in 1870, cattle feed in 1880, and table food in 1890 had the basic trajectory correct even if he was overconfident about how acceptable that transition was and how soon it occurred.[64] But the trajectory pointed to another reason the oil became conventional. Rather than undermining agricultural practices in ways that brought fury over margarine, cottonseed oil expanded the southern cotton growers' opportunities.

I couldn't say worries about the adulterating identity of cottonseed oil vanished after the 1910s—Liberty Hyde Bailey certainly considered it as corrupt as any number of other contaminants—but the broad public rancor that had previously put cottonseed oil into the same debates as other adulterants largely subsided. The new industrial marketplace of fats and oils was so integrated into the grocers' shelves and home kitchens that consumers relied on the label to ensure confidence. It was also the case that the health fears that pervaded the war on oleomargarine were less emphatic with cottonseed oil. In this case, the health concerns were less about the product itself, refined cottonseed oil, and more about the ways it was produced. That form of anxiety over the process of producing the accused adulterant would be even more pronounced with the third head of Berghaus's hydra, glucose, the subject of the next chapter.

6

Glucose in the Empire of Sugar

For the benefit of the glucose-eating public we propose to give the bottom facts regarding that mysterious substance. . . . The whole sequence is like this: corn, corn-starch, grape-sugar, glucose, molasses, sickness, pain and sudden death.

Wasp, 1883[1]

"What Fools These Mortals Be." Brooklyn's *Daily Eagle* headline on January 9, 1889, didn't mince words. It was a story about a con man. The paper tagged it with a reference to the character Puck from "A Midsummer Night's Dream." The mischievous fairy uses the line, calling mortals fools, in act 3, scene 2. Puck the sprite was an expert trickster. The *Daily Eagle* was uncovering a trickster's swindle. It was so audacious they compared it to none other than Alfred Paraf.

In early January of the new year, "Professor" Henry C. Friend's sugar con came to a crashing halt. He had a few irons in the fire at the time. One was a process for preparing sugar from corn and other starches developed earlier in the decade in Chicago. The sweetener led to the invention of corn syrup, which continues to vex health officials in its high-fructose form (HFCS). A second process was the one leading to his public downfall. He was peddling a process he claimed turned low-grade sugar into a highly refined, high-quality sweetener. In both cases, Friend was leveraging the flexibility of the idea of a "pure" sugar. That flexibility led others to process sugars from such a wide range of plants that the notion of a pure sugar—which for many people had been cane—lost its purported clarity. Friend found an opening to contrive new artificial sweeteners and benefit from the loss of a clear notion of purity.

Friend, who sometimes went by Freund and who wasn't a professor, swindled investors for several years with his evocatively named Electric Sugar Refining Company (fig. 6.1). The operation ran at 18 Hamilton

FIGURE 6.1. Stock certificate for Henry Freund's ill-fated scam, the Electric Sugar Refining Company. Source: Wikimedia Commons.

Avenue in present-day Red Hook, Brooklyn. It was a grand brick-facade factory, a "big building" that had formerly housed the Atlantic Flour Mills. In their comparison, the *Eagle* thought that Friend and Paraf were both men "of great intelligence," swindlers who were "quick to realize that the strain of avarice in human nature can be profitably played upon." It was clear, they said, "that there is little or no difference between the doings of the marplot [Paraf] and those of professor Friend."[2] To a weary public, Friend took the crown as yet another challenger to authenticity with "the most successful, audacious, and 'barefaced' swindle of the present generation." More than a local scam, his secret electrical refining process was "the bugbear of the sugar trust." It was ultimately his undoing.

The historical record is unclear about Friend's actual heritage and origins. The "Friend" versus "Freund" confusion was one he perpetuated in his signature, not just a result of poor reporting in the papers. His claims to be a German chemist were likely an intentional maneuver. They fit the cult of scientific personality he built through the "professor" title.

Friend's early alternative sugar efforts in Chicago yielded interest from investors. He was the president of a new company in 1881, the Grape and Cane Sugar Refining Company. That was the first of his two processes. But the spotlight of a lawsuit for embezzlement—he had "appropriated [stockholder] money for his own use"—ran him out of town in 1882.[3] Before his arrest, he joined with the similarly named Chicago Grape and

Cane Sugar Company to try to displace the power of what would become the Sugar Trust. A few years later, in 1888, he popped back up in Brooklyn.

Bystanders were apparently unaware of his criminal past. Nor did they know this was his second effort at a new sugar. They took great interest watching him rehabilitate the closed facility on Hamilton Avenue. He scaled up the factory and brought in investors from England and the United States. Of the ten thousand shares offered to capitalize the venture, he kept six thousand. Three thousand went to English investors. One thousand went to his company's president, an old grape-sugar associate from Chicago, William Cotterill. All this investment, hundreds of thousands of dollars, was bankrolled on the promise of a secret electrical process meant to change the sugar industry.

This is how it worked: Friend and his accomplices set up a room with a machine covered by cloth. The machine was supposedly an electrical processor that would act on the sugar to refine it from raw to high-quality pure. Since this "mysterious something carefully covered up" was a secret as yet unpatented—or not patentable, Friend told investors—he couldn't allow others to see it. "In order to convince skeptical sugar men and inquiring capitalists that the thing would work," one newspaper reported, "he invited a number of them to witness the operation of refining sugar." Visitors and potential investors would tour the room to see the machinery, which included crushers, chutes, sieves, belted wheels, wooden barrels, bins, and centrifuges. After the tour, the "investigators were turned out of the room and the door locked behind them." Friend then ran his "electrical" process and invited the men back in. When they returned, they found that "the clean barrels were full of the finest quality of hard sugar." His "presto-change performance" turned empty barrels into bins of gold in the form of highly refined sugar, all accomplished with the fantastical electrical process.[4]

What actually happened was this: Friend had the refined sugar already on hand but concealed out of sight. Nothing actually went on under that covered machine. There was no electrical process to speak of. Rather, once the room was empty, Friend and colleagues would simply dump the hidden refined sugar down chutes "through an arrangement of nine sieves," into the barrels. The raw sugar waiting for refinement was emptied through a side chute and whisked away to another room. It was audacious and, for a year, convincing.[5]

A decade earlier, Alfred Paraf had booked it to South America to evade the consequences of his fraud. Friend had no escape but the grave. He was known for his taste in expensive brandy and passed away in March

1888, reportedly of an illness brought on by alcoholism. (*Harper's Weekly* didn't believe it: "at least it is supposed he died.") The swindle was exposed nine months later after his wife Olive left the factory for her hometown in Michigan. Her fleeing left investors, most of them foreign, many of them from Liverpool, high and dry. The police set up a sting. They were ready to catch Olive, her family the Howards, and her coconspirators the Halsteads. They tracked her back to Michigan, staked out Detroit ports, and ensnared them in an ever-tightening dragnet. In mid-February, the New York Police arrested the electric refining "sharps." That June, Rev. Howard was sentenced to nine years and eight months of hard labor at Sing Sing. The women were let free for having been led astray by the immoral men.[6]

Henry C. Friend's scam was indeed downright Parafian. Or maybe Barnumesque. Here was a man tapping into the era of adulteration with a new sugar operation. His electric connection set him well within the Edisonian age, too, so much so that competitors fearing "the Friend Process" wrote Edison himself to ask how it was done. For his part, Edison had his office reply that electricity could bleach sugar, but he knew of no process to refine it.[7]

Like Paraf, Friend had been bold enough to approach the wealthiest, most powerful men in the business. For sugar, this was the Havemeyer family, Henry and Theodore, kingpins of the later Sugar Trust. Friend approached Theodore to invest in the new electric process in 1888. Havemeyer offered $2 million for it. Friend would not reveal the secret. Havemeyer said the deal was off. The larger sugar refiners eyed Friend's claims with suspicion for its challenge to their dominance. After news of the deception broke, West Coast sugar magnate Claus Spreckels, a fierce competitor of the Havemeyers, smiled to reporters to say he'd also had the chance to buy the secretive electric sugar refining process but declined. He knew it was a scam when he asked what happened to the dirt removed from the raw sugar and Friend could not answer.[8]

Spreckels, Havemeyer, Friend, and an industry of sweetener capitalists were well aware upstart sugar alternatives were a growth field. A worry for the sugar magnates as startling as the electrified refinery was Friend's other claim that he had a secret process for developing a new competitor, grape sugar. We don't use the term *grape sugar* anymore, and it wasn't just sugar derived from grapes. Grape sugar was the term for corn and starch sugars that produced a more dextrose-based sweetener than the sucrose-based cane of Caribbean and Pacific islands. Grape sugar went by many names, as it was derived from many products. Most common at the time was glucose.

Glucose and the Shock of the New

Few if any foods, commodities, or crops are written about more than sugar. If anything, cotton and sugar might be the dynamic duo on that score. The previous chapter on cottonseed oil sits in relation to the prevailing cotton production, distribution, and consumption patterns well canvassed in the history of agriculture, slavery, capitalism, and trade. Studies of sugar are voluminous for similar reasons. Talking about sugar always means talking about culture: plantations and slaves, colonialism and empire, taste, nutrition, diet, commodity economics, science, cultivation.[9] With a commodity so complicated and controversial, it is small wonder that the history of sugar as a sweetener would bring forth a history of alternative sweeteners.[10]

Like glucose.

Glucose was the third neck of the hydra-headed monster of adulteration. It gained that status as a manufactured and artificial version of presumably pure sugar. If margarine would confront agrarian patterns of dairying and land use while cottonseed oil would offend consumers for its commingling with other store-shelf products, glucose was problematic in the pure food crusades because of the complexity and obfuscation of its production process. It was outsize, industrial, and complex.

Then and in retrospect it was a strange beast. Generally speaking, glucose is a naturally occurring component of blood sugar, the bane of diabetics, and another "-ose" (a carbohydrate) in the list of dextrose, sucrose, maltose, and fructose. Technically speaking, glucose is one of two processed versions of sucrose. You get an appreciation for atoms and molecules when you're trying to figure out what makes one sweetener different from another. Glucose is six carbon, twelve hydrogen, and six oxygen molecules. Dextrose is too. Fructose is too. Sucrose is twice that. Glucose and fructose are monosaccharides. They are one assemblage of that carbon, hydrogen, and oxygen combination. Sucrose is a disaccharide. It has one part glucose and one part fructose. It is twelve carbon, twenty-four hydrogen, and eleven oxygen molecules. The two, the "di," share that last oxygen molecule, which is why eleven and not twelve. Our tongues only care so much. They taste reasonably the same, although, really, as sugar historian Deborah Warner put it, "fructose is somewhat sweeter than sucrose, and glucose is somewhat less sweet." An early report on this from the 1880s calculated that glucose was about three-fifths as sweet as sucrose.[11]

The part from chemistry you want to hold in your pocket is that it isn't the listed number of elements that matters so much as the ways they are arranged. As Warner explains, "with the addition of a molecule of water, sucrose can be split into two simple sugars, glucose and fructose [and] because their atoms are arranged somewhat differently, their properties are somewhat different." They may differ only slightly in sweetness on the tongue, but our bodies process them differently. It's the process that matters more than the bare identity.

Chemists first derived glucose from grapes in the later 1700s, leading to its initial generic name. In 1811, the German chemist Gustav Kirchhoff prepared it by applying a dilution of sulfuric acid to various starches. It was Kirchhoff's recipe that paved the way for the first half century of grape-sugar production in Europe. The name was common, though rarely thereafter was the sugar derived from actual grapes. Corn, sorghum, and beets provided the raw material for most glucose in the United States, with corn in the lead. Europeans derived most of their glucose from potatoes. An 1881 study found that it could be made from so many sources, including "potatoes, grain, rice, maize, moss, wood, fruits, honey, raisins, etc.," that it didn't even bother to finish the list.[12]

The *American Grocer,* a trade paper based in New York, provided an early overview of the production process. "The manufacture of grape sugar, or glucose," they wrote, "[operates] by the action of sulphuric acid upon the starch contained in Indian Corn." Their report came from Buffalo, where manufacturers erected a building seven stories tall "in the most elaborate manner to accommodate the enterprise." (Friend's electric sugar scam emulated that factory structure.) The factory men raised corn from rail cars to the top of the building. They then began a pulping process to create a mash. Through heating, filtering, and mixture with chemicals—sulfuric acid—the "pure, limpid thick syrup" sluiced through more filters and boilers at stepped-down temperatures to create a final product. "This is called glucose, or grape sugar by manufacturers." In its best form, it is "whiter than refined cane sugar, but less sweet."[13] When the Chicago Grape and Cane Sugar Company began operations in the early 1880s with Friend as an associate, it competed with twenty-eight other factories in the United States to produce what was then considered a unique sweetener.

In the Gilded Age, detractors saw it as a corruption of pure sugar. It was artificial instead of natural, the USDA noted, referring to the substance interchangeably as "Artificial or Dextro-glucose." A Massachusetts pure food law in 1884 used common antiadulteration language to criticize glucose as an egregious offense to stomach and wallet. State analysts found traces of sulfuric acid and arsenic in a sample sold openly as glucose. Of the three honey samples under inspection, two were actually glucose. Their analysis also paved the way for Berghaus's trinity, understanding glucose as part of a collective of adulterants. To wit, they also found twenty-one samples of butter in their report "more or less adulterated with oleomargarine." The lard, it turns out, was faithful to its claims, but thirty-two out of forty-nine bottles of olive oil were actually cottonseed oil despite labels maintaining them as "genuine."[14] The *New York Times* shared the same antagonism, putting glucose into the adulterating triumvirate in 1887. "Like the manufacturer of oleomargarine," they wrote, "the manufacturer of glucose lives by fraud." Its con was "as great a swindle [as deceiving] customers by selling cottonseed oil for cheese and lard [or] oleomargarine for butter."[15] A parody in San Francisco's *Wasp* told of the US Army giving Apaches in Arizona ham and sugar as a good will gesture. It was a ruse, though. Every American knew, the paper said, the ham was inedible and the sugar was mostly glucose and sand, a road that after "grape-sugar, glucose, molasses, sickness" ended in "pain and sudden death."[16]

Why was it such a big deal? As I have found when talking about the topic, most people today find glucose even more pedestrian than the already prosaic margarine or little-known cottonseed oil. It's but the name for a chemical constituent of the body's blood sugar.

Not so in the late nineteenth century.

When it was introduced, glucose (fake) and sugar (real) sat in a complicated relation to one another in large part because of the wide spectrum of sweeteners already in play. Cane sugar (sucrose) was the dominant form, the one at the center of global plantation empires. But there were many sources of sugar. Honey was a sweetener with a several-millennia-old history of human use. Molasses, maple sugar, and fruit sugars were nothing new. As Henry Friend knew, the sugar from sugar beets was also thriving by the second half of the nineteenth century. Plus, chemists at Johns Hopkins University had discovered saccharin in the mid-1880s. Saccharin's story is somewhat like glucose's in its status as an upstart alternative derived from laboratory insights. But glucose, among all of them, would be caught up in scams and swindles, elevating it to greater public scrutiny (fig. 6.2).[17]

A few things explain the situation. Naming was one of them. The confusion over the difference between sucrose, glucose, dextrose, grape sugar, and corn sugar as I write today was even more complicated in the 1800s. As with the plethora of synonyms for margarine, glucose wrestled with its nomenclature. The cloudiness over the terms was more than semantic. It opened the door for swindlers, accelerating the view that scientific terms were cloaking hazardous foods and creating possible distrust in the marketplace. Ellen Richards sought to clarify this. She acknowledged the "confusion… caused by the loose ways in which the term 'glucose' is

"OUR MUTUAL FRIEND."

FIGURE 6.2. Cover image from an 1885 issue of *Puck* highlighting the perception that the medical and mortuary professions were the end result of eating chemical sugars. It showed the ways people thought of glucose as one among many of those chemicals. And in its title it offered a call back to the 1865 Dickens novel of the same name, *Our Mutual Friend*, which centered on the health and purity of English life amid the pursuit of money at all costs. "Our Mutual Friend," *Puck* 16 (January 7, 1885): 1.

commonly used." Her view was that "formerly it was the designation of all the manufactured products, whether solid or viscous, but of late the term 'starch sugar,' or dextrose, covers the solid sugars and glucose means the syrup form, from the Greek *glukus*, meaning sweet." Harvey Wiley made the same point, admitting "the term 'glucose' is used by chemists, rather too loosely" to indicate the presence of "any sugar of the composition $C_6H_{12}O_6$." A report from the National Academy of Sciences (NAS) in 1884 similarly stated that manufacturers and purveyors used *grape sugar, glucose, beet sugar, dextrose, maltose, dextro-glucose, glycose, potato sugar*, and *fruit sugar* interchangeably.[18]

Another point of confusion was somewhat more diffuse. It followed from the ways glucose challenged the variety of notions of "pure" sugar at hand. With butter, lard, and olive oil, concepts of purity mapped onto concepts of nature in a way that positioned the three in a binary relationship with their artificial counterparts. To its advocates on the pure food side, butter was pure and natural, margarine was an artificial creation. To those who disliked cottonseed oil, olive oil and lard were pure products, cottonseed oil was artificial. With sugar, though, the argument for purity wasn't as simply binary. It wasn't just that cane and beet sugars were derived from nature—the argument butter, lard, and olive oil makers made to defend against their challengers—but that its process of cultivation and refinement met certain standards of sweetness and quality. That led pure food advocates to denounce glucose as artificial for two more reasons: one, because it was the product of too much artifice—from labs, factories, and chemical technologies—and two, because its process of cultivation and refinement could not meet the standards of sweetness and quality set by cane (sucrose).

This host of factors is difficult to parse out in retrospect because it was difficult for people to parse out at the time. As it was, that confusion only added to perceptions that glucose was a deceptive con. But of course antiadulteration advocates still tried to figure it out. By the 1880s the NAS investigated glucose's identity and proliferation, the USDA studied its agricultural derivation, the Treasury Department monitored its growth, the Bureau of Commerce kept track of its sales, and periodicals and newspapers—daily, urban, rural, and satirical—deliberated on its legitimacy. Glucose's place in the pure food debates arrived as part of an ongoing struggle to define and police the border of pure and adulterated sweeteners. Two of those sweeteners, cane and beet sugar, framed the proximate background from which glucose appeared.

Cane, Beets, and Purity

Preexisting arguments about the legitimacy of cane and beet sugars set the stage for debates about the legitimacy of glucose. The purity of cane sugar from South America and Caribbean and Pacific islands had long been a point of deep cultural importance by the time of grape sugar's incursion. Even the basic terms describing cane sugar carried dual meanings. In the nineteenth century, cane sugar's whiteness in its refined state referred to the agricultural process of cultivating, harvesting, processing, and packaging the sweetener. It was marketed and inspected for its degree of purity, a measure defined analytically by the percentage of sucrose. Akin to the story of white bread, the whiteness and refined status also referred to the class of eaters and the expectations of consumers who created the demand for such sugar. This showed that for some, the methods for whitening sugar meant more artifice and processing could be a sign of more purity, not less.[19]

That variable view of purity stands to reason. As the previous cases have shown, there was no singular sense of purity throughout the era of adulteration, no isolated definition to which people could defer. They had to make it. With sugar, they made it by arguing over its production process—its environmental and technical aspects—and the cultural values that went with that process. The arguments were not about whether sugar was processed but how. Thus, a highly refined sugar could be considered more pure than a less refined product so long as it followed acceptable principles of processing. This was Friend's downfall. His method was not legitimate.[20]

There were other reasons critics challenged sugar's (sucrose's) purity too. To the cultural point, cane sugar was inextricably linked to colonial politics. "From 1600 onwards," writes historian David Singerman, "the empires of Spain, Portugal, France, the Netherlands, and Britain contested control of much of the planet in the name of sugar. One by one, Brazil, Barbados, Jamaica, Louisiana, Saint-Domingue, Cuba, and Puerto Rico were the foci of capital flows, diplomatic intrigue, and outright battle." Pacific and Indian Ocean Islands were later conquered for similar ends—Hawaii, Java, Mauritius, and the Philippines among them. It wasn't incidental that these imperial efforts played a role in fostering slave-based labor regimes. All told, "of the twenty million Africans brought enslaved to the New World... sixteen million owed their perdition to the insatiable Western demand for sugar."[21] Such labor entanglements and modes of perdition made it difficult for critics to separate the cultural argument for moral purity from the agricultural (environmental) method for producing pure sugar.

The two, the cultural and the environmental, were often mashed together. Consider, for instance, that as early as the 1600s, slaves and other workers logged 210,000 *tons* of wood a year to burn in the kilns for the refineries. Cattle and horses required enormous amounts of feed too. In Barbados, sugar plantations so quickly denuded the soil and deforested the lands that even in the 1700s, growers developed practices to conserve resources. Manuring became more prevalent; colonists crafted "wood laws" to address overlogging.[22]

These were repressive cultivation strategies. Some people acknowledged and acted on that awareness. "Ethical consumption" advocates fighting for fair labor, fair food, and fair trade today—for coffee, produce, or sugar or against commodities like "blood" diamonds—echo the responses of sugar activists two hundred years ago. In preindustrial times, advocates paved the way for our modern concerns over morality and cultivation by making the case against cane a case for moral purity. They reasoned that Westerners could ease the violence of slavery by eliminating sugar from their diets. The author of *An Address to the People of Great Britain* made that point. He urged boycotts in 1791 with an admirably straightforward argument: "If we purchase the commodity we participate in the crime." A treatise in 1792, *No Rum!—No Sugar!*, told readers to avoid "slave sugar" until slavery was abolished.

Antebellum abolitionists in the United States sought to develop new crops, beet sugar being one of them, as part of antislavery activities. In that vein, a proprietor of the newly chartered Northampton Beet Sugar Company took his beets to the 1839 Massachusetts Anti-slavery Fair. Thousands took up a "free produce" movement led by Quakers in Pennsylvania to encourage abstinence from slave-produced sugar and cotton. The *Pennsylvania Farmer* advertised groceries "produced with free rather than slave labor."[23] Henry Ward Beecher, Harriet Beecher Stowe, and Frederick Douglass were members of the movement.[24]

For the cause of moral purity, one response to tropical cane was continental cane. Because of its climate, soil, and preexisting plantation model, Louisiana sugar plantations offered the most concentrated production site for sugar cane in the United States (fig. 6.3). The environmental needs for sugar and molasses relied on the temperate climate of the Gulf coastline. Cane grew in a variety of soil conditions. Temperature, moisture, and sunlight were the more important variables. Historically, then, most growers have planted cane between 30° north and 20° south latitude (roughly between northern Florida and central Brazil in the Western Hemisphere).[25]

The postbellum era offered a period when labor could ostensibly operate beyond the slavery of the antebellum regime. Investors at that time ramped up development in the southern half of Florida to compete with Louisiana's prominence (fig. 6.3). They spread cane to the Everglades in an effort to capitalize on the confluence of drainage-minded land developers and agriculturally astute plant pathologists. It was an early success. When some Florida cane fields "produced a phenomenal sixty to eighty tons of cane per acre . . . compared to an average of twenty-one tons in Cuba and thirty tons in Louisiana," Deborah Warner writes, one businessman "announced that the 'Alchemy of Muck and Sunshine' provided ideal conditions for the new enterprise."[26]

The story of Florida cane was remarkable for many reasons. As a reference point for fake sugars, it offered an example of seeking to escape

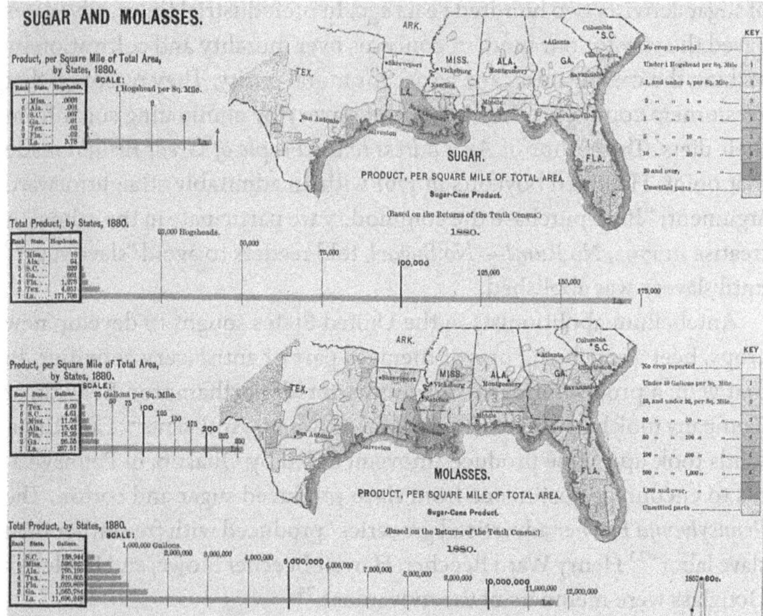

FIGURE 6.3. Plate 108 from Gannett's map of Sugar production, 1883, showing the presence of cane sugar mostly in Louisiana (based on 1880 census data). Source: David Rumsey Map Collection.

labor conditions that still existed outside the United States to cultivate a sweetener more geographically proximate and less morally compromised. This was purity in the sense of crafting an acceptable labor process for production. In this argument, "slave sugar" was impure because it was the product of enslaved peoples, not because of its percentage of sucrose. The sugar on the kitchen table was the result of an argument over pure foods, not the argument itself.

There were other options for seeking that method-based kind of purity. A good trivia question is who the largest producer of sugar was in the late 1800s. Few would guess Germany. But along with their French neighbors, the Germans had pioneered sugar beet production. Sugar beets were a more substantial response to cane from the Caribbean and Pacific than growing cane in the continental United States ever was. As historian Kathleen Mapes put it, "The rise of the sugar beet industry over the course of the nineteenth century represents one of the most important, if overlooked, developments in industrial agriculture."

Despite the early support of abolitionists, the United States was behind global competitors on this measure. Germany had the climate for it, the chemical insight to develop it, and the political drive to make it happen. In particular, they wanted to increase the sovereignty of their own sweetener sources and found that beets were good candidates. European beets were concentrated between the 45th and 55th parallels (basically northern Italy to Denmark). Germany produced over 600,000 tons in 1890 alone. All of Europe together produced 1.6 million tons of sugar, almost all of it from sugar beets. In 1884, half of all global sugar came from the sugar beet industry; by 1899, it was 65 percent.[27]

Europeans began their efforts earlier in the nineteenth century. After the US Civil War, government agents and private investors also advocated for beet sugar as a cane alternative in various regions across America. An early priority for the USDA (founded 1862) was to explore sweeteners. They hired a chemist from Purdue in 1882 specifically to investigate the possibilities. Harvey Wiley, as we've seen, would later rise to public fame for his staunch and rabid defense of the public good in the face of adulterated foods. In the early 1880s, he was dedicated to the more specific focus on alternative sweetener research. To quote Warner, "Wiley was generally considered the leading sugar expert on the federal payroll. He had studied sugar chemistry and polariscopic analysis in Germany in 1879, and he published polariscopic studies of sucrose and glucose after returning to his position as a professor of chemistry at Purdue University." He brought attention to the sugar purity question at a moment of public crisis in food confidence.[28]

Wiley and the USDA kept tabs on a fledgling domestic industry. By the 1880s growers were trying beet sugar in California, with limited coastal success. The prospects farther north in Oregon and Washington were much better when one of the nation's dominant sugar refiners, Claus Spreckels—the sugar tycoon who had wisely questioned Friend's electric con—took the opportunity to develop a beet empire.[29] Soon cultivators found their efforts take root in the Midwest between the 35th and 45th parallels (from about Tennessee to Michigan).

Beet sugar carried the same mixture of cultural and environmental senses of purity as cane, but it combined them differently. To that point, a sugar beet pioneer in the 1880s played up the civic duty of his work, calling it "profitable and patriotic."[30] For him, growing sugar beets was an effort in Americanizing and civilizing sweeteners. Michigan beet capitalists pressed the point further. Sharing the same terms soap makers were deploying in their advertisements, they took the view that theirs was a case for "civilized" sugar over "barbarian" cane. Although sugar and soap debates were generally part of different conversations—the one edible, the other inedible—they mutually benefitted from a broader cultural moment reducing kinds of people to civilized or savage. Beet-sugar advocates would advertise the racial dimensions of their product. They harked back to earlier arguments that the labor used to process cane tarnished its moral character. With Michigan's sugar beets, the whiteness of the cultivators offered a superior grade to the darkness of cane sugar cultivators.[31]

The racial, cultural, and labor elements comparing cane and beets exemplified the confusing assemblage of arguments about purity I noted above. The "free produce" and antislave sugar movements were premised on recognizing the humanity of enslaved peoples and resisting a racial hierarchy. Its proponents argued that a morally pure action was one that avoided Caribbean and Pacific cane. The early beet-sugar investors instead operated under the assumption of a racial hierarchy. Their point was to offer sugar untouched by dark hands. As part of a broader pure food movement, their argument for purity was related to environmental labor—how it was made—while intertwining cultural assumptions that whiteness meant purity.

And if that wasn't enough—purity because of labor, race, culture—the beet-sugar contingent added more complexity: they addressed one problem for purity advocates—namely, the labor dimensions of tropical cane—while introducing another—the murky industrial character of beet processing. They exacerbated the view that artifice and adulteration were aligned on one side of the pure/adulterated ledger. "The nature of cane sugar was intimately tied to its agricultural origins," David Singerman writes, "but the possibility of a consumable 'beet sugar' was Frankenstein's monster, inescapably the product of an early-nineteenth century German chemical laboratory." Having developed beet sugar's potential in their labs, chemists brought the laboratory efforts out to the fields. That relationship to artificial methods was problematic for pure food crusaders.[32] It added to a binary framing of artificial and adulterated over natural and pure. To them, to be artificial was to be impure. This would have been hard to generalize in all cases. After all, at the same time, other packaged and manufactured goods were lauded for their cleanliness, as with making bread untouched by human hands or tea sold in bags instead of loose tins. But for the sugar argument, those decrying beet sugar were decrying artifice.

The strike against beet sugar as industrial, artificial, and inauthentic would also become a strike against glucose. After all of this wrangling over the most legitimate way to produce sweeteners, a reader wouldn't be wrong to conclude that there really weren't any good ways to get sugar.

A "Safe and Most Profitable" Sweetener

That returns us to grape sugar, glucose, and why it was controversial enough to rise as the third neck of the hydra-headed adulteration monster. When it first became a public offense in the 1870s and 1880s, when glucose started to gain a name in the press, it joined an already vibrant debate over the legitimacy of various sweeteners. Producers were cagy, consumers were suspicious, and politicians were skeptical. Friend's electric sugar con unraveled in the late 1880s as a testament to the connection between fake sugars and audacious swindles. He left behind the debris of a sham factory and the legacy of a series of sweetener-based confidence schemes. Before Friend got the idea, a curious proposal in 1874 put the question of glucose into the spotlight of fake foods and insincere people. The proposal sought to fund a glucose concern in the United States. Its proponents wanted to capitalize on the demand for European potato-derived glucose.

The circular for the Grangers' Glucose Company was most certainly a scam. No records of the company's construction, production, or economic standing remain. New York publisher Edward Nieuwland, cutting a figure in a busy Manhattan, authored the four-page tract. He had big bushy sideburns that people said made him the spitting image of Commodore Vanderbilt. Stout, important, commanding. Before he came to New York, Nieuwland was a money broker in Europe. When he got to Gilded Age Manhattan, he began a career dotted with attention to currency, labor, capital, and farming, throwing his lot in with a new agrarian movement called the Patrons of Husbandry but known more commonly as The Grange. Its members were Grangers. Nieuwland ostensibly wrote his proposal to endorse a fellow Dutch émigré, Henry Becker, who was then producing glucose as "crystal syrup." It was a sign that food was so bad, so lacking in taste, that sugar-based table syrups were ubiquitous and necessary dressings. Of a range of versions, Becker and his glucose table syrup were "*better than any one else in the United States*" (italics in original).

In a textbook case of confidence scheming, Nieuwland assured potential investors that he "never recommended an investment in which I did not believe myself." It was easy money. While one manufacturer had been a "poor mechanic, now he is the possessor of many millions." Nieuwland was soliciting farmers' subscriptions up to $150,000 in shares of $50, $100, $500, and $1000 (fig. 6.4). Europeans were running eleven factories in five countries, Nieuwland wrote, and shipping it to eastern ports—New York, Boston, Philadelphia. They used potatoes as their starch feedstock because corn and wheat were too expensive in Europe. The Grangers' Glucose

The Grangers' Glucose Co.

A very Safe and most Profitable Investment for Farmers.

DIVIDEND, 46 PER CENT PER ANNUM THE LOWEST ESTIMATE.

Capital, $150,000, in Shares of $50, $100, $500, and $1000.

FIGURE 6.4. Header text for a circular seeking investment in Edward Nieuwland's Grangers' Glucose Company, 1874. Courtesy of the Wisconsin Historical Society.

Company would do better. It had two points to make. First, that it would bring the industry to the United States; second, that it would establish factories in the West where "corn and labor are much cheaper than in New York." Nieuwland was leveraging the conditions of the moment. In the shadow of the Panic of 1873, the Grangers could take advantage to build "a cheaper factory, buy cheaper machinery, and obtain cheaper labor, than before the panic."[33]

Nieuwland framed the circular in the language of trust and opportunity. He revealed few details about the actual production process except to say, as the header text summarized it, that it was "safe and most profitable." He also hedged against anticipated fears of adulteration, going so far as to reverse the common accusation that cornstarch was a degradation of natural sugar. "I can not refrain from calling attention to the fact that syrup and sugar, made of corn-starch, are perfectly pure and healthy, containing no acids or any impurities, which are often found in the sugars and syrup of sugar-cane." That wasn't true, but it didn't matter. The scheme was a classic example of Gilded Age hucksterism predicated on half-described details, vague testimonials, and the cloak of secrecy.[34]

Secrecy remained a point of conflict for the new alternative sweetener for some time. Secrecy led to suspicion, as with Henry Friend's Chicago and Brooklyn activities. At least beet sugar was grown and traded for what it was. Glucose was often veiled, feeding a preexisting public narrative about distrust in ways similar to cottonseed oil. The Grangers' Glucose Company may not have swindled as many people as Nieuwland had hoped, but it underscored the view of glucose as an unknown, untested, untrusted upstart.

Grocers, their customers, and their public advocates sought to lift the veil of mystery by asking, where was all of this glucose coming from? The Commissioner of Internal Revenue tasked the NAS with uncovering glucose's concealed origins in 1882. They addressed the question in two ways.

They first summarized an eighteen-step manufacturing process. Seven of those steps were chemical, aided by the addition of sulfuric acid, sulfate of lime, and sulfurous acid gas. The others were mechanical. These were high-pressure boiling and filtering processes. They then reviewed the twenty-nine existing glucose manufacturing plants in operation. They found facilities in nine states, with the bulk of operations in three. Illinois had twelve factories; New York had six; Iowa, four.

The 1884 report that resulted from the NAS's investigation only counted large urban manufacturers. That left out the dozens of scattered rural mills throughout the countryside of the Great Lakes region. At the time, the beet sugar and corn sugar industries were trying to keep up with an already vast empire of starch sugars across Europe. In 1871 there were nearly two thousand "beet root factories and distilleries" on the continent, most of them doing lucrative business in the sugar trade.[35] The more narrow register of grape-sugar factories, according to a letter to the *Chicago Tribune* from irascible correspondent "Zea Mays" (corn), counted ninety in Germany, ninety in France, and "many others in Austria and other European nations."[36]

The US glucose industry prospered during the twenty years following the NAS's investigation. In 1882, the first year the government kept such statistics on the commodity, producers shipped five million pounds to foreign ports. That increased more than sevenfold by 1890 to thirty-eight million. At the turn of the century, glucose exports quadrupled again to 185 million pounds in 1900 alone. Figures 6.5 and 6.6 show five-year averages of export flows. The period from 1881 to 1885 brought 46 million pounds of the product from three American ports to fifteen export nations and a handful of British colonies. Twenty years later, the exports accrued to 758 million pounds over the period from 1901 to 1905. By then, six American ports were shipping the product to the same number of export destinations.

Production ramped up as the economic merits of the alternative sweetener took hold, but onlookers saw something duplicitous about glucose. The government and grocers' press worked to explain the industry to overcome that perception of duplicity. This started with an eye to production, as production-side issues raised a host of fears for the new product. It continued with consumer-side attention to glucose's visibility and safety.

On the matter of production-side issues, one summary reviewed five different industry processes. Most of them were hidden in the same way Friend's electric sugar process was. An early patent holder for glucose production, Frederick Goessling, was clever enough to earn hundreds of

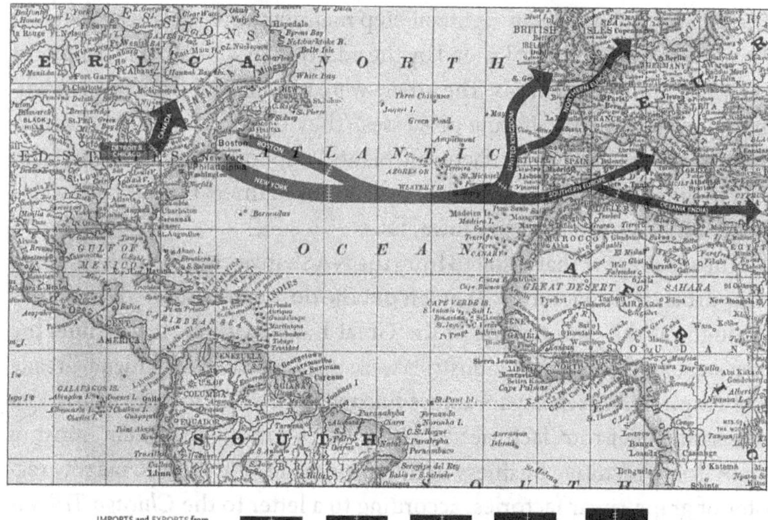

FIGURE 6.5. Map showing exports of glucose across a five-year span from the 1885 to 1889 from three US locations going to five global regions. The digital companion site https://purefood.lafayette.edu shows changes across twenty-five years of export patterns. Map created by Kristen Lopez.

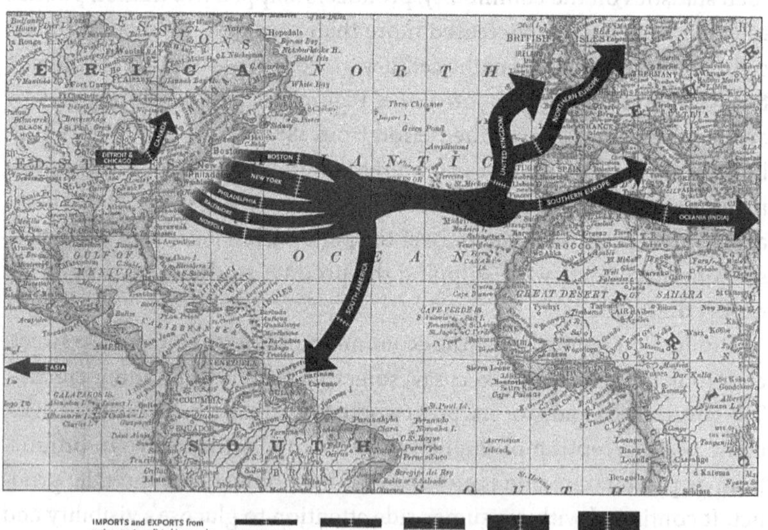

FIGURE 6.6. Map showing exports of glucose across a five-year span from 1905 to 1909 from six US locations going to six global regions. The digital companion site https://purefood.lafayette.edu shows changes across twenty-five years of export patterns. Map created by Kristen Lopez.

thousands of dollars in the first wave of production in the 1860s selling his expertise to investors. When he died, it all seemed like a scam: the patents were too vague for anyone else to replicate. Another investor, Cicero Hamlin, followed suit. He was a venture capitalist in Buffalo who bankrolled the Buffalo Grape Sugar Company in 1874, the same year the Grangers' Glucose Company sent out their proposal. Hamlin, Deborah Warner writes, "shunned everyone who might discover and reveal the details of his machinery, methods, or business practices." Even though he was a legitimate investor, his secrecy only fostered the perception of fraudulence. In her study, Warner found no less than a dozen patent lawsuits at the circuit court level in the early 1880s alone premised on forcing glucose makers to reconcile and explain the vague patent basis for their activities.[37]

Secrecy in the production process, like the range of confusing names, was a point of anxiety for a jaded public. The scale, complexity, and danger necessary to achieve glucose manufacturer's remarkable levels of production were others. The manufacturing process was intricate even in the 1870s. I have a degree in chemical engineering, yet reading the treatises on grape sugar, beet sugar, corn sugar, or any starch sugar is confusing and remarkably technical. The complex degree of processing details reveal a large footprint of materials, including water, chemicals, wood, metals, and buildings (fig. 6.7).

Earlier in the century, "slave sugar" activists opposed cane sugar cultivation because the process was exploitive in environmental and labor terms. Although the terms of labor and geography for glucose production differed, they, too, came in for criticism because of their disruptive environmental draw. Glucose factories left big impressions on the landscape. They used an immense amount of water. They hauled in tons of chemicals, literally, to convert the starch sugars into refined syrups and crystals. They processed thousands of bushels of corn. It was dangerous too. Inside huge wooden vats, boiling water diluted a mixture of hundreds of bushels of corn and doses of sulfuric acid. But the vats often blew out, "frequently causing great loss of life," as one court case put it. Brass clamps, oversize vats, copper coils, high-grade thermometers, stirrers, and judgments by eye about temperature control and mixing defined the vanguard grape sugar processes of the later 1800s.

Glucose production's voracious appetite for water and acid in particular was of concern to critics. A new generation of chemical companies provided the sulfuric acid, but cities, municipalities, and adjacent rivers provided the water. A new facility offered a snapshot of the environmental draw. The Chicago factory handled 12,000 bushels of corn a day at the outset, with plans to double its capacity to 25,000. A bushel weighed

LANDRY & LAUGA'S APPARATUS FOR THE MANUFACTURE OF GRAPE SUGAR.

FIGURE 6.7. Glucose manufacturing was complicated and industrial, as seen in this rendering of a Landry and Lauga apparatus, manufacturers from New Orleans. Workers in the scene show the relative scale of the boilers and equipment. Courtesy of Columbia University.

between fifty-five and seventy pounds, so on average the facility planned to handle about 1.5 million pounds of corn per day. The process typically used a three-to-one ratio of water to starch. This meant that each operating day one factory consumed 4.5 million pounds, or over half a million gallons, of water. To aid the process, chemists supplied three *tons* of sulfuric acid a day.[38]

For the sake of easy water access, it was common for a glucose company to incorporate in the east, often New Jersey, but build facilities in Iowa and Illinois. This was a problematic arrangement for challenging river uses and fostering an absentee landlord relationship. The absentee relationship led to legal disputes over raw material usage in ways that anticipated environmental regulations in the twentieth century. Against the glucose industry, plaintiffs worried that East Coast investors were poaching their supplies. In one instance of water overuse, a coal and iron foundry sued the Pope

Glucose Company over riparian rights. Manufacturers had built a damn on Illinois's Fox River to provide mill power, but Pope's glucose factory slurped up so much water from the river that the mill couldn't operate. Pope was ordered to reduce its use "on nights and Sundays or at hours and times that were unusual and unreasonable."[39] In another case, this one over water pollution, the State of Iowa brought suit against the Firmenich Manufacturing Company of Marshalltown, Iowa. Firmenich had moved from Buffalo for the cheap water. The state won the case for damages caused by polluting the previously "pure" Linn Creek. Waste from "the glucose manufactory," including "glucose, acids, sulphuric acid, sulphur and other poisonous substances," was damaging enough to make the creek "so impure that many of the fish in the stream died."[40]

Even before consumers could encounter it on their dinner tables, glucose was marked by danger and distrust. By looking at the production process, critics saw a range of factors deepening the perception that the manipulated starch sugar was a degradation of a more natural sugar. Sometimes that degradation occurred because the manufacturing methods worked outside norms of agricultural manipulation, as evidenced by chemical pollution, water use, factory scale, high-pressure equipment, absentee landlords, secrecy, and unclear naming of the product. Sometimes it was a more direct violation of nature's purity—as at Linn Creek. These violations were the consequences of glucose manufacturing's scale demands. A more "natural" sugar was one that resulted from less damaging environmental manipulation.

If the production of glucose centered critics' worries on secrecy, scale, and unnatural complexity, consuming it brought its own fears. This was all the more true because pure food advocates understood that the two sides of the producer-consumer relationship were connected.

Some of the consumer worries were about residual sulfuric acid in the sugar in their meals. This was similar to the worry pure food advocates had for residual charcoal and bone dust in cottonseed oil (chap. 5). For glucose, after the boiling, diluting, and separating processes, its makers would carry their product to a neutralizing vat. There, they added carbonate of lime, usually finely ground chalk or limestone. When mixed with the acid-containing mixture, this would form sulfate of lime (gypsum). The gypsum didn't fully dissolve in the water. Workers then separated the liquid sugar from the solid gypsum to leave with a neutralized product. That was the best-case scenario. Chemists almost always found residual sulfuric acid in the tested end product. Arthur Hassall, the leading purity chemist in Britain, reported that finding sulfite of lime was an easy way to tell whether the sugar had been adulterated.[41] If residual acids and sulfites weren't offensive enough, others thought problems of digestion could be. Although *Scientific American* considered it "perfectly harmless," the editor of the *Boston Journal of Chemistry* reported digestive problems. Glucose "gives rise to flatulency, colic and Nausea," its editor wrote. Elsewhere, consumers were reporting that glucose gave them "flatulency and painful affections of the bowels."[42]

Other worries connecting consumer and producer brought the argument back to storefronts and honest brokering. At the grocers', consumers were suspicious about glucose in a range of products. Grocers who "sell such compounds for honey, at the price of honey, [are] the refinement of swindling, and ought to be punished as such." A critic asked, "if your druggist sells you poison, the law compels him to label it; then, why not compel [glucose salesmen] to do the same?"[43]

Honey was a prominent target for adulteration with glucose, but it was hardly alone. "Brewers, tobacconists, druggists, confectioners, vinegar makers [all] use it freely," the *Chicago Tribune* reported. The "cordial makers of Copenhagen" were known to add the cheaper alternative.[44] Industries used it in "coloring" for liquors and "fruit jellies." At the time, candy and confectioneries were a $56 million industry and beekeepers were producing sixty-four million pounds of honey a year. They were large and profitable industries. The substitute sweetener was alluring to dishonest producers, who sought an easy way into that lucrative market, and worrisome to honest brokers, who feared for their product's pure standing.[45] In an era when, at $50 million a year, sugar tariffs provided the "single largest source of federal funds," the identity and supposed purity of a sweetener mattered a great deal.[46]

Grocers were, once again, caught in the middle, playing the role they played with other new commodities: they defended the healthfulness of the new product when they sought to curry favor with wholesalers and manufacturers, and they defended their integrity as honest brokers when they told customers they would always reveal the true identity of their sugar. One grocer captured the frustration of his profession before hearings of the Ways and Means Committee in 1878, justifying his anger over glucose. "Adulteration of sugar concerns the Committee of Ways and Means; it concerns the board of health; it concerns everybody."[47]

The *American Grocer* was happy to advertise sugar refiners, who were also boosters for the grocers' trade. Francis Thurber, the wholesaler who published the paper, certainly understood the need to curry favor. He worked with Theodore Havemeyer, who defended his business without reserve. Havemeyer tut-tutted confidently, "not one ounce of the adulterating matter or glucose enters my factory or is put in my sugars." The ad copy he placed in the *American Grocer* repeated common text: "Neither glucose, muriate of tin, muriatic acid, nor any other foreign substance whatever is or has been mixed" with his cane sugar. As figure 6.8 shows, the "sugars and syrups" they sold were "absolutely unadulterated."[48]

Grocers may have been equivocal about their self-interest, but the beekeepers were not. They supplied the grocers; they didn't want to be in a position where grocers blamed them for illicit products. They were upset about the glucose affront in no uncertain terms.

These were not trifling matters. Apiaries and hives underwrote a thriving industry in the late 1800s. "In no branch of rural work has the advancement been so great . . . during the last twenty years," one booster wrote, "as in bee-keeping."[49] But glucose was trouble, leading the *Beekeeper Magazine* to come back from a national convention in 1879 sounding warning bells. Most of the "finest comb-honey in New York" was "composed entirely of flavored glucose." Admittedly, the editors wrote, beekeepers were too smart to be afraid. Glucose could never compete with "sound, pure, virgin" honey. And at least it wasn't as bad as the dairy industry's margarine scourge, where "tallow butter [may] compete with real butter." But the

FIGURE 6.8. Advertisement for unadulterated refined sugars.
Source: *American Grocer*, December 15, 1881.

stakes were obvious. Any honest man "wants the genuine article, and will not cheat himself with a counterfeit, concocted of refuse and doctored with chemicals."[50]

Their fears gained further traction after British customs officials impounded a shipment of American "honey" that same year. Thurber and Company, who was otherwise busy telling its customers how honest they were, had shipped the tainted batch. Authorities in England held it upon arrival. Suspected of adulteration, a chemist measured polarity to find boxes of honey that were in truth 57 percent glucose. Although "there is a moral tribunal that requires every man to be a good citizen [and] not knowingly do anything to injure his neighbor nor his country," the *American Bee Journal* wrote, the impoundment showed that the glucose problem was real and present.[51]

Apiarists considered this a problem of corruption and deception and a challenge to their professional identity. If they could weed out the bad adulterating honey makers, they would preserve their customers' trust. They were right about grocers too. The Thurber impoundment confirmed fears that they would get the blame for others' contaminating schemes. *Judge*, an offshoot of *Puck*, took aim at Thurber to frame it that way (fig. 6.9). While ostensibly attacking the hypocrisy of his economically motivated antimonopolist stance, they took the opportunity to showcase glucose as a dominant feature of his ill-reputed store. Piling on, the *Beekeepers' Magazine* crafted a light rhyme to enunciate the issue: "How doth the little busy D. / work hard to make more money. / He gathers glucose every day, / and mixes it with honey."[52]

Critics distrusted adulterants for a number of reasons, as scores of cases demonstrated. The agricultural disruptions of margarine and the surreptitious commingling of cottonseed oil built up angst over dishonest brokers, mislabeled goods, and potentially unhealthy new products. Glucose appeared as the latest in a laundry list of sweeteners. Cane, beets, and honey were three of the biggest sugars already in the marketplace. Cane sugar had long been a lightning rod for arguments over the ways new foods connected distant consumers to unjust practices. The response to cane's place as a kitchen staple was an early version of demands for ethical consumption. Beet sugar spoke to those concerns, but not always because its promoters rejected the racial hierarchies that angered "free produce" advocates in the first place. Into this preexisting context of confusion over sweeteners and alternative sweeteners alike, critics distrusted glucose for many reasons. Not only did it inherit the suspicions of impropriety brought forth by sucrose before it but it also agitated principles of proper, natural sugar cultivation through its outsize, sometimes cloaked, artificial

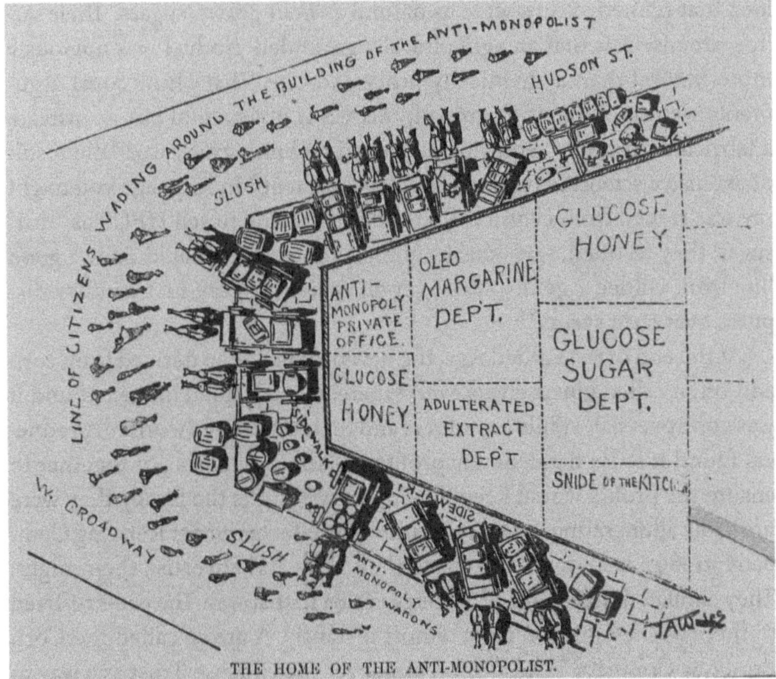

FIGURE 6.9. Cartoon mocking Francis Thurber's grocers' empire. While writing extensively about his antimonopoly beliefs in the public sphere, his own business was monopolizing trade to control the market, all the worse, critics claimed, because he dealt so much in known adulterations. Source: "The Anti-Monopolist," *Judge* 1 (April 1, 1882): 11.

identity. Secrecy, scale, and complexity raised fears at the grocer's counter. Those factors also made glucose seem like a scam. A "safe and most profitable" sweetener? Too many people were suspicious of the claims.

From Glucose to Corn Syrup

Starting in 2011 and lasting nearly five years, the lobby for sugar growers sued the Corn Refiners Association (CRA) for false advertising. The CRA had launched an ad campaign for high-fructose corn syrup (HFCS) claiming that all sugars were alike, or "sugar is sugar." Their online and television ads sought to blur the distinction between sweeteners to ask consumers to trust in corn. They wanted their product, Karo Syrup, to bear the label "corn sugar," not "corn syrup."

Cane and beet-sugar growers did not like that. Theirs was the "real" sugar; HFCS was not. *Saturday Night Live* parodied the ads, mocking the idea that refined syrups were as natural as field-grown sugars. Their satire's premise was that an agriculturally grounded product was obviously more natural than a chemically processed one. That's how court arguments went too. Plaintiffs argued, "HFCS is not natural but is, instead, a fabricated product requiring advanced technology." It was "the result of extensive scientific research and development." In language you might say was ripped from century-old headlines, trying to sell HFCS as "real" sugar, they testified, was "sneaky and dishonest." You could draw a good line from Gilded Age disputes over glucose to twenty-first-century disputes over corn syrup.[53]

At the end of the Gilded Age, the new product often derived from corn and often called glucose was everywhere, it was capital intensive, and it was controversial. While customers and their protectors worried, producers found it to be handsomely profitable. The capitalists got together to ensure the profits would keep flowing as sweetly as the honey they were probably adulterating. Investors formed the Glucose Sugar Refining Company in 1897 as a way to aggregate their risks and advertise their might. They emulated the model of the Sugar Trust in that way. The press referred to them as the Glucose Trust almost instantly. A group called the Corn Products Company (CPC) soon took over the Glucose Trust as a way to bury the glucose name and overcome its stigma. Then in turn the Corn Products Refining Company (CPRC) bought them out. If confusing naming conventions, concealed identities, and industrial imagery had dogged the early years of glucose, this quick succession of acronyms may not have helped, despite the CPRC's rebranding efforts.

Nevertheless, they kept at it. In 1903, the CPRC stewarded a new syrup to the market by attempting to escape the stain of adulteration through a multipronged marketing campaign. They placed ads in domestic-science papers and the womens' press. They took a page out of the meat by-products and oils books by selling cookbooks showing how to use the syrup in recipes. (Procter & Gamble used the tactic, too, with Crisco and its extended product line of edible consumer goods.) They also worked hard to say "corn syrup" instead of the tainted "glucose." The syrup designation was a way to avoid the glucose label.

The CPRC called their new product Karo Syrup. It was the most successful corn-based sweetener to follow the first generation of glucose. It found the ire of Harvey Wiley, who kept his zeal about specific ideas of all things pure and true and didn't want corn syrup to hide its glucose identity (fig. 6.10). But growing industry influence within the Roosevelt administration helped it overcome the USDA's initial angst and gain broader regulatory support within a few years, settling into a post-1906 Pure Food

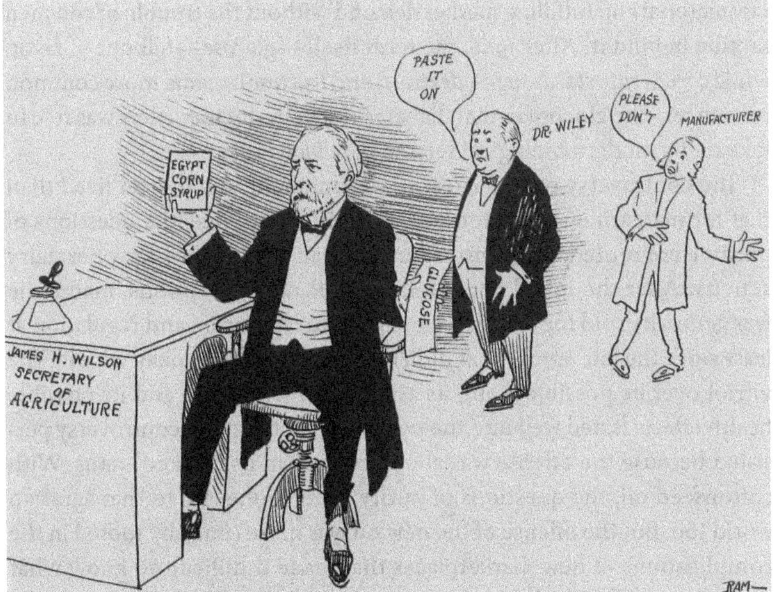

FIGURE 6.10. Cartoon from the Louisville *Post* (early 1900s) revealing the skepticism consumers had about corn syrup manufacturers and painting the USDA's James Wilson and Harvey Wiley as winning the fight over glucose through honest labeling. Courtesy of the Wiley Papers, Library of Congress.

and Drug Act environment. In 1908, with a signature from Roosevelt, the authors of Food Inspection Decision 87 ruled that glucose, as syrup derived from corn, could be labeled corn syrup.[54] This marked the end of *glucose* as a publicly reviled term.

Legally, the fight against glucose as an adulterant then died a quiet death, subsumed under a proliferating industry of processed sweeteners and lost beneath a new label. Culturally, the slavery connotations of earlier sugar lost public force as the range of alternatives increased and the distance from the antebellum period faded farther into the past. That muted the criticism about morally impure consumerism, but only so much given that the residue of those earlier moral commitments to proper food sources is apparent still in principles of ethical consumption. In the era of adulteration, the cultivation of starch sugars was an effort to control supplies and avoid the taint of Caribbean and South Pacific labor practices. The starch sugars of Europe filled 65 percent of the market by the early twentieth century; beet sugar and glucose from the United States reached hundreds of millions of pounds of exports. A change in name to "corn syrup" satisfied those export markets just as much, securing a stable raw material and fulfilling market demand without the trouble of con men lurking behind it. After 1908, the term itself—*glucose*—fell out of favor, while *corn syrup, starch sugar, dextrose,* and *fructose* became more common references. The glucose in that diabetes test your doctor orders was free to take on its modern biological reference as *blood sugar*.

The controversy over glucose as an adulterant was shorter lived than that for margarine and cottonseed oil. With margarine, the questions of purity were rooted in defense of an agrarian worldview of proper dairy activity. After the initial phase following Paraf's deceptions, margarine was generally sold for what it was, subject to legislation and regulation at least since the late 1870s. That didn't mean people were okay with it. The vitriol over its possible scam, its association with cons, and its possible health effects lasted well into the twentieth century. The controversy persisted because the offense was about more than its cloaked status. With cottonseed oil, the questions of purity were connected to that agrarian world too. But the offense of the new oil was more centrally rooted in the complications of new marketplaces that made it difficult to know what was in a product. Problems with cottonseed oil followed from its concealed use in a wide range of other products, olive oil and lard being the two principal ones. Arguments against its validity were thus wrapped up in arguments about the validity of wider by-product streams and complex industry product lines. The very presence of fake butter and fake oil as alternatives raised suspicions about their legitimacy.

With glucose, questions of purity were at once more diffuse, less agricultural, and less enduring. Secrecy, masked naming, dangerous manufacturing, industrial scale, and the residual effects of its production process cast glucose as an offense to consumers and an exemplar of a new product that was dishonest and distrustful. Fear of glucose was also fear of chemical intrusion, of labs over fields. That had been true of margarine as well, but at least margarine makers could defend themselves as part of another plausibly organic industry, that of cows and pigs. And margarine, for all of its downsides and despite cartoons claiming the contrary, wasn't enabled in the same way by chemical concoctions as much as glucose, with its sulfuric acid and sulfites.

The questions of purity were also less agricultural in glucose's case in the sense that it wasn't clear that going back to field-based sugar was a purer means of producing sweeteners. Cane had long been the end result of a morally and environmentally compromised labor system. Beet sugar was meant to provide an alternative to cane, but even that added new confusions while striving to avoid others. Beet sugar's racially acceptable Anglo-Saxon labor practices were more "pure" than the labor of dark-skinned slaves in the Caribbean and Pacific; beet sugar was more "pure" precisely because it did not perpetuate the repressive practices of imperial sugar colonies. Glucose avoided each part of that dynamic but created its own new problems as it sought to avoid the old ones. Its industrial labor was distant and complicated, large-scale and dangerous, artificial and untrustworthy.

No matter the labor relations, glucose had been a target of pure food crusaders from its earliest days. Berghaus and a host of pure food advocates put it in the crosshairs of their site, but ultimately its offenses were less fantastic and less persistent than its hydra-headed partners. After all, the cockroach in the candy from chapter 1 was a more visible and egregious offense to many. The blatant candy-cockroach adulteration was obvious and recognizable. It was the result of an everyday urban life that people understood. It was not difficult to argue that such things were neither right nor natural. New glucose and artificial sweeteners were subversive, complex, and unknown. Like Friend's Electric Sugar, they were the result of chemical factory production that was separate, concealed, and confusing. Like the Friend con, they were analyzed not just in laboratories but through long-standing metrics of trust and as a reaction to suspicions about authenticity.

III

The Analysis of Adulteration

✱ 7 ✱
Analysis as Border Patrol

What ho! Have you read the poor-food list
Subscribed and sworn by the state chemist.
There's aniline
In the soup tureen
And borax thick on the gay sardine

Oh, Man, self-proud your roast and stew
The animals don't envy you,
For note the ass,
He eats his grass
And knows it's all in the pure-food class!

JAMES FOLEY, poet laureate of North Dakota[1]

There was no substantial infrastructure of chemical professionals who identified, characterized, or prevented adulteration before the pure food crusades. There was one after. Chemists from private trade shops, the grocer's storefront, and civic bodies of governance sought to establish their authority to police the distinction between pure and natural on one side and adulterated and artificial on the other. They developed their role by generating commercial livelihoods and government positions. They did it by placing themselves on the stand as experts based on laboratory assays. And they got there by organizing to present a unified front for chemical testimony. They became border patrollers in the age of adulteration, arbitrating the boundary between pure and adulterated. How did they make it so?

Part of the momentum leading to impure foods in the first place was a shift from proximity and familiarity to distance and opacity. Part of it was the culture of distrust and deception that framed the call to fight for purity. In both cases, the *appearance* of foods no longer engendered trust from consumers to the same degree it had before. Or, perhaps better put, consumers wanted more than surface appearance. They sought instead to cut through surfaces to the depths of true identity. This impulse followed from the belief, long tested by a cultural search for authenticity, that the truth was inside.

Chemists could expose that truth. In that manner, the chemists were a consequence of the age; they found their footing amid the momentum, the culture, and the search. They didn't come from nothing. In turn, the ones detecting impure foods built their new status with technical accounting, a set of instruments in their tool kit, and the faith of newly quantified authority. They were the ones who could uncover "aniline / In the soup tureen / And borax thick on the gay sardine." Over the second half of the nineteenth century, they organized to insinuate themselves into debates about purity as those debates grew in prominence. While they were a consequence of the era of adulteration, they then became consequential, ushering in a new age—ours today—where we moderns understand the nature and purity of food through trust in analytical techniques under the guidance of technical expertise.

The late nineteenth century didn't introduce the first combination of analysis and purity, as earlier chapters have shown. There was a millennia-old record of political statutes intended to police adulteration through various means. Biblical passages and Plato's observations about acceptable and unacceptable foods make the point. There was also a long legacy of isolated attempts to define the problem and issue warnings through chemical observations. Fredric Accum's 1820 treatise stood out in that regard. But the framework of the pure food crusades produced a new generation of chemical professionals in the United States that grew in relation to the era's argument about artificial goods and artificial people. It was a new generation that brought chemical testimony closer to the center of government. By then, analysts staked out a position to draw distinctions between real and fake.[2]

A strange incident among many typified the point. It's one last margarine reference. This one's about the surprisingly dramatic conclusion to an otherwise humdrum rural fair in 1895. There, at the Meadville (Pennsylvania) State Agricultural Fair, a strange double cross occurred in the competition for best butter. The small town of Meadville is about forty miles south of Lake Erie in the far northwest corner of the state. That's four hundred miles from Chicago and 350 miles or so from the markets of Philadelphia and New York, about halfway between the major market centers of the nation. It also sits twenty-eight miles from Titusville, where Edwin Drake struck oil in 1859, overseeing a derrick boring into newfound petroleum deep underground.

In those far reaches of the state, amidst awards for cattle, crafts, and various crops, the fair hosted a competition for dairy prowess. As the *Meadville Messenger* reported, "Six premiums were offered for the best butter that should be presented" in the first week of February at this, the twenty-first annual meeting of the State Dairymen's Association. The *Messenger's* editor, Mr. A. J. Palm, entered three samples in the name of some local dairymen, two of which took prizes in the competition based on "flavor, texture, color and salt." The *Philadelphia Inquirer* reported the next week that Palm's one-pound sample took the second premium and his five-pound sample took third. So that was that, and a presumably trivial column inch in an urban paper would be the extent of the attention.[3]

But it wasn't.

In the *Messenger*, Palm quickly reported that he had played a prank. He did so because he "vigorously oppos[ed] the present State law on the subject of butterine"—namely, as we have seen in earlier chapters, that this artificial product should be taxed and separated from butter as an adulterated product. He was out to prove you couldn't tell the difference between the margarine (butterine) and, in the well-worn term he used, "the genuine article."[4]

Palm claimed he sent for three packages of fake butter from his contacts at Armour and Company in Chicago. He slipped the butterine into the butter competition banking on its apparent superficial similarities. The ruse worked. His samples made it past the respected judges. These included three experts in the State Dairymen's world. One was Professor H. J. Waters of the state agricultural college. The second was the Honorable J. C. McClintock, a prominent Jersey breeder whose renown came from scoring the highest of any Pennsylvanian at the recent World's Fair. The third judge was A. L. Wales, an experienced creamery man. They awarded Palm the second and third place positions. He had fooled the experts to prove his point that there was no discernible difference between butter and butterine.

This made the judges angry, being made the fools. They couldn't believe it was not butter. Their anger sent them scurrying to save face. After officials gave out the awards and Palm revealed the purported deception in the *Messenger*, the secretary of the dairymen's association sent the samples to chemists for analysis. As confirmed by a signed affidavit, the Wells Fargo Express Company delivered the sample's unbroken seal to William Frear, the lead chemist at the Pennsylvania State College Agricultural Experiment Station. Frear then analyzed the samples. He compared microscopic slides of the Palm samples and a known sample of butter made the old-fashioned way, from cows. Frear looked beneath surfaces to measure melting point, fat content, volatile fatty acids, and curd. His analysis revealed that the Palm deception was deceptive only to Palm—it was indeed butter, not margarine. The analyses showed that Palm had himself been double-crossed by distant contacts in Chicago. His fake butter samples were real butter, the ruse was up, and the judges were vindicated. With increasingly underscored verbiage, the dairymen subsequently published and circulated a pamphlet exonerating themselves: *It Was Butter* (fig. 7.1).[5]

The incident was confusing in its moment and reported and debated for months in the press, from the *New York Times* to small regional papers from Maine to Ohio. The light of history has done little to clarify it.[6]

Palm's sample may or may not have been butter. It's difficult to know why Armour would claim to send him butterine but end up sending butter, if Palm's contact was not an Armour agent after all, or if history has perhaps obscured a third character behind the scenes who may have put the wrong sample into competition. The judges may or may not have gauged its quality accurately. They had relied on a dual arsenal of experience and tests of taste, sight, smell, and feel that many, as chapter 3 canvassed, thought were under assault. The judgments of veracity in this case were soon augmented with analytical measurement from microscopes and laboratory equipment at the agricultural experiment station. Yet the head judge, Waters, was also a colleague of the ultimate arbiter at the experiment station, the chemist Frear, raising questions about possible conflicts of interest lying behind those results.

Agriculture fairs like that at Meadville were showcases of environmental activity. They presented all manner of end products from the agricultural life cycle. These included animals, crops, dairy products, and the crafts made from fibers and raw materials of the land. In their staging, the fairs claimed to be judging the quality of nature's bounty or, better put, judging the skills with which humans had acted on the land to produce that bounty. With that in mind, Palm made a political point about taxation, market structure, and consumer choice. He wanted to press an agricultural and sneakily conceptual point: What's the *real* difference between nature and artifice, and who has the power to say so?

There are many ways to read the Meadville affair. The base level points to a frustrated merchant avoiding taxes, that tried-and-true American theme threading back to protests over representation with tea in Boston Harbor. The local level points to clashes with new distribution networks in an area of the country growing for its source of newfangled petroleum. True, both. Yet Palm's prank was based as acutely on a dispute over reality, over what counts as one thing and not another. He was leveraging the abstract purpose of the Meadville fair—a showcase of environmental

It

Was

Butter

Chemical Analysis of the alleged Butterine awarded prizes at the State Dairymen's Association held at Meadville recently, shows that it was BUTTER.

FIGURE 7.1. Front cover of a pamphlet distributed by the Pennsylvania state dairyman's association in the month after the Meadville fair of early February 1895. Courtesy of the Warshaw Collection of Business Americana-Dairy, Archives Center, National Museum of American History, Smithsonian Institution.

activity—with the concrete example of fake butter. Or fake fake butter, as it were. He had a philosophical complaint, but it played out without recourse to jargon. It was an everyday philosophical problem characterizing the greater worry over distinctions between natural and artificial in the later nineteenth century. Easy answers didn't work out in Palm's case, but his motivation and the course of action that unfolded from his prank tapped into the prevailing concerns of the new age of manufactured food.

Palm's deception was different from Paraf's, or "Professor" Friend's, or the hucksters on Melville's and Twain's Mississippi steamboats. Palm was, in an odd phrase, honest about his sleight of hand. Con men in general are known for a lack of self-expression about their chicanery. But it was the reveal that motivated Palm. It was about pulling the curtain back on what he thought was a bogus distinction.

The case typified circumstances around pure food debates by the end of the century because the Meadville affair was ultimately resolved by deference to the state chemist. Such chemists and analysts have come up throughout this book as contributing voices to adulteration concerns. The chemist, the analyst, the lab technician, and the microscopist moved closer to center stage when public outcries of adulteration demanded a response.[7]

The age of adulteration provided a basis from which analytical chemists developed their modern identity as experts. Part I of this book attended to the cultural context of authenticity and trust to provide the proximate background against which the war on adulteration was fought. Part II drew out geographies of adulterants to show that the argument against adulteration was also an argument about changing environmental practices. Part III turns to the place of chemists and technical analysis amid those arguments. How did pure food arguments produce the new class of chemists? Where did they work? What did they do? Looking to the local and regional level helps address the questions. Broadening that view to the state and national scale helps, too, as in the next chapter.

"Not Strictly Pure, but Pure Enough"

Scientific analysts are complicated political actors in this story. Where one view considered the scientists' skills in manipulation and chemical rearrangement the very source of adulteration, another—like that of the judges in Meadville—saw chemists as salvation. They were cause of and solution to adulteration. A columnist at the time captured the tension. "If this is the age of improvement," he wrote, "it may be said also, with truth, that this is most conspicuously an age of adulteration.... Scientific investigation, chemical analysis and inventive genius have all contributed towards supplying what has been so eagerly sought after."[8] Chemist William Atwater at Wesleyan College agreed with the positive contributions. He thought cheaper alternatives were "one of the greatest boons that modern science has wrought for the benefit of the poorer classes."[9] Another chemist, C. W. Parsons, took the contrary view. Any alternative was adulteration, he thought, and "adulteration is synonymous with poison . . . the scythe of grim death itself." The same cause of–solution to dynamic played out with margarine, cottonseed oil, and glucose. Each case revealed the tension between producer-side issues about innovation and improvement and consumer-end worries about danger and risk. Chemists fell into that dynamic when people thought they were the font of or answer to adulteration. In *Puck*'s view, for instance, as with the lower-left corner of figure 7.2, chemists were paid-off coconspirators to the adulteration mob.

The producer side of that tension reveals the early history of industrial science and the birth of corporate research laboratories. Many an aspiring

FIGURE 7.2. One of *Puck*'s many efforts at highlighting the popular view that adulteration was rampant and unchecked, even by the chemists ostensibly paid to certify purity (*lower left*). The upper left boxes point to milk diluted with chalk; the upper right has "S. Windle" as sugar adulterated with sand. The middle right box correlates oleomargarine with unhealthy animal oils. The bottom right brings them all together as an unholy alliance with gravediggers. Source: "The Adulteration of Food," *Puck* 7 (May 5, 1880): 160.

analyst found or created positions at chemical works that produced raw materials useful for other industries. The sulfuric acid the glucose manufacturers depended on is an example. For reference, in the mid-1870s there were thirty-six private chemical works in New York City; by the turn of the century New York had 135 chemical manufacturers.[10] In parallel were those chemical manufacturers who operated in league with a range of corollary industries. Dye makers, for example, worked alongside textile makers. Recall that Alfred Paraf was the son of a dye maker, learning his chemical skills as part of a textile industry. Or there were oil and fats traders, like William Procter and James Gamble, who built ever-larger factories to produce and trade soaps and candles and an extended line of new consumer products. Procter & Gamble stands out in the history of industrial science for developing one of the first corporate research labs in the 1880s.[11]

Meat-packers and food processors did too. J. Ogden Armour, Palm's supposed coconspirator, celebrated the company scientist. "All the cunning of the chemist," he wrote, "has been called into service to save, to make the most of every scrap of material at hand, and to discover new ways in which some element of waste may be diverted from uselessness to use."[12] As long as chemists as company men were solving problems for industrialists, they were a good thing. That industry affiliation was problematic, however, when others saw it as a conflict of interest. Henry Mott had taken over the U.S. Dairy Company in the early 1870s after investors discovered Paraf's chicanery. He sought to testify in support of the chemical validity of margarine. But he was rebuked because he was too cozy with industry to hold credibility for consumers.[13]

The producer side is interesting, but the consumer protection side provides a better angle to explore the greater book themes of how people understood and responded to the fear of adulteration. Here came the chemists. The rest of the chapter explores the presence and growth of that analysis of adulteration. Doing so helps explain the new infrastructure of adulteration analysis that made it possible for the Meadville controversy to find its analytical end. Most telling about the new late-century stewardship of chemical analyses was the way the focus on adulteration from chemists confirmed a shift in consumer attention from attending to processes of agricultural production to fearing retail marketing. "Chemists could usually *identify* substances," historian Mitchell Okun has noted about this era, but "they could not truly *evaluate* them." By marking off their point of intervention away from the process of agricultural activity, the chemists were defining the problem of adulteration as one of identification, not derivation, moving the concept of purity into the analytical realm and out of the environmental one.[14]

Three main constellations of antiadulteration chemists show that shift happening. They focused on the consumer world, where grocers, cooks, and housewives lived and faced choices about food identity. One constellation worked as individual consultants, hanging their shingle along Main Street and advertising as an accountant or plumber might. These were entrepreneurs who sold their analytical services to those worried about foods, drugs, and drinks. A second kind of antiadulteration chemist parlayed technical credibility into the journalist trade, editing papers in urban settings, collating the problems of the day, working with grocers who often hired them to add credibility to their empire. And yet a third cluster of chemists worked as members of nascent governing bodies developing the virtues of disinterested public good at various scales: they worked with civic organizations or for municipalities, regional boards of health, and state commissions.[15] With such a flurry of new activity over the later decades of the nineteenth century, by the time Congress passed the Pure Food and Drug Act in 1906, the chemist, analyst, and public health scientist was visible, scrutinized, lauded, chastised, hired, fired, trusted, questioned, and typical. To be sure, there was no such thing as a pure scientist in the pure food crusades.

Those in the first group of chemists, the individual consultants, were thinly populated in the decade after the US Civil War. As food manufacturers, meat-packers, and urban grocers grew in size and number, though, so did the business for private analytical chemists. Their presence tracked the increasing concern over food identity in the urban sphere. In the mid-1870s, New York was home to nineteen private analysts (in addition to those working at the city's thirty-six private chemical works). A decade later thirty-six chemists advertised their services and, by the 1890s, about fifty.[16] Chicago, by comparison, had few chemists to hire outside the industry specialists working for meat-packers such as Armour and Swift. With a population close to four hundred thousand, only three advertised their analytical services in the mid-1870s. (For comparison's sake, you were just as likely to go into the chewing gum trade as chemistry, as there were also three gum suppliers in Chicago in 1874.) A decade later that number had grown to a dozen chemists working as private analysts in the city. Relatively speaking, Boston was a hotbed for chemical analysis. The city supported twenty-one separate chemical analysts in the mid-1870s, twenty-seven in the mid-1880s, and forty-six by the mid-1890s.[17]

Sometimes they called themselves "consulting" chemists. Sometimes they listed their businesses as a part of the dyestuffs or druggist trades. All of them made their money through the analysis of consumables, be they foods, drinks, or drugs. Given Accum's precedent, you couldn't say the association of chemistry and antiadulteration was new in the Gilded Age. And analytical chemists were doing more than adulteration analyses, as an affiliation with druggists and dye makers would suggest. But by the late 1800s their numbers had grown dramatically. By then they were a trade in a service economy built on the increased availability of and concern about produced goods. Here came the chemists.

Gideon Moore's storefront was a good example. He was a German-trained consulting chemist who ran an "Analytical Laboratory and Assay Office" in Lower Manhattan (fig. 7.3). He founded it in 1872, and by the 1880s, it was thriving. The shop sold assays for twenty categories of manufactured goods, food being a primary offering. Moore also offered his services as a court witness for a price. For the assays, prices ranged from loose change to several hundred dollars. He would measure the "percentage of pure cane sugar" for $1.50, "test for adulteration or foreign admixture" of soap for six dollars, do a complete workup of fertilizer samples for fifty dollars, or run a "complete chemical and mechanical analysis" of a soil sample for $250. (A similar soil test from the cooperative extension

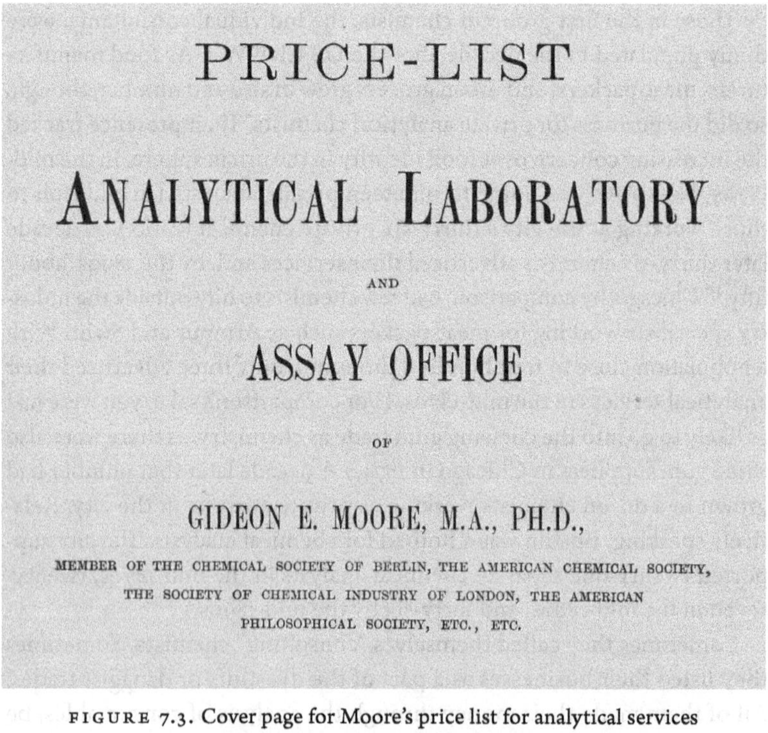

PRICE-LIST

ANALYTICAL LABORATORY

AND

ASSAY OFFICE

OF

GIDEON E. MOORE, M.A., PH.D.,

MEMBER OF THE CHEMICAL SOCIETY OF BERLIN, THE AMERICAN CHEMICAL SOCIETY, THE SOCIETY OF CHEMICAL INDUSTRY OF LONDON, THE AMERICAN PHILOSOPHICAL SOCIETY, ETC., ETC.

FIGURE 7.3. Cover page for Moore's price list for analytical services in Manhattan, 1888. Courtesy of the Science History Institute.

service today runs about twelve dollars.) There were fourteen separate tests for beer and another forty for wine. He did butter tests too. One by lactometer cost one dollar (more on that later), while a test by microscope was three dollars. He was a businessman in a trade flourishing amid demands for chemical identity.[18] At the close of the Civil War, Moore and his peers in the profession were a rarity in American cities. By the turn of the century they were common.

The second kind of chemist was more robust than the first. They were antiadulteration analysts affiliated with the grocers' empire. *American Grocer* (1869–1990), *Spice Mill* (1878–1963), and *American Analyst* (1885–1895), three serials speaking for that empire, open a view to this constellation of chemists. All three papers were engaged with chemical deliberations as a response to fears of adulteration. Each was anchored in the grocers' world of urban commercial interaction. Each showed chemical analysts shifting attention on adulteration from the agricultural process to consumer product choices at the market.

From a quiet postbellum start in 1869, *American Grocer*'s aim was "the protection of the public against the adulterations which are so extensively practised by unprincipled dealers." The editors promised to obtain samples from the "leading houses of common articles" and manufacturers making anything for family use. Their tools were "chemistry and the microscope," which would "afford the ready means of determining the true constituents of any article of diet." Its early years straddled the countertops of the grocer and the fields of the farmer. Weekly columns listed wholesale prices. Each number was replete with ads for farming equipment and agrarian wares. Every issue reported on the spectrum of adulteration already noted in this book—tea, milk, lard, sugar, olive oil, butter, beer, spices, bread, honey, and more.

I described *American Grocer* as part of the rise of the grocers' empire in chapter 3. I can add that it was also a premier site for commercial antiadulteration chemistry. A cursory view of its title might lead you to miss the analytical motive, but the chemist John Darby was first to helm the paper in an effort to bring together analysts, the antiadulteration movement, and the professionalizing goals of grocers. Darby published it in 1869 with financial backing from the grocery magnate Francis Thurber. He then handed the masthead over to Thurber in the 1870s to return to the consulting shop he operated nearby.[19] (It was also but four blocks and a half a mile from Gideon Moore's shop.) Like the paper itself, Thurber's career straddled farm and field. He was at once the president of the American Dairymen's

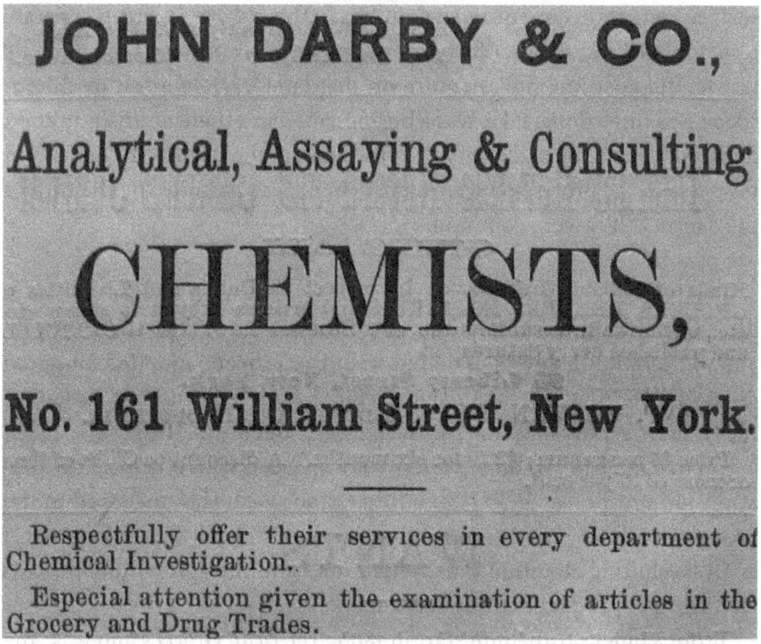

JOHN DARBY & CO.,

Analytical, Assaying & Consulting

CHEMISTS,

No. 161 William Street, New York.

Respectfully offer their services in every department of Chemical Investigation.

Especial attention given the examination of articles in the Grocery and Drug Trades.

FIGURE 7.4. Advertisement in *American Analyst*. Darby was also the publisher of *American Grocer*, a trade paper that began circulation on a weekly schedule in New York in 1869. He sold his analytical services to readers of *American Grocer* and *American Analyst*, from which this ad is taken. Courtesy of the Library of Congress.

Association—this was the national organization of which Pennsylvania's State Dairymen's Association, the one at Meadville, was a part—a powerful urban businessman, an avowed advocate for oleomargarine, and a personal friend of the board of health's Charles Chandler.

Still, Darby remained a presence in the paper. His frequent advertisements for analytical services set a pattern for others who would publicize their food detection trades (fig. 7.4). Joseph Geisler, an analytical and consulting chemist, considered "Analyses of Food Products a Specialty." J. F. Elsom, another "Expert Chemist and Analyst," placed ad copy in New York papers for services from as far away as Indiana. This was a common pattern for grocers' papers. George Heid, for example, advertised his St. Louis area services "on anything appertaining to analytical chemistry" in the *St. Louis Grocer and General Merchant*. Others advertised in *Chicago Grocer*, *Baltimore Grocer*, *Philadelphia Grocer*, and the like.

American Grocer was an industry leader as a hybrid grocer-analyst forum from the beginning. Darby and Thurber saw the tight connection between the dual purposes of analysis and grocer boosterism. Elevating grocers meant defending against charges of adulteration. Defending against the charges of adulteration meant building a place for the analyst to distinguish between genuine and fraudulent goods. Working for the paper provided a way for analysts to build that experience and credibility.

As I showed in chapter 3, the papers were eager to lay blame for adulteration on everyone but themselves, deploying the language and moral connotations of duplicity so common to the age. For *American Grocer*, it was the "wily Chinaman" who was out to create "trashy tea" by contaminating batches with exhausted leaves. It was the "cunning manipulative propensities" of the Japanese with their "lie tea."[20] Alas, they were equal opportunity jingoists. Europeans were guilty too. British shipments of tea were often rejected for the lead coating used as coloring matter. The Germans were worse, *American Grocer* claimed. They would dilute American kerosene into three grades, while adulterating flour with plaster of Paris, baryta, and potato flour and cutting lard with tallow and horse fat.

Then there was the new alternative sugar, glucose. "Seven-eighths of all the sugar sold in Chicago is glucose, a cheap substitute, made of rags in Germany."[21] For coffee and tea, analysts used microscopic tests to detect adulteration. They used polariscopes to identify fake sugar. They also determined that American-derived glucose, by contrast, was made safely from "corn starch and oil of vitriol." The American-made substitute was legitimate, that is. The imported one was an adulteration. "Honest American products are made unpopular by dishonest Germans." For the grocer, chemists validated these cultural observations through the clarity of their analysis.[22]

Three blocks north of Thurber's publishing house, Jabez Burns circulated the *Spice Mill* toward similar ends (fig. 7.5). He began publication in 1879. (The paper was in print until 1963.) You could again be misled by the title. Burns was a self-described "Machinist, Inventor, Manufacturer, and Dealer in all kinds of Coffee and Spice Mill Machinery," but he was really after adulterators. His paper would "deal in a spicy way with the spice of active manufacturing business life" while grinding "spices and other kindred lines of business" into fine powders. His goal was to reduce facts and figures to a fine powder to analyze them and put them to practical use. I have no comment on what appear to be witticisms, but let me say more about Burns's larger goal to enroll chemical analysis in a fight against adulterators.

The *Spice Mill* dealt with the full complement of adulteration, including candies, bread, olive oil, butter, sugar, cottonseed oil, and more. Each volume published columns on "the adulteration of food" and any bills against adulteration in Congress. Burns reported on "Terrific Adulteration

THE SPICE MILL,

DEVOTED TO THE INTERESTS OF MANUFACTURER AND CONSUMER.

Vol. I.] JANUARY, 1878. [No. I.

FIGURE 7.5. Masthead of the *Spice Mill*, Jabez Burns's coffee, spice, and antiadulteration paper. Courtesy of the Hagley Museum and Library.

in France," new British regulations on coffee and chicory, and the art of adulteration in his series "The Age of Adulteration." Another series called "Honesty and Purity" made explicit Burns's goal to fold the customer's concern for sincerity together with the producer's claims of authenticity.

Like *American Grocer*, the *Spice Mill* showcased not just the fact of adulteration but the place of the analyst and chemist in the war against it. Early volumes discussed German chemistry, British analysts, "the future of scientific discovery," analytical reports on suspected adulterants (many of them oleomargarine), different testing methods, and a column called simply "Analyst's Report." One report on cottonseed oil focused on the customs house analysts who detected the adulterations.[23] Others discussed the analytical methods for detecting grape sugar in deceptively labeled "pure sugar" lots.

The *Spice Mill* was anchored in the urban grocers' world. It offered another example of analysts focusing on the adulteration of consumer products at the market instead of production processes in the fields. Burns was a coffee and spice purveyor by profession, not an analyst. Like other grocers, he sought to enhance his professional status by crafting the lauded image of honesty and purity. Chemical analysts helped identify adulteration and vouch for honest goods.

Burns often reprinted columns from other local papers when he wanted to enlist chemical testimony to support a point. *American Analyst* was one of his favorite sources. In what was a very busy analysts' district, the chemist Henry Lassig edited *American Analyst* just another five blocks away from the *Spice Mill* in lower Manhattan. He, too, pursued a basic agenda of adulteration detection. *American Grocer* had brought together ostensibly different vocations of grocer and analyst in one office. The *Spice Mill* anchored itself in the merchants' world of coffee and spices, reaching out to bolster its claims with reports from analysts. Lassig was a chemist who further epitomized the role analysts sought to play in the food trade with his more explicitly scientific title *American Analyst.*

Lassig began editing a ten-year run of the *American Analyst* in 1885 with fellow chemist John McElrath (fig. 7.6). If the titles of *American Grocer* and *The Spice Mill* misled readers to think they were nonscientific papers, *American Analyst* offered the opposite misperception—by title, a casual reader could have mistaken it for arcane chemical commentary outside the pedestrian world of the grocers' empire. Its mission, though, was the more routine goal of "the suppression of adulteration." Lassig boasted that his was "the pioneer journal in this country to take up the fight against adulteration." He and McElrath would fight the "charlatans, adulterators, and handlers of poisonous food" to reveal "the utter frauds" of the grocers' kingdom. They would lift "the veil of chemical mystery in which tricksters . . . envelop their processes of manufacture." They would stand against "confidence operators" by using "the only method for counteracting their nefarious schemes."[24] They were neither shy nor humble in these pursuits.[25]

Through the "unceasing efforts of the *American Analyst,*" Lassig offered chemical detection as a tool to ferret out impure imports at customs houses too.[26] Like his publishing competitor *American Grocer,* he would blame Europeans for corruption. "Rigid anti-adulteration laws in most all European countries [have] compelled the shifty rascals to seek 'pastures new.'" It would be unfair, he wrote, "to punish the innocent purchaser or transporter" if the problem was allowing goods to come into the country.

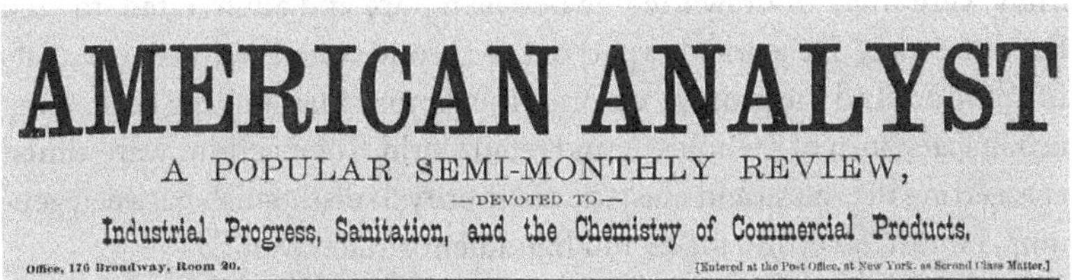

AMERICAN ANALYST
A POPULAR SEMI-MONTHLY REVIEW,
—DEVOTED TO—
Industrial Progress, Sanitation, and the Chemistry of Commercial Products.
Office, 176 Broadway, Room 20. [Entered at the Post Office, at New York, as Second Class Matter.]
VOL. II.—NEW SERIES. NEW YORK, MAY 15, 1886. No. 10.

FIGURE 7.6. Masthead for the *American Analyst*, published semiweekly out of New York beginning in 1885 and edited by the chemists Henry Lassig and J. McElrath. This version is from the second volume, 1886. Later volumes changed the subtitle to *A Semi-Monthly Journal of Pure Food, The Suppression of Adulteration*. Courtesy of the Science History Institute.

Customs agents were key actors in this scenario. Theirs was a more complicated task, as they had to differentiate grades, varieties, and strict chemical compositions arriving in barrels of uneven quality in the ports of New York. Since tariffs were such a critical contribution to federal coffers, cane sugar importers sought to influence the results of port-side analyses to lower their tax by manipulating the analytical gatekeeping in those customs labs. In a situation where "the craftiness of overseas sugar-makers enabled them to subvert the appraisal of American customs officials," *American Analyst* said, distinguishing between one grade of product versus another had significant economic importance. I was speaking metaphorically when I said chemists marked off a border between pure and adulterated with their analyses, but here the borders were physically real. Imports entering the country showed the geographical boundaries chemists were there to protect. Some things were supposed to be inside, others were not; chemists could tell the difference.[27]

American Analyst joined *American Grocer* and the *Spice Mill* in blaming the "other" for causing or allowing adulteration. In turn, the manufacturer was to blame. Foolish customers were to blame. Native Americans were to blame. South Asian Indians were to blame. The "Hebrew" was to blame.[28] Italians were to blame. Lassig told the story of an Italian milkman accused of fraud who quite literally had tricks up his sleeve. The "cowman" held a rubber tube of water up his shirtsleeve to be dispensed into the milk just before closing the lid on the jar and selling it to an unsuspecting customer.[29] Tests for specific gravity and consistency determined that the milk was cut; judgments of character found that the Italian immigrant was the culprit.

In search of differentiating the "genuine, honest, healthful character" of food from impure dangers, activists shouted, "who can tell the difference!" Lassig's point at the *American Analyst* was that chemists could. They were there to draw the line between pure and adulterated, to "tell the difference." He worked as part of this second constellation of chemists forming around the grocers' empire. They found their footing addressing urban fears born of a wholesale and retail world. Their actions were aimed at asserting the moral and epistemic authority to distinguish between genuine, honest, healthful foods and fraudulent adulteration.[30]

If one alignment for consumer-side chemists was the private analytical chemist and another was the grocer-aligned trade chemist, a third one took elements from both to build a presence at the governmental, quasi-governmental, and civic organizational levels of chemical detection. The

Gilded Age was similarly the age when cities, states, and regions built governmental boards of public health. New York City (Manhattan) founded its Metropolitan Board of Health in 1866. Massachusetts founded its state Board of Health in 1869. New Jersey founded theirs in 1877, the same year as Illinois. Like the federal chemists in the next chapter, analysts on those boards worked to get beyond metrics of community-based trust that had given cultural import to their efforts for years. By this period, as one historian wrote, they "led a charge that supposed that regulation by reputation was inadequate and that methods to guarantee agricultural success were far beyond the farmers' ken."[31]

In addition to boards of health, civic groups organized to stand up for beleaguered consumers by endorsing and encouraging chemical analysis. The Women's Christian Temperance Union (WCTU), founded in 1874, worked endlessly during the later 1800s to fight not just against alcohol but for pure foods, drugs, and drinks. Historian Loraine Goodwin characterizes the WCTU for its prominence at the forefront of the antiadulteration movement. In state chapters, members of the WCTU "sponsored lectures and debates featuring public officials, physicians, and lawyers" to promote the cause of purity. They worked with state chemists to test supplies of food supplements. At one point in Massachusetts alone they distributed hundreds of pamphlets and held dozens of pure food meetings. Some of their members would testify that adulterated foods were "unnatural," others called them "demoralizing." Throughout these efforts, WCTU members worked in association with chemists and analysts and built ties to leaders of the domestic-science movement, Ellen Richards being one of them.[32]

The new American Society of Public Analysts (ASPA) also fostered antiadulteration analysis for public protection. They first convened in 1884 to address problems like tainted and adulterated milk, false ingredients, and broad contamination. Central to their larger mission was "the dissemination . . . of knowledge relating to the detection and prevention of adulteration and falsification of food, food products and drugs." Given the stubborn problems of tainted milk in particular, their inaugural meeting of October 1884 devoted a significant amount of time to debates over tuberculosis, milk supplies, and oversight. Once assembled, J. H. Raymond, Brooklyn's health inspector, led the meeting to make clear that "a harmony of action" between various health governance districts could help in the broader portfolio of "the detection and punishment of frauds in foods."[33]

The ASPA's formation offered a way for Gilded Age chemists to establish authority at local and regional scales. Watching them meet was like watching analytical authority happen. Those at the inaugural meeting represented boards of health from New York (i.e., Manhattan before the boroughs consolidated), Brooklyn, Albany, and Paterson, New Jersey. The premise, as Raymond put it, was "to see if some kind of association can be formed by which there shall be not only similarity of methods employed but also a blacklist made, so when a violator is found here that fact shall be communicated to the Boards of Health and Food Inspectors of the surrounding cities and country." They were trying to bind together different regional governance boards.

Raymond was proud of his success in Brooklyn identifying contamination, but he knew that success was ephemeral. The inspectors were playing a kind of Whac-A-Mole of suspected products, tamping them down in one area only to see them pop up in another. "If we drive a man who is a violator from Brooklyn, he goes to New York or New Jersey, and if a violator is caught in New York he comes to Brooklyn or New Jersey." Raymond's associate, the chemist E. H. Bartley, had been dealing with milk sent from Brooklyn to New York, trying to catch it before it crossed the river. He was chasing down adulterators right up to the docks, watching in frustration as they took off in boats across the river, barely escaping just outside his jurisdictional grasp.[34] If the analysts could join forces, they could address the problems more effectively.

At the board of health's meeting hall in Brooklyn, the forty men agreed they should organize and coordinate their analyses. But they also understood the problems facing them. One was the jurisdictional coordination Raymond and Bartley were talking about. Another was the role the analysts could play in preventing adulteration if they didn't know enough about the process. Adulterated milk provided the topic that night. They talked at length about the conditions that produced contaminated milk, conditions they understood from years in the inspection business, which led to their shared assumption that the adulteration was difficult to prove. The 1850s swill milk controversies hadn't gone away, they'd just become more challenging to detect. "Unless we go to the stable and see the cow eating," the inspectors said, they couldn't tell if the cows were healthy. They had to understand the conditions of milk production to truly ferret out adulterators.

Like air flow, for example. It must've been a fun night when forty chemists talked at length on the topic of adulteration and airflow. If you've ever railed against the minutia of committee meetings, these guys would look familiar. "Improper ventilation is a source of trouble," one of them said. "The air space," he thought, "should be fixed by law." Some in the group figured "600 cubic feet of air space" ought to do. Dr. Bell, the editor of the public health journal *Sanitarian*, disagreed. "When we consider the breathing capacity of a cow I should say air space should be from 1,200 to 1,500 cubic feet."

Either way, Dr. Elwyn Waller of New York was flustered that they were getting pulled into the weeds. He drew the conversation to a more general level, adding, "It seems to me it is the duty of all Health authorities to take the broad ground. An animal that is diseased certainly cannot produce healthy secretions, or the tissues must be abnormal." He insisted they consider the broader conditions—a dairy ecology—of feed, air circulation in stables and dairy stalls, water supply, the other cow's they were in contact with, and so on.[35]

Then Raymond explained that even if they could understand the living conditions, the milk could still be corrupted between the stable and the consumer. Some cases were ostensibly simpler than swill milk or tainted country milk; sometimes it was simple watering down, and that could happen anywhere. In fact it was common practice. "There is an odd milk dealer," Raymond told his colleagues, "who came to me and said it was the commonest thing to add water. He did it and everyman did. He said they add 4 or 5 quarts to 40 and that it was a business necessity." So even if they monitored the final product—the premise of new food-health statutes—and even if they knew how healthy the cow's quarters were, it was still unclear whether this would stop the larger problem of trust in milkmen and trust in process.

What the chemists *could* do, they all agreed, was develop instrument standards and shared methods. What they really wanted was to measure products with "mathematical demonstrations," as Raymond said. Standards and methods would provide the basis for their credibility in fighting adulteration. That approach had implications, the chemists knew. It placed their evaluation of purity outside the agricultural processes that previously anchored it. This was a lesson they'd drawn from their collective experiences monitoring and addressing adulteration in the New York metropolitan area and from their positions as consultants in public controversies over the problem. Some of the chemists at the ASPA meeting in 1884, for instance, had been part of a court case the decade before, *The People v. Daniel Schrumpf* (1876). That experience paved the way for the later discussion about standards and methods.

In the old *Schrumpf* case, Elwyn Waller offered testimony against the small-time milkman Schrumpf. The expert testimony drew from mixed methods. Some were older organoleptic tests of sight, taste, and smell at the farm and with the cow. Those were part of the consideration of the "broad ground" of the problem.

Waller would get frustrated with his colleagues for getting lost in the details in 1884. In 1876, though, he was the one in the weeds. For instance, Waller's first review of the suspected product was to swish it around a drinking glass, testing "whether the milk adhere[d] to the side of the glass as milk generally does." Schrumpf's lawyer challenged the legitimacy of such a fickle test. Waller held his ground. He showed how he augmented the direct sensory tests with others mediated through specialized instruments. Court journalists reported that "On the table in front of the bar was an imposing array of glass tubes, bottles, measures, and other vessels, a miniature drying oven, and two steaming water baths, heated by gas drawn through a rubber pipe."[36] The display was intended to make the apparatus of chemical detection visible and empowering, similar to today when a scientist in a movie wears a lab coat to denote cultural authority. The table of instruments was a performance.[37]

The Schrumpf case is notable in the history of adulteration analysis for a few reasons. One is that it enunciated the difficulties of evaluating what should seem like an easy thing: is this pure milk or not? Tasting the milk seemed to make sense; checking out the stables in which the cows that made it was reasonable; measuring its specific gravity would also seem important. This wasn't a case of intricate factory innovation but a problem of purportedly normal life. And even that was difficult to arbitrate. Another reason is that in order to judge milk's identity, the court case assembled and then relied on trained chemists who lived and worked outside the context of dairy production. They were public advocates, third-party inspectors serving as mediators between the producers and the consumers. They were available to play that role by the Gilded Age in ways that hadn't existed before. But the most relevant reason is that the case offered a snapshot of a transition, providing an image of the analysts just before they came into their own in the fight for pure foods. They were still building their credibility. They were still developing their role of public advocacy and mediation. All the material pieces were in place, the tubes, the bottles, the glassworks that all ended in *-meter*, but the ways those instruments added up to certainty and veracity was still unclear.

On that point, the chemists in all three constellations of analysts—the consulting chemist, the antiadulteration press chemist, and the quasi-governmental public chemist—were in a bind in the 1870s that they would climb out of by the end of the century. In public, they defended the legitimacy of their instruments and their authority on the stand, basing the legitimacy on experience and methodological rigor. But internally the men didn't actually agree on that validity. The simple lactometer, for example, was a point of much dispute. Schrumpf's lawyer knew and exploited that fact. Professor Robert Doremus was an analytical chemist at the City College of New York. He called the lactometer "that perverse and mendacious instrument."[38] Chandler had written in the *American Chemist* in 1871 that the lactometer was "a very unreliable guide." A British study found that "the lactometer test is a total fallacy."[39] The analyst-for-hire Gideon Moore used it, but not with the same degree of confidence he placed in other instruments. He offered analytical detection using the lactometer at his shop in lower Manhattan, but it was a third the cost of a more thorough microscopic analysis, one dollar versus three dollars.[40]

When analysts founded the ASPA eight years later, they were still arguing about it. This time it was behind closed doors and, by then, the experience of a crucial extra decade of adulteration and analysis helped. They talked about policing the foods across jurisdictions. They weren't naive about market conditions either. They understood the business demands that may have pressured milkmen to dilute their products. Yet *American Grocer* was still reporting that "the lactometer is useless, and plays into the hands rather of the fraudulent dairyman than of the consumer." So when they organized they took as their charge the need to develop standards and methods to increase the public efficacy of their analyses. *American Grocer* told readers "the only method known by which [purity] could be satisfactorily tested was complete analysis."[41]

That's where they operated. For analysts, it begged the question, how complete is a "complete" analysis? When could you be satisfied that enough was enough? In the 1880s, the ASPA answered in a directional, not absolute, sense. *More* measurements and *more* instruments were better than fewer. In fact, the longer they talked about it, the longer the instrument list grew. A chemist from New Jersey noted the lactometer but also wanted the lactodensimeter to get its due. Then there was the creamometer, "an instrument for measuring the amount of cream which is thrown up to the surface of the milk." And the pycnometer, "a little glass bottle provided with a glass stopper, which stopper is perforated so that we may fill it to a fraction of a drop [and] holds a thousand grains of pure distilled water." Others praised the centesimal galactometer, Soleil's saccharometer, the centrifuge, the microscope, and something called a "chameleon test." This last one was a way to test milk "on the fact that [the contaminant] potassium permanganate is equally discolored by [the adulterants] casein and albumen."[42] The premise of the listing of the instruments was that a complete analysis was one with more measurements.

The ASPA represented a new kind of organization dedicated to drawing the line between pure and adulterated. For them and others, the specific instruments became less important than the fact that they used instruments. Site inspections didn't go away—investigating the sources of the problem—but they became a form of intervention complementing the more expedient, more efficient, more controlled instrumental tests in the city and for the grocers' customer. Marking off their point of intervention away from the difficult to track process meant the chemists saw adulteration as a problem of identification, not derivation. That was a big move.

Whereas an assembly of public health analysts arguing about adulteration detection methods a generation earlier was a rare thing, by the Gilded Age chemists were busy working to identify the best ways to intervene, not whether or not they should or could. They developed their role by generating commercial livelihoods and government positions. They did it by placing themselves on the stand as experts based on laboratory assays, by organizing to present a unified front for chemical testimony. In so doing, they embedded a new assumption into the detection of adulteration that close inspection, scrupulous testing, and monitoring made foods safe. It was a supposition based on the logic that the measure of safety was the degree of testing rather than an agreement that the thing being tested matched a definition of natural or pure. Lassig provided a sense of this when he admitted in *American Analyst* that analysts could find products "pure, and not strictly pure, but pure enough."[43]

Pure enough. The work to be done was the work of distinctions and differences. When had you crossed the line over to adulteration? Analysts were patrolling that border with the shared premise that there was no preordained, natural (as in nonhuman) standard that defined purity. It had to be built and defended. It had to be made true.

Their answers weren't absolute, they were negotiated. On what terms were they negotiating? Nature had variations, so that didn't provide an objectively self-evident guide. "Nature does not allow herself to be confined in narrow bounds, but delights in pleasant variations," a spokesman said in testimony before a hearing about contamination. Natural butter was various because cow's milk varied. Yet as chapter 4 showed, the Schrumpf case argued, and the members of the nascent ASPA understood, "the very name of milk has been used by different persons in quite a different sense." Some thought of it as "the unchanged product as drawn from the cow." Others thought "a little more or less fat" made it skimmer or fatter and both were still milk. Analysts answered by deploying instruments to determine whether the milk, the butter, and the food were pure enough. They were negotiating on the terms of analytical measurements, not agricultural practice.[44]

The chemist was in position to do this by the end of the nineteenth century at the local and regional level because of dedicated organizing to address a recognizable public problem. Their detective work was shaped and informed by the complex conditions of adulteration and impurity. It wasn't just the fact of adulteration that explains their activity. It wasn't just the perception, valid or not, of dishonest foreigners that aided their work. It wasn't only the difficulty of managing imports—whether across the width of the Atlantic or the narrowness of the East River. It wasn't only the legitimacy of various types of technical evidence that fostered more work to refine methods and build consistency and reliability. It was all of those things.

The Meadville Affair, Redux

By the Meadville controversy of 1895, it was possible to defer to the state chemist in ways that hadn't been possible before. Palm's prank may have been ostensibly about taxes, but it was fundamentally about distinctions. Ellen Richards had written about the difficulties of dairy distinctions in ways that Palm understood and tried to leverage, accenting the belief that one was "real" and one fake. "Rarely has there been a fraud so difficult to detect," she wrote, "since not only the apparent but the real differences between genuine and artificial butter are but slight."[45] *American Grocer* had reported on "butter and oleo tests" to tell readers that chemists were devising new techniques to differentiate the two. A set of "microscopical" studies could do it. Jabez Burns wrote consistently in the *Spice Mill* with praise for analysis as a way to cut through differences that were confusing to the naked eye. Lassig printed a joke about distinctions in *American Analyst*, retelling a ham-fisted tale about two calves that looked the same. Butty was "a bright yellow-haired cherub" who went for a little promenade with his parents, Mr. and Mrs. Cow, only to stumble upon a twin at the grocery store. The twin, Oly, was "a poor little orphan" with no parents yet was the spitting image of Butty. The cow family took an interest and noticed the striking resemblance. Oly then followed his new family home. When they got there, Mrs. Cow looked at them both but couldn't tell who was who. Unable to distinguish them, Mr. Cow asked her why it even mattered. "If there isn't any difference, then what's the difference?"[46]

Palm asked the same thing, touching on key adulteration-era themes of distance, trust, and distinction. The swift response to Palm's prank at Meadville brought into stark relief the moral as well as agricultural boundaries he had crossed. It spotlighted the challenges he brought to prevailing trust systems and proper codes of market conduct and showed that more than just physical distance explained adulteration debates. As Paraf and his historical entourage of fellow con men so capably demonstrated, and as Lassig's Butty and Oly were supposed to show in a more mundane way, appearances were still in play.

At the Meadville Fair, the episode turned on the existence of scientific agents ready to intervene and determine the identity of the product by looking beyond appearances. Local judges with credibility garnered from experience and involvement in the dairy trade had gauged that the sample was legitimate. They used taste, sight, smell, feel, and the familiarity born of daily practice. They were part of the traditional norms governing face-to-face conduct that were then under duress.[47]

But Palm said he'd ordered his entry from afar. In the glow of his success in duping the experts and upending those norms, he was happy to cooperate with officials to use further analysis to reveal his stunt. With the spectacle of affidavits and hired couriers, the dairyman's association sent the samples to a state authority. That state chemist was invested with that authority by a larger governing body and given credibility for the very fact of disinterest and separation from the scene. He deployed the kinds of laboratory equipment by then available to show instead that the samples were indeed "natural" butter, thus explaining the genesis of the "It Was Butter" pamphlet published to alleviate questions about the integrity of the judges and justify their decision to award the prize to Palm's entry. The state chemist was placed at that slice between surface and interior, between deceptive appearances and what they argued were authentic substances.

At the intersection of private analysis, the grocers' industry, local boards of governance, and experiment station activities sat William Frear, final arbiter for the Meadville samples. Frear was the vice director of the experiment station in State College, Pennsylvania, its chief Chemist, and a professor of agricultural chemistry at Penn State. The station's mission was to do "conscientious and thorough work along a few lines, steadfastly adhered to, rather than by attempting superficial work in many directions." Deploying the language of neutrality soon to become so central to modern science, "state authorities who were disinterested parties" in the broader agricultural industry peopled the station. Frear and his colleagues were public servants, chemists who'd published on milk and butter, known entities in the midst of a complex manufacturing and agrarian world.[48]

Figure 7.7 shows the analysis summarizing four parameters compared against a butter standard and oleomargarine standard. Of those, only volatile fatty acids would show a difference between butter and oleomargarine. The Meadville sample aligned with "pure butter" on that count, an

	Melting Point Deg. Centigrade	Saponification Equivalent.	Volatile Fatty Acids.	Curd.
Pure Butter . .	29–35	236.5–260.7	13.5–36.4	.19–4.78
Meadville Sample	34.1	256.7	25.6	.9
Butterine . .	34–40	274–290	0.4–8.6	.74–1.82
Oleomargarine .	34 40	274–290	0.7	.62

FIGURE 7.7. Published results from analysis of the Meadville sample by William Frear, chemist at the Pennsylvania Agricultural Experiment Station. The chart was included in the pamphlet (fig. 7.1) distributed by the State Dairymen's Association. Courtesy of Warshaw Collection of Business Americana-Dairy, Archives Center, National Museum of American History, Smithsonian Institution.

order of magnitude higher than its butterine comparisons. Apparently it was indeed butter.

While such analytical possibilities were a consequence of a new generation of instrumental attention, they then had consequences for future generations. One of them was that distinctions, differences, and variations were to be pushed to the side of final products, not agricultural processes. It shifted the concept of natural from the "product as drawn from the cow" to the product as measured against an experimental standard. In that position, the chemists were substitutes for hands-on, farm-based experience. They preferred laboratory assays to dairy stable inspections for a practical reason, no doubt: they simply didn't have the time or means to police the stables with microscope in hand for every dairy sending milk, cheese, or butter to the city. They also preferred the assays because they were based on ostensibly repeatable, methodologically agreeable practices, ones they could define and defend. As groups such as the ASPA and others showed, the later 1800s were an active period for defining and defending. Their findings were "pure enough."

Reporting from Meadville, newspapers would write, "The evidence in the hands of the committee on awards . . . as to the genuineness of sample, accuracy of analysis, and correctness of the conclusion that alleged packages of butterine were undoubtedly butter, is indisputable." The rural paper *Agricola* wrote that the "sample of the butter whose genuineness was in doubt from the original package in which the so-called butterine was shipped" had been properly shown to be "*pure butter*, instead of an artificial product." The *New York Times* repeated an interpretation that considered the state tests confirmation of the need for "dairy tests by experts." This was not just trust in numbers but trust in analysis; the reports considered the genuine character of the measurements a means to parse the genuineness of the product.[49]

In retrospect, Meadville sits neatly between two epochal moments in the age of adulteration and food policy history. The US Congress's 1886 Oleomargarine Act brought the first federal scope to the pure food question by drawing on input from public health officials and relying on chemical analytical characteristics. The 1906 Pure Food and Drug Act made the language and goals of purity explicit even if the underlying terms of authenticity were implicit. Temporally splitting the difference, Meadville represented the extent to which nascent scientific practices had become credible ways to make sense of the new marketplace of food identity. Although the interests, ideologies, and backgrounds at play varied depending on the specific situation and particular chemists, together their work presented a picture of vibrant activity, intense debate, and deliberate investigation of adulterated products. In parallel with that work was the work of state and federal agents dedicated to the same goals of analytical authority to police purity. The next chapter examines their efforts.

8

Food & the Government Chemist

It is not so much a campaign for health as it is for honesty.

HARVEY WILEY, 1905

The genuine and unadulterated article is driven to the wall by the artistic counterfeit.

UPTON SINCLAIR, 1906[1]

Harvey Wiley loved to sing. He loved to drink too. If he could do both, all the better. He met with colleagues to found yet another new organization in September 1884. This one was the Association of Official Agricultural Chemists (AOAC). When chemists in New York formed the American Society of Public Analysts (ASPA) the next month, they did little more than spend a brisk Friday night hanging out in a government building in Brooklyn. For the more nationally attentive AOAC, the occasion called for a beer-battered feast. A banquet hall in Philadelphia hosted the government chemists. The USDA had called them together to secure uniformity of fertilizer analyses.

At their two-day meeting, they lined tables to swing steins of beer and belt out tunes from a printed songbook. One was "Guano." Wiley called it "a scream on fertilizers." They sang another jingle to the tune of a Gilbert and Sullivan song:

> When the easy-going chemist's analyzing, analyzing,
> He loves to take his time to do it well;
> His *otium cum dig* is quite surprising, surprising,
> And when he will get through no tongue can tell.[2]

The men in attendance were academically trained chemists. Most had spent some time in German universities. As if those Germanic connections were not visible enough, many of the songs in their banquet songbook were written in German. Others were dotted with Latin phrases and inside jokes. (*Otium cum dig,* I now know, means leisure with dignity.) In "Lac Loquitur" (the milk speaks), the chemists sang about one of the bugbears of the day, contaminated milk. The image of a buxom dairymaid adorned their printed lyrics sheet. The lyrics told the story of milk that was once the symbol of a happy land but was now a masquerade, deceiving those who bought it.

> Wonders stand on every hand,
> we strut around in new clothes,
> It's common talk that I am chalk
> and honey only glucose.[3]

Chalk masqueraded as milk, glucose masqueraded as honey, adulteration masqueraded as purity. The agricultural chemists were motivated to find the difference. Congress had founded the USDA only twenty-two years earlier. The congressmen and chemists sought to bring coordination and analytical attention to the demands of agriculture. It was the production side of the farm-to-table life cycle that mattered to the USDA, as is still the case today. Fertilizers and related soil analyses took up a good deal of their time. Because of that early analytical attention, fertilizers were one of the first products in history labeled with certified analyses. Like foods, they had only recently started to become manufactured at chemical works.[4] By the Civil War, fertilizer providers were for the first time manufacturing their product away from the farm and shipping it back for application. The new manufacturing presented new problems of veracity, authenticity, and purity.

The USDA gathered chemists to form the AOAC at a key moment in the era of adulteration. They came to Philadelphia as members of the late nineteenth-century world of shaken confidence bent on catching huckster fertilizer men and dislodging dishonest brokers. They gathered to maintain the character of genuine products and authentic farm aids, and they sought to do so with the consideration of governmental oversight at the national level. The fact of adulteration was not in dispute; the best means to address it was.

The chemists of the USDA's Division of Chemistry took charge of analyses, a point they sang out loud when they formed the new AOAC. It was a key moment for the analysis of adulteration because, as they acknowledged in song, they were attending not just to the fertilizer for growing the grass, wheat, cane, or cows but the foods and drinks that were a result.

The agricultural chemists had mainly attended to production-side issues such as the purity of fertilizer, the integrity of soil, and the validity of crop choices, but by the 1880s their new organization recognized that purity required analytical attention to the consumer-side too.

Twenty years later, when congress passed the Pure Food and Drug Act in 1906 they vested authority for patrolling the authenticity of foods with the USDA's Division of Chemistry, which by then (from 1901) had been promoted to the level of a Bureau of Chemistry.[5] When the agency spawned by the new law adopted its current name in 1930, the Food and Drug Administration (FDA), the transition from farm to fork was complete in a semantic, organizational, and analytical sense. The separation between producer and consumer was so distinct that it required two separate federal agencies to police the health and veracity of both. The move from the USDA to the new FDA pronounced a shift in food oversight from regulating the process to regulating the product.

The 1906 act was a signature achievement of the Progressive Era. Along with the Meat Inspection Act passed at the same time, it has long stood as a prominent example of the ways the United States built federal agencies and their bureaus—bureaucracies—to manage American governance and, perhaps more pointedly, the identity of what it meant to be American. With the decades-long surge in immigrants fundamentally transforming the demographics and the workforce of the nation, historians point to the wave of bureaucratic actions as ways of shaping the complexities of new peoples, new urban densities, and new marketplaces. Pure food advocates were part of that larger administrative shift.

Wiley was present at the creation. He was there to found and sing along with the AOAC; he was there to lead the Division of Chemistry; he was there to commission and coauthor the USDA's doorstop of a bulletin, *Foods and Food Adulterants*; he was there to testify in one fin de siècle congressional hearing after another; and he was there to take credit for antiadulteration legislation for the remainder of his life. As food historian Bee Wilson put it, "no one ever accused Harvey Washington Wiley of false modesty."[6] He collected and celebrated accolades about his work for decades.

Figure 8.1 may have been most on the nose: Wiley was the knight in shining armor in this crusade against impure foods, his weapon chemical analysis. Early histories of the pure food crusades mostly took him at his word, playing to his vanity, while later studies make it clear that Wiley was not alone. The previous chapters of this book subscribe to that later view. Wiley's place in so many debates over adulteration, however, helps show how antiadulteration efforts grew from the local, municipal, and regional

FIGURE 8.1. Cartoon of the pure food shield protecting Wiley as he challenges "Ye Knight of Impure Food," taking the crusade imagery of the pure food fights and making it literal. Early 1900s. Courtesy of the Wiley Papers, Library of Congress.

scale reviewed in the previous chapter to the state and federal scale examined below, illustrating the public visibility of scientific purity at the turn of the century. His work provides a thread that characterizes how formerly environmental concepts of purity, dynamic and agricultural, moved into the realm of chemical analysis, static and analytical. It was, for Wiley and the AOAC, an effort to take command of patrolling the border between pure and adulterated within the halls of government chemists.

Born in 1844, Wiley arrived in Washington as a nearly forty-year-old chemist from Indiana. It was the early 1880s. He stayed in the city until his death in 1930, at which time he was buried in Arlington National Cemetery. He was raised in a strict religious household—his was one of those austere "born in a log cabin" childhoods—and brought a kind of zealotry for honest dealings to his lifelong work. As an early biographer would say, Wiley had a "sense of duty" and "feeling of obligation to humanity" that grew from an upbringing that instilled in him a drive "to do God's work."[7] He fought in the Civil War. He earned a medical degree after that war. He studied chemistry at Harvard and taught at the Indiana Medical College and Butler University. Then, in 1874, he took a position at the brand new Purdue University.

Pure and honest foods were at the center of Wiley's career from the start. Early in his days as an Indiana chemist, he studied the purity of sweeteners. In the decades to come, he would find himself investigating the full cabinet of suspected adulterants. He was an evangelist for honest brokering in those efforts, motivated early on by a duty to fight hucksters and charlatans in the fields of his home state. He had equivocated on career choices for a while, unsure whether he should go into the medical profession or pursue further chemical studies. A timely trip to Germany in

1878 fostered his commitment to chemistry and, specifically, to the cause of identifying impurity through analysis. While there, he listened to lectures by the chemists Helmholtz and Virchow on the adulteration of food. He also stayed in Paris to meet more chemists before visiting labs in Bonn and Leipzig. He watched the famous Robert Bunsen (of burner fame) at work in his Heidelberg lab.[8]

Wiley learned to appreciate specialist equipment while in Germany like an aspiring connoisseur on a wine tour. He wasn't interested in generic instruments, to be sure, but precision tools. A "Schmidt and Haensch polariscope" caught his eye; "the Vogele spectra" was particularly impressive. As part of his instrumental sightseeing, he visited experts whenever he could. A stop to see Germany's official chemist of the Imperial Health Office first led him to learn about the polariscope.[9] When he got back to Indiana, he sought the best one available for his office. He had Purdue buy the one the French government used in its customs houses.[10]

Wiley was the kind of guy to be charmed by a polariscope. His work on sweeteners brought him to the instrument as a better way to examine sugar. French scientists had pioneered the technique decades before. The scope had captivated American agents seeking a more effective way to patrol sugar imports since at least the 1840s. With the French instrument in hand, Wiley began to conduct a set of sweetener tests, first on maple sugar from trees and later on cane, beet, and sorghum sugars. He reported on that research at a meeting of the American Association for the Advancement of Science (AAAS). He then published his first professional research papers in the new *Journal of the American Chemical Society* (1880) and the second volume of *Science* (1881). That work led him to consider new sweetener options. At one point, he tested a concoction of cornstarch and sulfuric acid that made a new kind of syrup.[11]

That was glucose. The product intrigued Wiley. He petitioned for funding to study a glucose manufacturer in Peoria, Illinois.[12] He wanted firsthand knowledge of the process by which corn, water, and sulfuric acid mixed together to become an alternative sweetener. He considered moving to Chicago in 1880 to work as a chemist in the sugar business. "Six months later," as an early biographer put it, "he tried to mobilize capital to buy a defunct beet-sugar factory in Boston and to convert it to the manufacture of glucose; this project failed, however, as did another attempt to organize a company to produce glucose in Lafayette."[13] He was clearly quite familiar with the alternative sugar.

Dabbling with alternative sugars also brought the first public rebuke of Wiley's work. Beekeepers accused him of advocating for "artificial" honey, though ultimately they misunderstood his tack. After more studies, and with the failures of the two business projects behind him, Wiley grew to oppose what he considered dishonest sugar. To confirm his view, he wrote a series of articles about glucose and adulteration for *Popular Science Monthly*.[14] Whereas beekeepers were suspicious of his motives at first, soon, Wiley bragged, they became his "most enthusiastic supporters." His work led him to consider other professional avenues in the sugar business, advocating for new alternatives to cane and safe alternatives to glucose. Sweeteners were a matter of immense public value. After all, "the consumption of sugar," Wiley thought, "is a measure of progress in civilization."[15]

State officials recognized his expertise in 1881, tapping him to serve as the Indiana state chemist in charge of analyzing sorghum, sugar, and fertilizer purity.[16] He wrote in Purdue's annual report that year about concerns over a problem far wider and more rampant than fake sugar. His studies of syrup and grape sugar led him to conclude that adulteration writ large was a vast public problem of authenticity in the food marketplace. Waves of new food manufacturers were not being honest brokers. By the early 1880s, Wiley had come to recognize the broader narrative already being written in England, France, Germany, and other European states, one prevalent in the grocers' empire across the urban United States and subject to scrutiny and organization in municipalities from domestic economy treatises and through manufacturers' forums. He wrote for the Indiana State Board of Health in 1882 that it "is high time that the demand for honest food should be heard in terms of making no denial."[17]

It was also time for Wiley to move on professionally. His mobility at Purdue was limited. One reason was personal. His ambition for the presidency of the college was seemingly out of reach. His bosses considered him uncouth, a bachelor without the proper social mores. Most damming, apparently, he was the teacher who had the audacity to ride a bicycle. One day he was called to the president's office. He assumed he was getting a raise, but instead the trustees rebuked him for being "dressed up like a monkey and astride a cartwheel riding along [the] streets."[18] Such a demeanor, they said, "was beneath the dignity of a professor."[19] It didn't help that he also had the nerve to play baseball with his students.

He also had a more intellectual reason for leaving Indiana. Research on sorghum, glucose, and cane sugar had brought Wiley to forums at the state and national level, where he saw that the problems of adulteration required more than local counteractions. He met the USDA's commissioner George Loring in St. Louis at one such sorghum convention. (That was a thing, sorghum conventions.) Soon after, Loring offered him a job in Washington. Given his friction with the Purdue administration, his business failings, and his growing determination to confront adulteration with "no denial," he took the chance to move to a federal position. The USDA hired him in 1883 to lead efforts in sugar and fertilizer analysis.

The next year, Wiley helped organize the AOAC. By that time, he had come to recognize the challenge of adulteration as a challenge of scale. His view was that in the USDA and through the AOAC, "all the problems relating to the adulteration of food, the general method of agricultural analysis, and all other matters which concern, in common, the analyst and the public, could be included among the duties and privileges of this body."[20]

The USDA served as the AOAC's patron as the AOAC managed the strains of working along the entire span of the farm-to-fork life cycle, attending to fertilizer and soil analysis on the land along with food measurements beyond it. Wiley helped organize their first banquet at the AAAS meeting in Philadelphia. He sang along with the others from the book of tunes made just for the occasion. Two years later he served as the organization's president.[21] Secretary of Agriculture Loring and his new hire from Indiana were not breaking new ground in their effort to coordinate and aggregate the analysts. It was the same effort the American Society of Public Analysts (ASPA) was making concurrently in New York.[22] Loring's and Wiley's contribution was to address the problems beyond that municipal scale and within the particular cultural dynamics of the changing American food system.

It was a busy time, the 1880s.[23] The mounting alphabet soup of acronyms was one sign. USDA, AAAS, ASPA, AOAC. They bespoke organizational growth and an appeal for better governance. The public conversation about authenticity and sincerity, con men and veracity, the fake surface and the real core, had by then grown to give substance to the Gilded Age as a moniker. Were all things merely gold plated to cover the rot of their true interior? Agriculturally and economically, each of the main examples in part 2 of this book accelerated in visibility and public concern around the same time. Wiley had already staked a claim in the argument that glucose was troublesome. Cottonseed oil had become a viable substitute for olive oil in the decade before, but paired with its use in the lard industry, the 1880s was its true coming out decade. And margarine, fake butter, was dominating the news about the contested value of natural and artificial. We in the twenty-first century often look back to ask when "modern" food began to take shape. Wiley's view from the

1880s USDA provides a reasonable reference point. He came to work at the USDA with an eye to hucksters amid a moment of great agricultural and food-system change.

Wiley stood witness to dramatic changes in food production, distribution, consumption, and marketing. He also participated in the response to fears over those changes from his perch in Washington at the labs of the USDA. How did they do it? Organizing the AOAC was one way to respond, one that assumed the role of agricultural chemists was salient in the fight against adulteration. A more specific response was a series of government projects that illustrated the approach agricultural chemists took to gain control of the fight against adulteration: divisional bulletins from the head offices of the USDA, research at the state level by the Office of Experiment Stations, and a set of trials that reporters called the Poison Squad. Each worked with the support of the AOAC's chemists. Each was a salvo in the war on adulteration. Each was unsatisfied to say only that adulteration was a problem, going further to say how and by whom it should be addressed. They did this by arguing that purity should be understood through analytical methods, and they did it with the view that experts and technical analysts were the ones to referee those methods and the subsequent understanding of pure foods. Trust the government chemist, in other words, for reputable analysis and genuine character. Focus the angst over dishonest brokers and deceptive producers on Wiley and friends. Their work was, to quote Wiley, "not so much a campaign for health as it is for honesty."[24]

The USDA's Bulletin 13 first provided the space for Wiley to make his case to a national audience. On the heels of the 1886 oleomargarine hearings (chap. 4), he began direction of the thirteenth bulletin, *Foods and Food Adulterants*. It would eventually run to ten volumes, published in fits and starts over a sixteen-year period and dealing at length with, among other things, oleomargarine, cottonseed oil, and glucose. Part 1 (1887) was the margarine one, focusing on "imitation dairy products"; part 4 (1889) attended to lard and its adulterants, with cottonseed oil at the center of the report; part 6 (1892) took sugar and sweeteners as its subject.

The Bulletin was a big deal. Significant prepublication buzz showed a wary public looking to it for guidance. Although the USDA was not in the business of publishing best sellers, Wiley spent most of the spring of 1887 fielding queries from all corners of the country about the new Bulletin's imminent publication. Chemists from MIT wrote to ask for a copy. So did agents in New York and New Jersey. So did editors of grocers' papers.

Henry Lassig at *American Analyst* insisted he be sent a personal copy. High school principals urged Wiley to share a version as soon as he had one. An inspector at the Philadelphia customs office asked for one. A representative from Armour and Company wrote more than once to explain that his company had not been sent a copy as promised and would Wiley please send one as soon as possible? They felt left out.[25]

It's hard to imagine how Wiley had time to serve as president of the AOAC, run his labs at the USDA, prepare the publication of the agency's thirteenth bulletin, and handle the copious amounts of correspondence flooding into his office on adulterated butter alone. There was so much pestering that Wiley wrote the public printer in April asking him to "inform me of any delay in the print of Bulletin No. 13." The *Rural New Yorker* ran its twelve-part series about adulteration in the spring and summer issues that same year. They published Berghaus's hydra-headed monster in May. In June, Bulletin 13 finally rolled off the presses at the Government Printing Office.[26]

A government document such as Bulletin 13 appears dry and trivial to a twenty-first century reader. In its efforts to say how and by whom adulteration should be addressed, the commentary was mainly dispassionate. The first part, on imitation butter, combined an impressive review of the available literature, a summary of lab experiments commissioned just for the report, and a presentation of over a dozen different analytical methods. Wiley drafted the report but had Edgar Richards, a staff chemist, lead its final authorship. Richards summarized forty-nine separate registered margarine patents. He included testimony about the value of the substitute butter from Gustavus Swift and Philip Armour. He noted the views of margarine's "wholesomeness" from the leading lights of chemistry, a cast of characters that should be familiar by chapter 8 of this book—Charles Chandler and Henry Mott in New York; Samuel Johnson at Yale; Henry Morton in New Jersey; and Wilbur Atwater in Connecticut. He reviewed tables, charts, and a series of plates of images produced through a microscope. The summaries together made the point that the USDA had a handle on the adulteration of butter, that they were there to take care of it.

As they made their case, the report's authors spoke of a close affiliation between pure and natural. "Pure fresh butter" referred to butter "prepared in the ordinary manner." A genuine product was the opposite of an artificial one, as Richards explained that the "natural" parts of butter were those that occurred outside human intervention and before artifice. In the USDA's assessment, when manufacturers added things away from the farm or stockyard they were moving beyond nature and into artifice. This was a distinction based on assessing the proper degree of manipulation—

just enough and it was still natural; too much and it edged into the terrain of artifice. To decide that difference, the report explained what "pure butter is *supposed* to contain" as a product of nature (italics in original). The distance between what was supposed to be there and what tests found defined the degree of adulteration. This all happened in a lab away from the source of production.[27]

In their labs, the bureau adopted the same strategy as the ASPA's more regionally focused chemists and the locally attentive consulting chemists in the grocers' empire. Namely, they made the case that methods and standards of analysis would anchor their credibility. Thus would Bulletin 13's part 1 present over a dozen different analytical methods, including a combination of physical and chemical adulteration-detecting techniques. One involved measuring the polarity of light through a sample. To make their point, Wiley and his staff included twenty-four color plates of microscope slides showing beef fat, butterine, and lard magnified from forty to 160 times. The plates confirmed that seeing was believing but that seeing with the naked eye was insufficient. To gauge purity, part 1 of the Bulletin argued, was to follow the trained chemists with specialist equipment.[28]

In each new part of the Bulletin, Wiley and his chemists would build their case that fighting dishonesty along the food chain should be the domain of specialists (fig. 8.2). In part 1 they illustrated the use of sophisticated tools to make this point. By part 4 they did this by enunciating the distinction between "ordinary" and specialist approaches. Wiley served

FIGURE 8.2. Harvey Wiley and associates at the USDA pose outside their office, mid-1880s. Wiley is third from the right. Courtesy of the Wiley Papers, Library of Congress.

as lead author for part 4's comparable study of "Lard and Its Adulterants" (1889), dealing with olive oil adulterated with cottonseed oil. Here Wiley presented analytical assessments from laboratory instruments, tables of quantified data, and plates of microscopic images as ways to supersede the tests of the "ordinary" customer. "In external appearances to an unskilled person adulterated lards are not appreciably different from the pure article," he wrote. "An expert, however, is generally able to tell, by taste, odor, touch, and grain, a mixed lard from a pure one. There is usually enough lard in the adulterated article to give it the taste and odor of a genuine one."[29] He augmented his direct sensory assessment, as he had with butter, with a set of analytical methods. Wiley brought together qualitative and quantitative metrics to bolster the expert's role in adjudication.

Part 6 (1892) dealt with sugar to continue the running themes of the Bulletin on methods and specialization. Sweeteners were an area of career-length experience for Wiley. It might not be surprising, then, that he took it for granted sugar purity should be understood through chemical analysis. He cut to the chase in this part of the Bulletin. It was "wiser to devote less time to the methods of detecting adulteration which for the most part are simple operations and well understood," he explained, "and to give greater attention to the extent of the practice of adulteration." The report included a bibliography of 106 different studies of adulterated honey, hundreds of summaries of molasses and "sirup," and pages of tables on confections analyses. It was the work of an exasperated commentator, piling evidence upon evidence to show that there was clearly a problem here. Chemists across the country found "pure sugar" samples that actually contained glucose. Of 250 separate candy samples, 173 of them had glucose.[30] To Wiley's mind, the barrage of details served to further separate ordinary and specialist techniques. Ordinary grocers and customers might have taken the "pure sugar" claim at its word, but the specialist knew better.

Each part advanced the case for specialist control of the problem of adulteration (the who) and instrumental analysis (the how), working at a level of federal governance to forward the same argument running at local and regional scales. In keeping with the late 1800s framework of confidence and trust, the arguments were couched in the language of honesty and propriety. Purity, to the Bulletin's authors, followed from the attempts to verify honest behavior. The margarine from part 1 might not be unhealthy, but it should always be sold for what it was. The lard and olive oil of part 4 might be cut with cottonseed oil, but Wiley and friends showed that analysts could detect the difference. For the chemists, cultural attachments to right behavior—what did it mean to be a good and decent person?—became subsumed within the halls of government labs.

In part 6, for example, Wiley dealt with honey dealers who opposed glucose not because it made their product less healthy but because it was un-American. A honeybee dealer in Chicago confessed to Wiley that in "1870 I tried some French and German glucose, using it as a part substitute for sugar," but only because there weren't good domestic supplies. "When good glucose was made in this country," he wrote in 1890, "I became patriotic and used only goods of home manufacture."[31] The "pure confection" to him was not natural, as in from the land, but one that was culturally correct, as in from America. Wiley and the USDA could detect it either way.

Bulletin 13 was a major effort in Wiley's work to address pure foods with federal attention. He did this as a function of the agency's Division of Chemistry and by leveraging AOAC methods. The USDA and the AOAC coordinated another thrust of food analysis in parallel with the Bulletin. It came from an experiment station model sprouting up as part of the proliferation of land grant colleges. This was the USDA's new Office of Experiment Stations (OES). As historian Jessica Mudry writes, the OES further fostered "the marriage of science and agriculture and perpetuated and inculcated the structures and language of science, the rigidity of experimentation and method, and the need for objective and quantifiable results."[32] In the war on adulteration, the OES helped advance a philosophy of purity anchored in techniques of measurement and buoyed by principles of instrumental analysis. It, too, was less about how food was produced and more about how you could detect the contents of that food at the market. It, too, nudged the concept of purity away from the process and toward the product.

The OES's efforts began as a commitment by Congress and the USDA to bring agronomists, chemists, bureaucrats, and farmers together to address problems in agriculture. Yale chemist Samuel Johnson founded the nation's first experiment station in 1875 in New Haven. Johnson's mentor was the preeminent agricultural chemist of the century, the German Justus von Liebig. Liebig's leadership was instrumental in the experiment station model in the German states, where there were fifteen stations in 1862 and about eighty by the 1880s. The sites brought a consistency of method and control to agricultural research. For Liebig and his benefactors, they were undergirded by chemical principles and intended to bind scientific protocols to matters of public significance. Johnson and his former student Wilbur Atwater imported the model of experiment stations from those German precedents. In the United States, that form of agricultural research took an important step forward in governmental visibility when Congress passed the Hatch Experiment Station Act.[33]

President Cleveland signed the bill into law in March 1887, the preposterously busy season when Wiley was preparing Bulletin 13's first print run. The Hatch Act sat with Bulletin 13, the AOAC, and the ASPA as yet another addition to establishing institutional attention to agricultural problems. On the heels of the act's passage, the USDA created the OES to coordinate the work at each state's station. Administratively, it would operate in parallel with Wiley's Division of Chemistry. The idea was that each of the state's land grant colleges would house an experiment station. When the judges of Meadville called on William Frear to arbitrate their fake butter decision in 1895 (chap. 7), they sent it to his office in the Pennsylvania experiment station at Penn State University.[34] When Wiley sought analytical results about various adulterated products—butter, lard, sugar—he leaned on his colleagues at OES stations across the country.

The two forums for addressing agricultural and food analyses, Bulletin 13 and the OES, were tied together in purpose and key personnel as well as in timing. They couldn't have come about any other time, as even ten years earlier the personnel, administration, and public demand for attention to adulteration had not yet coalesced. Both were housed within the USDA, and both had coordinated experimentation and analyses as their mission. Chemists supervised both of them. And both had the touch of German influence.

The secretary of agriculture plucked Wilbur Atwater from the member rolls of the AOAC to serve as the first chief of the OES. His early career ran remarkably parallel with Wiley's. Like Wiley, Atwater traveled to and studied in Germany. Like Wiley, Atwater brought a penchant for agricultural chemistry and food analysis to bear on his own research program. He and Wiley inhabited the same world of chemistry, agriculture, and food in the later decades of the 1800s. They knew each other. They didn't get along.[35]

In the world of food analysis, most of the key actors were socially acquainted and academically related. Part of their antagonism was institutional. Wiley's Division of Chemistry now had to compete with Atwater's OES for funding, a point of conflict Wiley bemoaned to his bosses. Part of the antagonism was conceptual. The two men had different philosophies about what counted as a pure food.

Atwater advanced a more biological and nutritional view. He saw the human body as a machine that could be analyzed using the same instrumental principles of a lab. Wiley took a more chemical view, preferring to avoid the bodily processing of foods as part of his analyses, though, as we will see below, the differences between the two philosophies began to blur by the early twentieth century.

Before that, Atwater had proposed studying the nutrition of foods at a meeting of the AOAC in 1886. Unfortunately for him, his fellow members were less enthused. In 1885, in one of the first studies by the AOAC, they had focused on animal nutrition, comparing the contents of food eaten and waste excreted. They found no difference. As one chemist reported at the AOAC's annual meeting, they thus concluded that analysis was better suited to identifying the contents of food rather than the ways the body took up that food during digestion. It was a statement of principle that one could treat foods in a chemical sense without confusing the matter with the body's biological role in eating the food.[36]

Atwater thought differently. Historians often refer to him as the father of nutritional science in large part because of his different views of food, diet, and physiology. He was among the first to apply a broad-based quantitative and instrumental ethos to the problem of human nutrition. His philosophy was born of the arguments over food health and identity given shape by the war on adulteration.[37]

It's worth noting that before Atwater's time, health advocates were more commonly vitalists. They sought health and fitness as the result of a balance of vital, life-giving forces. Such health advocates also associated food choices more directly with moral criteria. This was similar to the long-standing framework for assessments of purity. A pure food was one produced by an honest broker. In similar fashion, a nutritional food was one that drew from appropriate moral lifestyles. That moral connotation was evident in the spate of utopian leaders who combined communal choices with eating habits. Sylvester Graham and his eponymous cracker were forerunners of this from earlier in the century. Later century advocates like John Harvey Kellogg, he of the cereal Kelloggs, and Horace Fletcher, "the Great Masticator" known for a protocol of slow but rigorous chewing, worked in the same vein. They all "judged diet by moral and aesthetic criteria," writes Nick Cullather, "rather than the objective, numerical standards of an industrial age."[38]

Atwater didn't. He subscribed to the notion that the body was "a de-animated machine." This was partly Liebig's influence. The Prussian chemist had already set the model for Johnson and Atwater on experiment stations. He was also a prominent authority on the then current understanding of the chemistry of food, pioneering studies and publishing widely on the subject in midcentury. He even launched a famous meat extract product line to benefit from his chemical claims, Leibig's Meat

Extract.[39] Liebig brought a more mechanistic approach to his work than others brought to theirs.

Through his work with the OES, Atwater applied that mechanistic approach to social problems in a variety of studies of diet, nutrition, and poverty. He published work with Ellen Richards in the 1891 collection *The Science of Nutrition*. In 1895 and 1896, he pioneered a dietary survey of Chicago working with the famous settlement activists and suffragists Jane Addams and Caroline Hunt of Hull House. He then took the idea of quantified food to the next experimental level in ways that not only pushed concepts of purity toward products and away from processes but, once focused on products, pushed further into mechanistic and quantified form.[40]

"The work of rendering food into hard figures began just after breakfast on Monday, March 23, 1896, when Wilbur O. Atwater sealed a graduate student into an airtight chamber in the basement of Judd Hall on the Wesleyan University campus," as Cullather captured the scene. It took place at the site of the Connecticut Agricultural Experiment Station. Atwater's support for the Hatch Act a decade earlier was partially motivated by his desire to build the room-size instrument. The experiment brought college boys into the room to measure every ounce of input and every ounce of output.[41]

That first student "took measured quantities of bread, baked beans, Hamburg steak, milk, and mashed potatoes through an airlock during rest periods, which alternated with intervals of weight-lifting and mental exertion."[42] The same approach had not yielded substantive results when the AOAC tried it with studies of animals in the 1880s. But by the 1890s, Atwater's eagerness for quantifying that which had only been qualified had found support institutionally, conceptually, and publicly.

When Wiley published Bulletin 13, it was a public phenomenon, everyone clawing for a copy. Atwater's room calorimeter was a public spectacle too, though it was looked on more as a Barnum-like peculiarity of chemists and strange instruments. The small Connecticut town was bombarded with requests by onlookers who wanted to be part of it, to protest it, and to see it. This included a champion bicyclist whose fitness, he thought, made him the perfect specimen for Atwater's needs. Reporters came to Middletown by the droves to write about the 170 cubic feet copper-lined wooden box (fig. 8.3). Atwater and his patrons at the USDA took the public's eagerness to see the experiment as a sign that it was good use of government chemistry.

Atwater's calorimeter was crucial in the development of modern nutrition, especially in the ways it defined the calorie as a measure of food character. As a component in the fight for pure food, his work at the OES

Fig. 1.—Respiration calorimeter.

FIGURE 8.3. Early rendering of Atwater's calorimeter, a small box in the middle of a room, 1897. Source: Wilbur Olin Atwater, F. G. Benedict, and Charles D. Woods, *Report of Preliminary Investigations on the Metabolism of Nitrogen and Carbon in the Human Organism* (Washington, DC: Government Printing Office, 1897), 11.

and with the AOAC was crucial in helping secure an analytical idea of purity. Despite their different views of how to assess food identity, Atwater and Wiley shared a philosophy of purity that assumed technologies of measurement were more relevant to know food than the experience of production or the process of agricultural work. Here the chemists extracted food from its agricultural and culinary context, moved it outside the realm of the grocers' empire, and placed it in the hands of technicians. Food was part of the machinery of digestion for Atwater, where "the calorie," Cullather writes, "represented food as uniform, composed of interchangeable parts, and comparable across time and between nations and races."[43] Atwater's experiments were a specific example of the general trend to assess purity and identity outside the space of agricultural production at scales of state and federal authority.

Despite his intra-agency friction with Atwater and earlier disagreement over the biological (nutritional) and chemical contours of food, Wiley was captivated by the basic conceit of the room calorimeter. The core idea guiding the experiment was that human bodies could be calibrated as instruments and tested in labs. Wiley and his AOAC peers had resisted this more bodily measurement in the 1880s; they found it appealing by the early 1900s. Atwater's success had something to do with this. By the turn of the century, Wiley and the AOAC thought they could control those bodies to suit analytical aims. Wiley would expand Atwater's idea into what became his most publicly popular effort, the Poison Squad.[44]

Along with the bulletins and the OES work, the Poison Squad also sought to pull control of food definitions inside the corridors of federal government labs. Bulletin 13 showed the ways federal oversight could manage adulteration through reputable analysis; the OES showed how to judge food identity through quantified metrics of bodily interaction; and the Poison Squad combined the two. Let's see how it worked.

Wiley returned from a European trip in 1902 to propose the new project. He sought and quickly received a $5,000 congressional appropriation to study the effects of preservatives on able-bodied men.[45] Bulletin 13 had evolved by the 1890s away from a smaller gaze that sought to determine *whether* adulteration was a problem toward a wider-reaching view about the best way to handle the fact *of* the problem. By the turn of the century, adulteration was not a matter of facts but a problem of methods. For Wiley and staff, if the dose made the poison, how could you measure the dose? To find out, they would test a chemist's shelf of additives, including borax, boric acid, copper sulfate, potassium nitrate, saccharin, salicylic acid, salicylates, sulfuric acid, sulfites, benzoic acid, benzoates, and formaldehydes.[46]

The press called the proposed experiments the Poison Squad, a name Wiley eventually though reluctantly embraced. He didn't want to limit himself to a 170-cubic-foot chamber, though. He used a room, not a box. His Poison Squad was to be stocked with "young, robust fellows, with maximum resistance to deleterious effects of adulterated foods," those capable of withstanding increasingly dosed portions of possibly poisonous preservatives in their meals.[47] It was an outrageous idea. Twelve "hale and hearty young men" would spend weeks in the basement of a USDA office building on the west end of the Washington Mall, within earshot of the White House and in the afternoon shadow of the Washington Monument.[48] "If they should show signs of injury after they were fed such substances for a period of time, the deduction would naturally follow that children and older persons, more susceptible than they, would be greater sufferers from similar causes."[49]

If the room calorimeter had captivated the public's imagination, the Poison Squad made that earlier attention seem quaint. Reporters wrote with fascination each time a new set of trials began. The men in the experiments were "heroic volunteers"; their "scientific martyrdom" had them escaping a "death diet"; women across the country wrote to ask for the recipe after they read the men had "pink cheeks" from eating one of the trial meals.[50] Advertisers used references to the experiment to show how pure their products were. Ceylon and India Green Tea boasted that their tea was so good, "Professor Wiley will not use it on his poison squad," because there'd be no point. Wiley was only using products "poisoned by the use of deleterious substances."[51] At 1903's annual banquet of the Gridiron Club, an elite social club in DC whose members included ambassadors, senators, governors, and titans of industry—J. P. Morgan was there—a skit had actors posing as a "poison squad." One of them pretended to be Wiley monitoring the diners' heads with stethoscopes and X-ray machines. They were looking for "borax on the brain."[52] The rest of the night was stocked with rhymes and songs and easy banter.

The borax joke landed because a contamination scandal five years earlier had added a new layer to adulteration fears. The "embalmed beef" affair of 1898–1899 arose when troops in the Spanish-American War were sickened by their rations of tinned meat. Rumors spread quickly that bad preservatives were the cause. Wiley investigated to find that the main suspects, borax, boric acid, and sulfites, were not the culprits. Poorly sealed cans and a heated climate seemed to have caused the problems. But fear had taken root, and the memory of the event stuck borax and boric acid in the minds of those afraid of tainted meat. The reference was well known enough to rib J. P. Morgan with it at the robber barons' fete. Those guys.[53]

I'm intrigued by the Poison Squad because it advanced an approach to adulteration that was proudly quantifiable, capably under the control of federal administration, and, frankly, on the sketchy side in terms of scientific ethics. To be sure, the *Washington Bee* reported that government clerks were "delighted at the chance of securing their table board for six months absolutely free of cost to them," happy for three free meals a day.[54] But the protocol of intentionally poisoning research subjects offered a morally sullied example of experimentation and, to endorse historian Ann Vileisis's take, followed "a research method that prudence would never allow today."[55] The very listing of chemicals, preservatives, and additives had the effect of titillating readers and placing the presumed adulterants in a skeptical light. The *Washington Post* mocked the squad's holiday meal in 1902. It would be "apple sauce, borax, soup, borax, turkey, borax, borax, stringed beans."[56]

Wiley thought the "robust fellows" were perfect studies despite the criticism. He made a methodological assumption that had him reverse engineering the results: if borax or sulfites sickened the strongest, then children, infants and the infirmed would also suffer.[57] Onlookers saw the flaws of the approach as easily as Wiley boasted about it. The *Washington Post* was skeptical: "There is only one puzzling feature of Prof. Wiley's delightful experiment, and this we may put in the form of an interrogatory: How does Prof. Wiley propose to prove anything?" The *New York Evening Post* pointed out what might seem obvious. "Apparently," they wrote, "it has not occurred to Dr. Wiley that experiments on such healthy robust young men will not be of much service."[58] They would call it "scientific poisoning" to lightly mock Wiley while admitting its scientific pretensions. At one point, someone floated the idea that it would be better to conduct tests on babies and the infirm, those most susceptible to dangerous foods. The USDA actually considered it for a moment before rejecting the suggestion that intentionally poisoning babies was a good idea. A century later, Bee Wilson wrote of Wiley that "by his own admission, he didn't know just how dangerous these preservatives were before the trials began; otherwise, why would he need to do the experiment?"[59]

Here's what mattered, though, in the ways analysis fit into the history of purity and adulteration. Even if the criticisms were easy to make, they accepted Wiley's premise. Observers bought into the analytical worldview, seeing that the issues were less about the fact of adulteration and more about the means to detect it. Their critique aimed at the validity of the methods, not whether or not the problem of purity should be addressed with analyses.

It's interesting, too, that even as Wiley, the AOAC, the OES, and the USDA sought to move the policing of purity into federal labs, their work was still interpreted through the lenses of trust, sincerity, and deception. The Poison Squad and the USDA's broader efforts struck observers as false and duplicitous. Critics thought Wiley was not forthright; others were more explicit and said he was a huckster or shill.[60] Four months into the Poison Squad's experiment, *New York Evening Post* writer Henry Finck suggested Wiley was eating up federal dollars on a dubious lark. He wrote skeptically that it "seems to require a very long time to decide whether his 'brigade of poison eaters' . . . are really eating poison or harmless preservatives."[61] Reflecting a decade later, Finck was sure "Dr. Wiley was a fraud," a shill for Big Borax. Popular comics and cartoons played into the dynamic, such as fig. 8.4, with ambiguous views of Wiley fighting industry but perhaps being part of it too. There was a trust problem, as the battle for purity remained set against the backdrop of confidence and deception.[62]

When it wound down after five years, the Poison Squad had shown several things. It was a journalist's treat, hitting a public nerve and placing

FIGURE 8.4. "Old Doc. Wiley's Sure Cure for all Adulterations Fake Foods Quack Remedies," a clipping circa early 1900s that Wiley saved showcasing his centrality in the government's fight to protect the consumer. Courtesy of the Wiley Papers, Library of Congress.

Wiley in the national light he had long sought with scores of profiles, papers, bulletins, and articles. It pushed forward earlier bodily experiments that paved the way for larger-scale digestive research. It showed that the political and physical scale of chemical analysis could be managed at the highest governmental levels. And for all those reasons it has since been fodder for almost every historical account of the ramp up to pure food legislation success in 1906.

It also showed public visibility of scientific assessments of purity at the turn of the century. To Wiley, this was for better or for worse. For better, Wiley's purpose, the *St. Paul Globe* wrote, was "to demonstrate the truth or falsity of scientific theories on the subject of adulterated foods."[63] Nearly a century after Accum's publications on the chemistry of adulteration, the problem of purity had grown from being the subject of niche publications and assessments of trust to a subject of broad governmental science where consumers were asked to invest confidence in federal agents over local purveyors. For worse was the criticism that followed a public spotlight. Wiley's efforts were a victim of the success of a half century of increasing chemical and analytical activity. Public debate and journalistic critique over scientific protocol were possible.

Criticism of the scientific integrity of Wiley's Poison Squad missed the same point that critique of Bulletin 13 had. Wiley was after the virtues of quantification, not ethically rigorous experiments. His point was to demonstrate that the question of purity should be handled in a lab, beyond the farm, and under the control of expert clinicians. The purpose of the Poison Squad was not to prevent impurity. It was to establish a commitment to the bureaucratic and technical management of the pure food question. Bulletin 13, the OES, and the Poison Squad gave substance to a broad set of initiatives that brought the control of food governance inside the federal lab. The three-headed monster in the fields of American farms would eventually be subdued not by the spears of agrarian grangers but by the microscopes of analytical chemists.

As all of this happened, there was Wiley giving testimony to a House committee on the problem of adulteration. This was 1899. Congressmen grilled him on how he knew what the standard was for various foods. They asked him about all the highlights in this book—about olive oil, milk, butter, sugar, and alternative sweeteners. But what could he say about the *value* of that food? How did he know what counted as "good," pure food? A century later, one of the planks of the local food movement has been to consider a "good food" movement. The moral connotation of goodness is intentionally provocative, helping advocates for good food develop metrics that extend beyond narrow industrial calculations of efficiency and productivity. Wiley sat in front of congressmen a century earlier addressing those same issues at a moment when the centrality and dominance of those industrial metrics were still in the making.[64]

In his first answer Wiley said that if it *tasted* the same, it was up to consumers to decide whether they liked a food. Senator Mason of Illinois pressed him, because olive oil was sneakily cut with cottonseed oil, and wasn't that terrible? Wiley said he personally preferred "the flavor of olive oil," but that was just his "taste."[65] If others preferred other flavors, so be it. Then they talked beer. Critics suspected that brewers were using new fake sugars. Mason asked whether Wiley could tell good beer from bad. Wiley said sure, he "can tell the difference... upon analysis." Pushing more, Mason asked, "can you tell by taste?" Wiley could, yes, but "the ordinary consumer of beer"? Here Wiley's answer differed from his first as he admitted that taste was not so straightforward a measure. "Unless he is a connoisseur," he told Mason, "it would be almost impossible for him to tell except by chemical analysis, which requires apparatus and chemical skill to be of any value."[66]

That was just it. Why couldn't people use taste anymore? Why couldn't they tell the value of food from appearance or smell? Why did they need chemical apparatuses? Wiley wanted to assuage concerns by saying taste mattered, yet his own testimony acknowledged that taste was fickle and complex and difficult, that "aesthetics should be considered" too, that it could take a connoisseur. In different hearings fifteen years earlier, a senator had complained, "The four senses which God has given us... are completely baffled."[67] That struck people as wrong, cheating, and deceptive. Wiley was among many, his contemporary Ellen Richards included, to amplify the point by considering artifice and adulteration synonymously: "Only the most careful chemical examination can reveal the difference [between] a natural and artificial" product, he said. "The person dining ordinarily at the table... could not."[68] Mason bemoaned the point; Wiley enunciated it to justify analysis.

Two years later, Wiley again. It was yet another congressional debate about pure foods. Wiley took the stand for two full days, leaving fifty pages of testimony and, for good measure, offering an additional fifty page statement of his own. He commanded the room. His comments continued the contrast developed in earlier hearings between the "ordinary customer" who might reject a food or drink because it didn't "appeal to the eye" and the expert who could tell from analysis if it was truly pure. Wiley's friend and close colleague William Frear, who resolved Meadville's margarine controversy back in 1895, stood with him, testifying to the need for chemical authority in "the punishment of cheats and swindlers."[69] As congressmen wondered aloud why they couldn't simply adopt English antiadulteration laws, Frear pointed out the scalar differences. English law was hampered by district jurisdictions, with no uniform agreement and standards only "fixed by local authority." Hepburn asked about the AOAC's efforts in seeking nonlocalized methods. Frear happily championed the

AOAC's work—he was a member too—as crucial background for a new federal law.

Two more years later, 1904, same thing.[70] And then finally, another two years later, committee hearings had much the same tenor with the advantage of a best-selling muckraker's exposé to back them up.[71] The Commerce Committee called their hearings to order on February 13, 1906. They ran for fifteen days. Doubleday, Page published Upton Sinclair's *The Jungle* later that week, on February 18. It sold twenty-five thousand copies in six weeks, one hundred thousand by the end of the summer, and a million before the year was over. Four separate officials sent Roosevelt the book. A young Winston Churchill reviewed a copy when the British edition came out.[72]

That popularity often leads us to read *The Jungle* backward. I know I did in high school. Many accounts of the pure food crusades include it as an explanation of the law's passage. That attribution is problematic in one way, given that Sinclair's work was the product of long-standing agitation against adulteration. He added a heavy straw to break the camel's back rather than presenting imagery that was shocking for its novelty. The attribution is also problematic because *The Jungle* was more directly helpful in finalizing passage of the Meat Inspection Act, signed concurrently with the Pure Food one. It was the meatpacking horrors that most attracted Teddy Roosevelt's attention.

Sinclair the journalist tapped into rhetoric, references, and imagery that had been in the public eye for decades, advancing the view that artifice and adulteration shared a common morally suspect foundation. At one point his main character, Jurgis, is in jail among the dregs of society, including "counterfeiters and forgers, bigamists, shoplifters, confidence-men, petty thieves, and pickpockets." Sinclair the socialist wrote in sympathy for the "honest merchant" for whom "the genuine and unadulterated article is driven to the wall by the artistic counterfeit." To him, the problem was capitalism, the challenge was market pressure, and the fact was that "imitation and adulteration are the essence of competition." For Sinclair, the Beef Trust could not be trusted. His cultural characterization of people as honest and true fit the material characterization of food as genuine and pure.[73]

Wiley was once again the most prominent voice in the 1906 hearings. He had to be tired of the routine. Of eighteen witnesses, his testimony filled more space than the other seventeen combined. He wasn't offering anything new.[74] By the early twentieth century, the general argument was about good chemists and bad chemists ("alchemists," like the Chevalier Paraf) instead of whether or not chemistry played a role in public governance. Years of work in domestic economy and scientific cookery, the analytical suggestions by chemists such as Ellen Richards, and the public advocacy of groups such as the Women's Christian Temperance Union helped elevate that analytical visibility. At the same time, chemists had built up an infrastructure of analysis—methods, standards, instruments—at the metropolitan level, in storefronts, within the grocers' empire, in the courts, at state experiment stations, and at the docks through customs and trade officials. Wiley benefitted from and pushed along those larger domestic efforts, characterizing the new analytical authority of the pure food crusades.[75] As summer began, the bills moved to a full debate. President Roosevelt signed both the Pure Food and Drug Act and the Meat Inspection Act on the last day of the congressional session, June 30, 1906.

Earlier attempts at passing legislation had failed 189 times. This one took. One reason the earlier efforts failed was because the consequences of passage were too cumbersome to accept. Some of those consequences have cast a long shadow. Some were visible at the time.

As the era of adulteration moved into the twentieth century, courts, manufacturers, and the grocers' empire understood what was at stake when analysts shaded the idea of purity to the product end of the life cycle. One consequence was the replacement of environmental and moral concerns with technical criteria. One court case, for instance, pitted an importer of Fairbank & Company's "American lard" against an inspector who had found 30 percent cottonseed oil in the sample. Was it pure or adulterated? The lawyer's questioning irritated a chemist witness for the defense, telling the lawyer he couldn't possibly understand the testimony. "I think you would require to attend a course of lectures on chemistry before you did." Later, the lawyer cornered him, asking whether adulteration was the *right* thing to do, to which the witness said flatly no, "it is not right." The lawyer seized on that: "You are not the analyst of right and wrong, but only of fat and stearine." The courtroom laughed as the lawyer made his point: the chemist had conflated an analysis of purity with an assessment of honesty and integrity. The lawyer didn't think they were the same. Neither did the court.[76] But Wiley and his cohort of acronyms did. He included the story in Bulletin 13 because its rejection of chemical authority over right and good was no longer valid. He thought by then they *were* analysts of right and wrong. He told the story to show that by the early twentieth century the technical had subsumed the moral.

Another consequence was the overcentralization of purity's definition in labs that tested end products. The manufacturing sector had long been involved in campaigns for pure food and brought criticisms to bear about the assumptions underlying regulation, not that federal authorities were seeking it. As economic historians have shown, their influence in the pure food debates was sensibly enough meant to ensure that businesses would benefit.[77] Heinz, for instance, had for years fashioned a business model predicated on an image of purity, from the moral demeanor of his employees to the claims on his label.[78] A pure food law would help his cause. Armour and Swift thought the same. They understood the value of the pure label and sought the sanctioning offered by government certification. They also figured an era of lightly regulated marketplaces would give them sufficient influence on inspection and enforcement. But their view, too, shifted the idea of purity to the product end of the life cycle. To this day, the bottle of Heinz Ketchup on my shelf tells me it is "grown, not made" without explaining how the seven ingredients besides tomatoes (including high-fructose corn syrup) got there. I'm asked to take their word for it.

Grocers brought criticisms too, of course. They feared losing their authority, understanding that the knowledge embedded in chemists' expertise was of a specific sort. History shows they were right to do so. With the twentieth century transition to self-service groceries, the profession of the grocer was reduced to the corner stores and mom-and-pop shops local food advocates today are seeking to reestablish.[79] Grocers also understood that the government chemists' approach offered a specific kind of knowledge about purity, one among many ways to understand it. *American Grocer*'s Francis Barrett objected that proposed pure food legislation "leaves the matter entirely in the hands of professional men, leaving out altogether any practical men." Those with firsthand experience working with food had their own knowledge and familiarity to bring. "I certainly think a practical man connected with stock yards is better adapted to formulate a standard for meat products than five experts selected by reason of their attainments in physiological chemistry and hygiene."[80] Some politicians made the same point. During the final pure food hearings, a congressman from Georgia jumped in to voice opposition. "I believe there are millions of old women, white and black," he said, "who know more about good victuals and good eating than my friend Doctor Wiley and his apothecary shop."[81]

Grocers, manufacturers, courts, and consumers were clearly aware of and sometimes worried about losing control of the policing of purity a century ago. In the decades and century to follow, all of them recognized more consequences of the new form of analytical policing. A loss of environmental relations was one. The analytical view, for instance, was a long way from the ship deck of Melville's *Fidele,* the lake lands of the upper Mississippi Valley, incursions from British grainland development across

Canada, the cotton fields of the postbellum South, and cornfields of the Ohio Valley. It was less attentive to farmers considering when, where, and how they sent their cows to graze for the sake of richer milk—decisions they would make when they assessed the rain and turf and seasonal movement of their cattle or the pattern of manuring and reinvigorating soil to optimize the color of the butter their milk would produce. Nor were agricultural complexities of, say, the cotton harvest or the labor demands of separating seeds from the plant, of pressing those seeds and extracting their oil, of much notice. Nor was the draw on rivers in Iowa or Indiana for water used in the process of turning cornstarch into alternative sweeteners something to factor in. And nor did they emphasize the differences in climatic conditions, soil properties, moral commitments to labor, or harvest dynamics between beets, sorghum, cane, and corn. The environmental work of producing pure foods became subordinate to the products that resulted from that work.

Furthermore, analytical approaches tended to box out a discussion anchored in the household stovetops of kitchens and cooks and families, though if and when they did it was because chemists sought to redefine those areas as spaces to be managed like a bomb calorimeter. In that way, the gendered politics of control remained a sometimes subtle, sometimes overt element in the march for technical control of purity. The gambit of the USDA, Wiley, and associates was to take control of food identity out of the hands of the housewife and into the hands of the analyst. This was to exchange the kitchen countertop for the lab's countertop. Or, if not that, and as with Ellen Richards, it was to turn the kitchen counter into a lab. That wasn't merely a locational difference. It was a commitment away from one type of knowledge, embodied in the labor of procuring, preparing, cooking, baking, and managing the household to another type of knowledge embodied by the labor of instrument operations.

Observing the changing ways of knowing could tempt a reader to think that adulterated foods were simply the consequence of manufacturing processes that prevented the firsthand knowledge of earlier generations. There's a kernel of truth there. The increased distance between producer and consumer has and does disrupt familiarity, trust, and knowledge. Yet adulteration had been a problem of civic life for millennia, even when food sources were closer at hand. Plus, some measures to protect the consumer against adulteration led to packaging and machine handling that took consumers even farther from the food's organic identity, not closer.

Sometimes more packaging was better, that is, a vote of confidence in the purity of the product. Today we might prefer handcrafted artisanal breads. But a century ago, breads made by machines untouched by human hands assured customers that they were clean and untainted. Customers divorced from the processes of wheat production and baking practices, plus those fearing the dirtiness of the baker and his assistants, were gratified to find their loaves in clean bags indicating sanitary production. Given the legacy of milk contamination, it wasn't uncommon to laud the technologies of milk storage and distribution as progress, showing that sometimes consumers considered more technology—more artifice—a means for purity rather than a hindrance. Packaged tea followed the same trend. Tea in loose cans was a danger in the grocer's market. Who knew if someone had sullied the container or contaminated it with leaves or debris? Lipton and Tetley pioneered bags of tea in part to assuage customer fears of contamination and to ensure purity.[82]

What was going on? At stake in situations close and far was trust and faith in the food provider. The rise of branding, packaging, ads, and company identity was a means of assuring customers that they could trust the source. For a study of purity and adulteration, the point is that such measures and responses were solutions to problems caused by changing agricultural practices and engendered by prevailing dictums of distrust, skepticism, and fears of frauds. Addressing adulteration could have meant addressing changing degrees of agricultural manipulation and modes of distrust and skepticism. That would have been a way of aiming at the source of the problem, not its symptoms. But with the new political viability and scope of the sciences of analysis, pure food advocates changed the debate to one about trusting the label, not the farmer.[83]

This has been consequential for how we think about food and nature. Consider that as modern consumers in a food marketplace, our connections to the land tend to be mediated through the label on the shelf, where there is usually an image of a farm or the pastoral countryside. This form of interaction was being put in place by the early twentieth century, when human interaction with the land was mediated less by the experiences of the agricultural process than by the experience of selecting a box, can, bag, or tin of food from a grocery store shelf. The sciences of analysis helped the cause of defining purity by narrowing the debate and providing focus, but they created new problems that we are dealing with still, namely, just that distrust, skepticism, unfamiliarity, separation, and confusion over what we are eating. In the main, then, this is less a story about scientific analysis being helpful or detrimental and more a story about what conditions led analysis to become a solution. We are part of that story still as we wonder how we know what to eat and who we trust to tell us. Despite the enormity of agricultural, manufacturing, political, and demographic changes since his time, one senator's question in 1886 still feels close at hand: "Is it not fair that I should know what I'm eating?"[84]

A story circulated in the papers in the early 1900s. Alfred Paraf was back in the news. He had been dead for over a decade and out of the public spotlight for nearly two. At the turn of the century, he was part of a wave of reflection on the swindles and frauds of the age. They still called him "the educated swindler" and "the Alsatian alchemist." They lumped him in with an account of famous "Wall Street Swindles." The USDA was publishing more parts of Bulletin 13, Atwater was running bomb calorimeter experiments, the AOAC was meeting to agree on methods, municipal analysts were fashioning local authority, grocers were organizing pure food conventions, A. J. Palm was lashing out at his fellow Meadville dairymen for real taxes on fake butter, Wiley was about to plan the Poison Squad, and readers of the daily news were looking back to wonder over the lore of "a scientific and commercial impostor named Paraf." He was "one of the most successful imposters," the *Brooklyn Eagle* reiterated, "who ever operated in this country." They reminded readers of the early cons that led him to a "career he afterward pursued to his eventual disgrace and downfall."[85]

His appearance on the public stage in the 1870s had preceded and then exacerbated widespread anxiety over the upstart adulterant oleomargarine. A year after his death, the US Congress passed its first full-scale federal legislation against adulteration with the 1886 Oleomargarine Act. Some considered margarine a con too. With the encouragement of the grocers, housewives, farmers, doctors, manufacturers, and chemists who have appeared throughout this book, from that point forward the era aimed for greater regulatory action, more legislative attention, and further evaluation of the veracity of foods and the people making and selling them. They sought the alignment of goodness, propriety, and truth in their food.

Their search for that alignment ended up shifting the concepts of natural and pure. The two had been bound together into the mid-nineteenth century: a pure food was one from nature. In reference to food, nature was that which humans manipulated through culturally agreed on practices. In an agrarian world, nature wasn't simply the nonhuman world. It was the life cycle of animals, plants and crops, the seasonal patterns of ecological change, the living soil, the healthy results of centuries-old methods of cultivation. That sense of nature also led to environmentally acceptable ways of processing and distribution, such as milking in dairy sheds and refining lard from butchered animals and extracting sweeteners from beehives or trees or vegetables. And those agreed on practices required attention beyond the fields too, to cooking, taste, and tradition. There were right ways to do these things and wrong ways. Adulteration was a violation of the right ways. It was a way of cheating on nature.

The view of pure and genuine food was different in the early twentieth century from what it had been in the mid-nineteenth century. Whereas a pure food had been one from nature into the mid-nineteenth century—understood through the metrics of community life and a sense of provenance (knowing the food providers, knowing where the food came from)—by the early twentieth century a pure food was one whose analytical certification said it contained the ingredients the seller claimed. In this way, an artificial product could be pure. So long as its contents were certified to be what the seller said, it fit the concept of pure.

An enthusiasm for technical analysis similarly took the core concept away from an evaluation based on values such as taste, palatability, or appetite. It narrowed the concept so that gastronomic, culinary, and gustatory considerations—values local food advocates are trying to reinvigorate today—were largely outside of or secondary to the evaluation. What is more, the emphasis of analysis on end products moved the study of purity away from considering the moral, economic, and agricultural circumstances that led them to those ends. The analyst's perspective suggested that purity was technical, not agricultural or culinary, that assessments of purity were about market-based products, not agriculturally dynamic processes. The industrial food system we live with now has grown in the shadow of those changes. The changes were as much about how we understand nature and how we argue for some sense of proper agriculture as they were about the specific adulterants that challenge our anxieties over health and wealth.

I said in the prologue that there were three eternally braided questions when it came to food and nature: what to eat, who says so, and how do they know? That last one, the epistemic demand asking how we even know what we're talking about, is often subtle or invisible or ignored. During the era of adulteration, scores of hearings, treatises, and advertisements deployed the term *pure* on the assumption that its meaning was self-evident. After the fact, too many studies of the era of adulteration have seen it as a march of progress away from contaminated foods to clean and pure ones without noting how the very terms of the debate were being reconfigured. Too often we assume that there is a preordained, a-cultural concept of pure and natural simply waiting in the wings to be recovered and reinstated. This would be to purify the past, too, as if things used to be ideal and got corrupted slowly and painfully. It's tempting, an Edenic vision of lost purity waiting to be regained. It's also wrong.

The story of purity and adulteration is less about the fact of impure foods, dishonest brokers, and fears of adulteration and con men—all of those were indeed valid and present in the 1800s, as they were before then and as they are today. It is more about what people meant when they deployed those terms. So many people spoke to the need and claim for purity. So many presented him- or herself as genuinely interested in the genuine. But how did they and their customers know? How did farmers and wholesalers and grocers and cooks and housewives know? The battle against adulteration Richards's and Wiley's generation fought was waged on the shifting sands of the very concept of purity. What changed over the course of the pure food crusades was as much the concept of purity under debate as it was the presence of adulterated foods. More to the point, the case those generations made was that some people had better access to controlling—and producing—knowledge of that change than others. Wiley was speaking to those matters on Capitol Hill. He wanted the government chemist to control the answer.

Doctor Wiley had his own denouement in this story. His tenure at the agency ended in 1912 (fig. 8.5). By then he was a sixty-eight-year-old chemist who had worked in Washington for twenty-nine years. He had seen the transformation of food policy from local and municipal origins to federal statutes; he had witnessed and helped usher in a set of organizations, methods, and instrumental practices to manage that food; and he had overseen and authored thousands of pages of documented analyses from every corner of the food system. When he arrived in Washington in the 1880s, his office employed six people. When he left, the Bureau of Chemistry was home to over six hundred employees. When he arrived, the food system was anchored in agrarian communities and rural vitality. When he left in the 1910s, the urban, immigrant, and industrial changes before him were profound. The changes were visible enough to warrant a first generation of appeals to get back to the land and return to nature, a point the horticulturalist Liberty Hyde Bailey made when he cataloged the roster of adulterants common to urban shelves in a book about rebuilding connections to the land.[86]

For Wiley, internecine struggles between the bureau, the secretary of agriculture and the Republican administrations of Roosevelt and Taft led to poor enforcement of the new food laws in ways some had feared during the previous decades' debates. That frustration pushed Wiley to resign. He then heard and declined an offer to run as Woodrow Wilson's vice

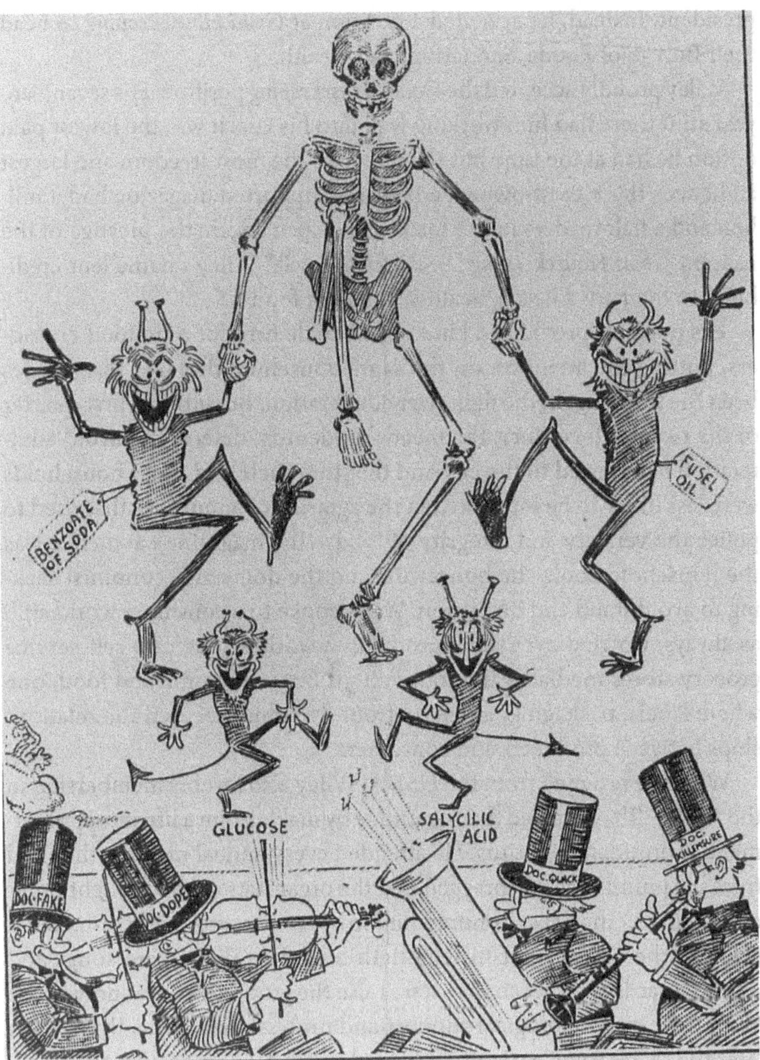

DR. WILEY HAS RESIGNED!

—Plaschke in the Louisville *Post.*

FIGURE 8.5. Death, chemicals, and even glucose dance with glee at notice of Wiley's resignation from the USDA in 1909, serenaded by the quartet of "Doc. Fake," "Doc. Dope," "Doc. Quack," and "Doc. Killemsure." Courtesy of the Wiley Papers, Library of Congress.

president. Instead, he accepted a position at *Good Housekeeping* to head their Bureau of Foods, Sanitation, and Health.

Wiley proudly accepted the *Good Housekeeping* position. His seventeen-year stint there had him working well into his 80s. It was the lowest paid option he had at the time but the one with the most freedom and largest audience. The Hearst-owned, advertiser-supported magazine had a million and a half readers in the 1920s. He helped secure the prestige of the nascent *Good Housekeeping* "Seal of Approval." Wiley's name lent credibility to their ads for safe, healthy, and pure foods.

His position provided a kind of full-circle turn for pure food crusaders, putting the attention on the same household that had for decades been the front line in the fight over adulteration. But into the first quarter of the twentieth century, the means to identify, detect, and avoid adulteration had moved to the city and the store shelf, and those households were less likely to be integrated in the agrarian communities that used to police the veracity and integrity of foods. The magazine's audience was the household cook, the housewife, and the domestic economist seeking to avoid fraud and deception. Wiley spoke to a consumer world still, as always, worried over sham products, a world where new self-service grocery stores mediated the transactions between home and food, one where labels, packaging, and storefront dynamics secured the relationships between producers and consumers.

When he resigned from the USDA, Wiley also lost his membership in the AOAC. They showed their gratitude by making him a lifetime appointment as president emeritus. He attended every annual meeting they had from 1884 until 1928. He bragged that the organization had brought order out of chaos. They also honored him in 1924 with another beer-battered feast on the occasion of his eightieth birthday. There were songs and rhyme, awards and speeches, as it was also the group's fortieth anniversary. With their methodological support and professional cache, three years later the secretary of agriculture helped spin off a new agency, the Food, Drug, and Insecticide Administration. They shortened the name in 1930 to the Food and Drug Administration. Its onion-like regulations encased the analytical commitments of the Wiley generation in its core. The FDA has evolved and changed in almost a century since that time just as the material conditions and political structures of modernity have changed, but the contested modern assumption of what and who defines a pure food remains there at the center. Purity had become a scientific concept.

EPILOGUE
The Persistence of Adulteration

None of us is pure, and purity is a dreary pursuit best left to the Puritans.

R. SOLNIT[1]

A friend wrote to tell me about fake green olives. When you write a book about adulterated, contaminated, and fake foods, you get an extensive list of examples from everyday news. I started a running tab of these a few years ago before quitting after it topped one hundred. Spinach, calamari, whiskey, pomegranate juice, olive oil, honey, coffee, almond milk, parmesan cheese, wine, chocolate, cantaloupe, and cereal, to name a quick baker's dozen.

My gradually accrued catalog was full of echoes from a century ago. The arguments in that list were all premised on some incident where people questioned whether what they were eating was what they thought they were eating. The cases turned on matters of trust, legitimacy, and authenticity. One story explained that researchers found a compound common in yoga mats, azodicarbonamide, in five hundred food products, including bread. The reporter wrote that it was another case from the prevalent "*what's-really-in-your-food* campaigns" of our day. Although it was not clear whether the compound's presence was necessarily bad—does it make you sick or not?—it gained a bad reputation through guilt by chemical association. The compound fostered the "ick" factor similarly prevalent in an infamous "pink slime" expose in 2012. Pink slime is a colloquial term for "low-cost processed beef trimmings." The nickname caused alarm once people found out fast food restaurants used it in their burgers. The shock of the reveal was less about the health consequences—though ick—and more about the inherent suspicion that it just wasn't right. An unspoken assumption about what counted as "right" sat inside the public rebuke.[2]

Other examples were just as resonant with a bygone time. The parmesan cheese incident in my list seemed classic when it made news in 2016.

It was about "the cheese police" busting down the doors of a Pennsylvania company to find them doctoring the so-called 100 percent Pure Parmesan with wood pulp. There was the honey case too. Bees feeding on effluent streams from an M&M candy factory were producing honey that was green, red, and blue. The case confused people, because did that count as adulteration or not? Around the same time, a seemingly more egregious and straightforward exposé outed a meat factory selling pig intestines masquerading as calamari. It turns out that case was more a story of urban legend and whimsy than a health assault, but its terms—trust, deception, confidence—made it a lively tale. The popular appeal of the story banked on the availability of shock and fear from a skeptical consumer audience, one who would ask, is that right? Do they really do that?[3] All of the cases—cheese, slime, honey, calamari, etc.—touched on perceptions of cheating, poisoning, and confusion, and all of them operated under terms common both today and a century ago.

The fake green olives were perhaps most telling as a comment on the contemporary balance of food, trust, and authenticity. My friend, I was told, pulled the lid off of a slow cooker of chicken and vegetable stew that had been simmering all day. She let the burst of steam wisp away and stirred the cooker with a wooden spoon, revealing a weird sheen on top. After turning on the overhead kitchen lights for a clearer view, she saw that the whole stew looked green. Spooning out pieces of chicken revealed bright green smudges on their sides. The only green items in the entire dish were olives from the Italian butcher down the street. Scanning the list of ingredients on the plastic olive container explained little because they had been bulk packed and likely pulled from a giant vat in a back room of the shop. After working backward, it seemed clear that the olives had been dyed green at some point before they were packed in bulk from a mysterious hidden vat and that the long bath in the crockpot had released the dye into the stew. Puzzling through the confusion of a green-sludge stew led her to conclude that "the guy who sold these olives was definitely a huckster, and these olives were definitely adulterated." After obviously not feeding it to her toddler daughter, she was still unable to shake the idea of a long conveyor belt of olives running beneath a line of spray nozzles dousing them in green paint.

These stories are reminiscent, evocative, and proximate. The issues they speak to are familiar and pressing as versions of a scenario that is long standing. They remind us that adulteration was a problem in the past, adulteration is a problem today, and adulteration will be a problem of public concern tomorrow. Its presumed antithesis, purity, is a dreary, if not constant, pursuit.

The continuing presence of adulterated foods is one way to see how the concerns from the era of adulteration have carried forward to today. This seems the easy, uncomplicated historical case: perceptions of adulteration have not gone away, they have only taken on new forms and dimensions. In that story of historical continuity, food purity and health is a game of Whac-A-Mole. Knock it down in one place, see it pop up in another. In the 1880s, analysts in Brooklyn met in part to tamp down shifting domains of adulteration. If they found it in Brooklyn it would skip over to Manhattan. If they found it in Manhattan, it would jump ship to New Jersey. If they found it in New Jersey, it would refashion itself in upstate New York. Or in ever-farther factories in Iowa or Illinois.

This has happened over time as well, not just then. We knock down one hazard as others pop up. The phenomenon evokes an ethical axiom of technological development the philosopher Paul Virilio wrote about some time ago, that when you invent the train, you invent the train wreck; when you invent the plane, you invent the plane crash. In kind, when you invent agriculture, you invent agricultural contamination; when you invent processed and manufactured food, you invent yet more adulterated and corrupt food practices. Economists and historians of capitalism alike note that fraud and markets are and have always been wedded. They arrive together. As long as there has been food, there has been food adulteration.[4]

The bare *fact of* adulteration provides a thread of historical continuity. But my sense is that that kind of persistence is less meaningful than the ways the pure food crusades led to transformations in the long history of adulteration. The story of this book is how our fears of adulteration and deception changed markedly over the course of the later nineteenth century into a modern form. In one manner, the ways people thought about adulteration's causes changed—its derivation, its sources. In another, the ways people responded changed, especially at broader political and governing levels.

A number of factors in this book help explain the first dynamic, the new ways adulteration came about. These include cultural debates over sincerity and trust, globalizing marketplace networks, and the rise of a consumer class separate from and outside of an agricultural producer class. Confusing new environmental and large-scale manufacturing processes also shaped the new moral landscape of adulteration. The environmental aspects of the pure food crusades spoke to changes in the very concept of nature that came about with the introduction of a new era of food production, processing, and distribution. Natural and pure had been conjoined, bundled in the same conceptual schema into the mid-nineteenth century. Artificial and adulterated sat in opposition to those notions. In fact, they were defined less by specific attribute and more by contrast with the former concepts. The same was often true of natural and pure. With respect to food, they were understood less as absolute concepts and more as ways of understanding what was right and proper *in contrast to* artificial and adulterated.

This could be problematic, certainly. Sometimes people thought artificial things worked more effectively than their supposed natural counterparts, leading them to consider artifice right and proper—cleaner standards for bakeries, healthier milk, packaged tea that helped escape the problems of contaminated tins, water purity that followed from better filtering and technological (artificial) aids. But such present views of a past era can't change the fact that people at the time challenged the validity of other artificial products because they were an affront to some concept of natural and pure. For members of the era of adulteration, the line to be drawn between pure and adulterated was a line of acceptable manipulation of agricultural practices. Too much intervention was adulteration; the right amount was pure.

This changed by the early twentieth century. By that time there could be purely artificial foods—Purity brand margarine, for instance. The line drawing was about whether the end product's identity matched the claims of the seller more than whether the seller was part of an acceptable process of agricultural manipulation. Analysts marked the boundary by evaluating the contents on the store shelf.

Environmental aspects helped characterize what shifted during the pure food crusades in another way: the very production of suspected adulterants followed from newly complex manufacturing processes, which were new ways to manipulate agrarian nature. New trade routes, production schemes, resource uses, processors, packagers, refineries, factories, and distribution systems all reconfigured the environments of agriculture. Margarine's integration of milk, animal fats, and chemicals drew together meatpacking and dairy shed ecologies from the lake lands of the upper Midwest to the stockyards of American cities to the traffic of imports from the Mediterranean and India. Cottonseed oil's integration of cotton fields, crushing facilities, olive groves, and pig or cow fat influenced the environmental relations of the US South and the world, tying the orchards of the Mediterranean to the rail depots of the US black belt and the trade routes of the North Atlantic. Glucose industrialists organized resource-intensive operations, drew millions of gallons of water supplies, corralled thousands of bushels of corn, input tons of chemicals, and polluted streams and rivers in ways that were opaque and intricate.

The changing circumstances of the period influenced the other main difference wrought by the era of adulteration: the new set of responses. Into the mid-nineteenth century, the dominant mode of policing the identity of food came from regional structures of enforcement based on common modes of social interaction following from well-entrenched practices of community norms. Thereafter, chemical and public health actions grew to new degrees of public visibility, viability, and authority. As the twentieth century dawned, new structures of scientific analysis and detection guided the reaction to adulteration at the level of federal governance. Analysis could help with the different directions of perceived adulterations, cheating and poisoning. If grocers lied to customers across the counter, were their customers angry because they were cheated economically, because their health was compromised, or both? One response (the cheating one) fit evolving business regulations of the marketplace; they were problems of commerce and economics. The other (poisoning) fit new structures of public health management; they were problems of bodily health and medical practice. No matter the direction, analytical attention was meant to identify the true substance of a product.

The changes that occurred during this first phase of manufacturing food history led to a now century-old structure of response to adulteration. Some features of that structure have only become more entrenched over the past century. Since that time we have accelerated toward more instrumentalism, for one thing, seeking in regulatory policy to displace home-based bodily senses with lab-based quantifying instruments. The personal experience of food knowledge will never go away, but it continues to be politically outpaced by the more authorized instrumental techniques embedded in policy and regulation.[5] We have also sought to dig deeper as time goes on, to look further inside. This approach began with a fight between surfaces and interiors in the nineteenth century. Does the truth lie inside? Is outer character—the way we present ourselves, our face value—a sign of that inner truth? In the twentieth century, the trend to look further inside continued with more refined detection strategies for food. These aimed initially at mechanical properties and microscopes, but they moved further beyond the naked eye to uncover chemical composition, then biological views of molecular identity, then with even further specification to particular genetic receptors, and so on. We keep digging deeper inside, believing that's where authentic reality resides.

That pursuit of material identity points to an additional notable trend, the further development of a consumer society in distinction to an agriculturally productive one. More and more people have come to interact with the products of agricultural life through storefronts; more people have come to know their food through labels, brands, and packages. Less than 2 percent of the US workforce is made up of farmers today. That transition took shape in parallel with the tilt from majority rural to majority urban living by 1920, when for the first time it was 51 percent urban, 49 percent rural, a ratio that became 80/20 by 2000.[6] Fewer agrarians and more urbanites means that the ways we respond to anxiety over food have increasingly come through consumer-based metrics. And that has nudged the discussion of purity toward food (the product) and away from agriculture (the process).

The overall payout for these trends has been to fashion our relationships with food and agriculture more through purchasing choices and less from the ways we understand our lives as connected to the land and with one another. You could read the local food movement of the past generation as a call to redevelop those relationships. Few people imagine we will return to some previous era, where the demographic trends of the past century would be reversed. That is folly. But others, myself included, imagine ways forward that can reengage our relationships with food and agriculture to build trust and confidence in finding that we are getting the real thing. That real thing isn't sitting out there waiting to be found. The real thing is produced through our relationships of trust and confidence.

Concerns of the pure food crusades have carried forward to today in another way too. I can attest that when you write a book about purity and adulteration, you not only get a pages-long compendium of current examples of food hucksters; you also get a slew of questions about new challenges to notions of nature and purity. These are often morally charged issues about "doing the right thing" to fashion an appropriate food system.

My running list of modern adulteration carried scores of examples of straightforwardly contaminated food, from wood pulp in cheese to yoga mat particulates in bread and low-grade cocoa in high-grade chocolate bars. But it had just as many stories about genetically modified organisms (GMOs) and genetically engineered foods. Those examples strike me as the more relevant and harder case in terms of policy and environmental and food ethics. They elicit the same questions as pure foods did. They also follow the structure of debate set out in the nineteenth century. Granted, GMOs are not the same thing as water in milk or cotton oil in olive oil. Nor are they the same as residual sulfuric acid in cornstarch or traces of various chemicals in any number of foods. Yet GMO debates are surprisingly comparable in structure to earlier pure food debates. Are GMOs natural? And if they are, does that mean they are okay? And if they aren't, does that mean they should not be part of the future of food?

There isn't much chatter that GMOs are an adulteration in the same language used a century ago, but there has been a great deal of angst over the legitimacy of genetically modifying so-called natural systems, their rightness and propriety.[7] Trying to answer the nature question is thorny. It requires us to defer to some notion of naturalness that, to this point, has not been clarified. Do people mean nature in the moral sense, as in "the way it is supposed to be"? Do they mean natural as in outside and before human manipulation?[8] The fact that the USDA and FDA lack standards for labeling "natural" food is illustrative. It confirms that nobody has yet been able to make the case that there is a commonly applied definition within the borders of the United States let alone as part of a global marketplace. Before the nineteenth century, pure and natural were not the same thing—but they were conjoined, they lived together. Judging whether a food was pure was akin to claiming it was natural. But as we've seen, by the twentieth century there could be a purely artificial product. To some, GMOs are unnatural. We should not do things that way. It is not as God intended. To others, GMOs are acceptable manipulations. They are contrived and artificial, sure, but that doesn't make them bad.

In this, too, the persistent cheated-versus-poisoned framework rears its head like a hydra-headed monster. Do consumers know when they are eating genetically modified foods (cheated)? Do GMOs make them sick (poisoned)? Here comes Senator Blair from 1886, wondering, "Is it not fair that I should know what I'm eating?" When concerned eaters ask those questions, they are imposing cultural assumptions about proper environmental action and claiming that genetic modification has crossed a line. So it is that GMOs are not necessarily the same type of manipulation as the adulterations that agitated the nineteenth century, but all of them have and still confront cultural angst over confidence, trust, sincerity, and authenticity.[9]

If the GMO debate follows with moral questions that are strikingly similar to those from the pure food debates, it also falls prey to the same limitations of that older framework. For one thing, the cheated-versus-poisoned framework places both options at the consumer-product end of the food life cycle. Proponents assume that the role of scientific analysis is to determine whether GMOs make people sick and whether it is present in a product. On the former, there is no substantial body of evidence that genetically modified foods make us sick; on the latter, there is considerable industry resistance to telling consumers their products are genetically modified, considering the rancor over state laws proposing to label GMOs. Both of those matters are being addressed through analysis. That possibility comes from a century of analytical progress. It also leads the question of GMO legitimacy further away from broader considerations of environmental practice.

Rather than understanding the argument through a binary of cheating or poisoning, it is more faithful to the ecological and cultural dimensions of food systems to see cheating and poisoning bundled on the consumption side of an argument and environmental and ecological practices extending to the production side. In that approach, we also need to recognize another element from the pure food crusades that has not gone away, namely, the fundamental awareness that the debate at hand is one of boundary making. It's about drawing boundaries between those practices that people deem acceptable and those we deem inappropriate. While so much has changed over the past century, we remain indebted to a food system that requires our trust, that asks for confidence, and that leans on assessments of appropriate environmental activity.

I study science, technology, and agriculture as an environmental historian because it provides the hard ethical case about human intervention. There's no place in a study of food, land, or animals to presume a utopian or Edenic prehuman state of affairs to which we should return. Agriculture is always "the intentional alteration of a natural ecosystem," to quote the environmental ethicist Paul Thompson. All foods are the result of human manipulation. Some are lightly manipulated, as I pluck a berry from a bush; others are heavily manipulated, like refining corn into highly processed syrup. Because of this, we will always have to differentiate between what is proper intervention and what is not, just as there will always "be some place to draw a line between an acceptable environmental impact and an unacceptable one."[10]

Thompson notes as well a "naïve food ethic that focuses intently on food choices and their economic impacts." Such an ethic is the outcome of assuming that "agriculture is just another sector of the industrial economy," no different than a widget factory, no closer to a grounded environmental activity than making a box of bolts. When agriculture is treated as an adjunct of industrial economy rather than an environmental process with embodied cultural practices of cultivating, cooking, eating, and knowing food, it assumes the relations we have are those restricted to a shopper at the store. That's problematic.[11]

A study of the era of adulteration shows that debating the merits of food identity and safety at the end of the food life cycle—the consumer option at the store—gives up the majority of the work we call agricultural and ethical. In turn, to argue over GMOs and chemical additives today requires more attention to the environmental circumstances producing them and the cultural values at play, not just the product resulting from those processes.

We don't have to make the questionable presumption that the past was better, that we should eat like our grandparents, to say that better connections to the agricultural process can make for a healthier future. Just as the causes of adulteration are cultural, environmental, and scientific, so, too, should our responses be. The better questions are, what circumstances lead us to judge the quality and identity of our food? and who has responsibility—whom do we trust—to patrol that safety-danger border? These issues should drive an agenda for food and agricultural reform. They are not mere calls to ask different questions. They require different, more environmentally demanding forms of engagement with food and agricultural policy, with how scientists and engineers imagine what interventions they make in their studies and innovations, and with what things consumers pay attention to in their quest for the genuine article.

Acknowledgments

I read backward. I read books and magazines backward. I flip to the end and work my way forward, checking either the back pages (magazines) or the acknowledgments and bibliographies (books) first. I'm not entirely sure why. There's something about the boundaries of a book, that might be it. I can see the cover, I know what they're trying to sell me, but I also want to know where it's going, where it ends, what might happen after. Since that's never possible in any other part of life—knowing its bounds or where it's going—I guess it's a comfort of reading. This means I'm often digging into acknowledgments first, which is what you're reading, though I don't know whether it was the first or last thing you found.

There's a corollary to this irregular reading style. It leaves me wondering not so much what was in the long middle between front and back, because I'll eventually get to that, I'll read the thing. It makes me wonder instead about how it got there, about who and what made it possible. Rather than last, then, as is somehow strangely customary in acknowledgments, I want to thank Chris, Harper, and Alex first for listening to random and out-of-context comments about hucksters and con men and foods and chemists, for supporting everything along the way, for providing the backdrop against which all the writing happened over the last many years. That goes for my parents, too, to whom the book is dedicated, and of course my sister and brother-in-law and niece, Shosh, Blair, Leanna, and my extended family, the Rodriguezs.

Then there are the accrued debts. I used to scoff cavalierly at the listing of the great many names in book acknowledgments. I was so naive. There are so many people along the way who have helped, encouraged, and supported this book that now my fear is missing someone. As with my first project, I start with Wyatt Galusky, who has read and reread this material enough times to have parts memorized, I'm sure. Then I want to acknowledge the number of students at Lafayette who helped collect, analyze, and process data. I'm grateful to Matt Plishka, Jay Kasakove, Andie Mitchell, and Claudia Umana for working through a trove of files, putting the basics of the book's digital companion in place, and assisting with documentation and translation.

I had a good deal of institutional support along the way too. As I was trying to figure out what this book was about, NSF Scholars Award #0924932 got me on my way. I also had backing from the Smithsonian's National Museum of American History (NMAH) through a research fellowship, where Jeffrey Stine, Deborah Warner, Pete Daniel, and Marcel LaFollette all offered guidance and encouragement. The same goes for Jody Roberts, Ron Brashear, and Carin Berkowitz at the Science History Institute and Roger Horowitz at the Hagley Museum and Library, who supported research with fellowships and archival grants. I'm similarly grateful to librarians at the New York Public Library, Columbia University Archives, New-York Historical Society, Library of Congress, Smithsonian's NMAH Archive Center, the Library Company of Philadelphia, National Archives and Records Administration, University of Virginia Special Collections, Bancroft Library, MIT Archives, Massachusetts Historical Society, Wisconsin History Society, and Lafayette's Skillman Library.

I began this work at the University of Virginia (UVA) and completed the bulk of it at my new home at Lafayette College. At UVA, I'm grateful for help in early stages from colleagues and friends. Thanks in particular to Laura Kolar Silson, Katy Shively Meier, Tom Finger, Jaime Allison, and Ed Russell; the cohort of friends in the UVA Food Collaborative and Charlottesville food and sustainable ag scene, especially Paul Freedman, Tanya Denckla Cobb, Kendall Singleton, and Emily Manley; and Josh Yates and the Institute for Advanced Studies in Culture.

At Lafayette, here in Easton, the list is extensive. It begins by acknowledging support from the provost's office through an R. K. Mellon Research Fellowship, various grants from the Academic Research Committee, two Andrew Mellon Digital Humanities research grants, funding for research students through the EXCEL program and backing for research assistants. It then extends to colleagues in my home programs: Julia Nicodemus, Kristen Sanford Bernhardt, and Dave Veshosky in Engineering Studies; colleagues from the environmental programs, especially Andrea Armstrong, Dave Brandes, Dru Germanoski, and Kira Lawrence; and those in the history department and beyond, where Paul Barclay, Emily Musil Church, Chris Philips, Rebekah Pite, and Jeremy Zallen all offered feedback on parts of the work. That doesn't even yet count the extensive collaboration with librarians, including Kylie Bailin, Terese Heidenwolf, Ben Jahre, Ana Luhrs, Diane Shaw, and Lijuan Xu; the digital scholarship group of Emily McGinn, James Griffin, Eric Luhrs, Adam Malantonio, Paul Miller, and Charlotte Nunes; and Karen Haduck in the Interlibrary Loan Office. John Clark's work on GIS stands out for a several year commitment to developing a unique and labor-intensive process of map creation. Kristen Lopez's graphic design work on the export maps cinched together an array of different ideas in just the right way. And of course, without the purportedly capable team at Various Breads and Butters—Simon Tonev, Michelle Polton-Simon, Renan Dincer, Ben Gordon, Jen Giovanniello, Will Gordon, Ian Morse, Claire Swanson, Thomas Williams, and again Andie—it's hard to imagine how this book would've been finished, or it might've been finished years earlier, whichever.

This compendium is probably an expanding circle of connections over time and place, because those from afar—from conferences, invited talks, dinners and drinks, workshops and random conversations virtual and face-to-face—are another source of extensive debt. Nearby colleagues from across the Lehigh Valley include Kelly Allen, who invited me into his NEH "Food in the Public Square" project and kept asking questions that put the future of just and sustainable food systems at the front of my mind as I was finishing the book; others did the same as we have worked on the Veggies in the Community project in Easton, including Larry Malinconico, Sophia Feller, Sarah Edmonds, Lisa Miskelly, and another half dozen students. Beyond the valley, and in addition to the aforementioned Wyatt and Jody, there's Kelly Joyce, Shobita Parthasarathy, Melanie Kiechle, Kara Schlichting, Neil Maher, Raechel Lutz, Mark Barrow, Steven Stoll, Jackson Lears, John Warner, James McWilliams, Julie Guthman, Carolyn Thomas, the California food-studies group and their Food Anxieties workshop, Aaron Sachs, Susanne Freidberg, Anna Zeide, Michael Kideckel, Alan Marcus, Mark Hersey, Jim Geisen and the CHASES group at Mississippi State, Michael Egan, Hayley Goodchild, Adam Sowards, Peter Alagona, Laura Sayre, Adam Rome, Catherine McNeur, David Singerman, Matt Wisnioski, Holly McPeak, and Jessica and Mitchell Kane. I'm also grateful for audience feedback (beyond the many at Lafayette) about various parts of this book at the American Association for the Advancement of Science, Drexel University, the Hagley Museum and Library, Johns Hopkins University's Center for a Livable Future, Massachusetts Historical Society, Mississippi State University, New York Metropolitan Environmental History Seminar, Rutgers University, University of Delaware, University of Scranton, University of Pennsylvania, University of Virginia, US Department of Health and Human Services, and Virginia Tech.

At the University of Chicago Press, I thank Karen Darling for her support and encouragement and for the anonymous reviewers who provided thoughtful feedback on the draft manuscript. Before that, parts of the introduction appeared in the *Appendix*, to Amy Kohout and crew's credit; the inklings of chapter 1 got their start at *The Morning News*, thanks to Andrew Womack, Rosecrans Baldwin, Kate Ortega, and Nozlee Samadzadeh; Chris Otter included what would become parts of chapter 7 in a special issue of *Endeavour* he edited; and Chris Monks was kind in hosting a short series about the research for the book at *McSweeney's*.

That feels like a lot. This took awhile. Thank you to everyone.

Notes

Chapter One

1. *Washington Bee*, April 29, 1899, and *New York Times*, May 1, 1885.

2. *Los Angeles Herald*, April 26, 1891; *New Zealand Herald*, April 15, 1885; *Brooklyn Eagle*, as reprinted and quoted in the *Providence Journal*, January 9, 1889.

3. Various histories of adulteration and the pure food crusades speak to these issues, although mostly they attend to the responses to adulterated food, not its environmental basis. In recent decades, these include Okun, *Fair Play in the Marketplace*; Young, *Pure Food*; Goodwin, *Pure Food, Drink, and Drug Crusaders*; Coppin and High, *The Politics of Purity*; French and Philips, *Cheated Not Poisoned?*; Ferrières, *Sacred Cow, Mad Cow*; Wilson, *Swindled*; and Blum, *Poison Squad*.

4. *San Francisco Chronicle*, November 7, 1873; *Washington Post*, September 27, 1880.

5. The secondary literature on Paraf is paltry. Just two substantial peer-reviewed sources deal with his work, and those focus on his chemical accomplishments in the dye industry. See Travis, "From Manchester to Massachusetts via Mulhouse," and Reinhardt and Travis, *Heinrich Caro and the Creation of Modern Chemical Industry*. I thank Tony Travis for further leads on Paraf and Ana Luhrs for assistance with Latin American newspaper databases. Details about his life in this and subsequent paragraphs comes from the newspaper sources cited in notes below.

6. *Kansas City Journal*, February 19, 1899.

7. *El Comercio* (Lima, Peru), October 3, 1877.

8. *California Mail Bag*, October 23, 1873.

9. *Brooklyn Eagle*, November 17, 1877.

10. Mège-Mouriès filed his French Patent 86,480 on July 15, 1869. It was granted on October 2, 1869. He filed the British Patent 2,157 two days later, on July 17, 1869, and it was accepted the next New Year's Day, January 1, 1870. The US patent was filed by Paraf in February 1873 and assigned on April 8, 1873 as Patent 137,564; Mège filed his American version, Patent 146,012, on December 13, 1873. It was quickly granted two weeks later, on December 30, 1873.

11. *California Mail Bag*, October 23, 1873, v.

12. Letter from Henry Mott to *Scientific American*, January 6, 1877, 9. It was housed at 6 New Church St., Brooklyn, New York.

13. *Los Angeles Herald*, August 5, 1874.

14. It was "the golden age of adulteration," to quote French observer Paul LaFarge from 1880. A British commenter chimed in similarly in 1885 to call it "the

era of adulteration." (cf. *British Monthly Homeopathic Review* 29 [March 2, 1885]: 158). A report from the *Colonial and Indian Exhibition* from 1887 said the same (273). The *New York Times* reported that the era was still going strong two decades later (December 31, 1904) as from testimony at the annual AAAS meeting.

15. Freidberg, *Fresh*, 6. Freidberg's study explores the point more fully with attention to the relationships between new cold-chain technologies meant to secure an idea of freshness—chilled warehouses, ice transport, refrigerated rail cars—and the ways people "understood the value of perishable food" (5). The freshness concerns sat in league with debates over purity. Just as with concepts of purity, "over millennia people developed all kinds of ways to keep perishable goods. But most methods were home-based or artisanal. . . . In the late nineteenth century, the business of stopping spoilage [as with that for debating purity] became less visible and more controversial" (4–5). Smith-Howard, *Pure and Modern Milk*, adds important commentary on the changing nature of food freshness and purity with the example of milk. Also see Petrick, "Arbiters of Taste" for a food-studies analysis of the cold chain, taste, and purity.

16. Bailey, *Holy Earth*, 101. For works on character and culture in the nineteenth century, consult Haltunnen, *Confidence Men and Painted Women*; Orvell, *Real Thing*; Fabian, *Card Sharps, Dream Books, and Bucket Shops*; Lears, *Fables of Abundance*; Cook, *Arts of Deception*; and Salazar, *Bodies of Reform*. See Worster, *Nature's Economy*, for a broader discussion of concepts of nature (and ecology) even before the era of adulteration.

17. Burnett, *Plenty and Want*, 99.

18. See Cronon, *Nature's Metropolis*; Freidberg, *Fresh*; Smith-Howard, *Pure and Modern Milk* for further exploration of similar points.

19. For a good entry point to a discussion about networks and chains, see Hamilton, "Analyzing Commodity Chains."

20. LaBerge, *Mission and Method*, 199.

21. Hamlin, *Science of Impurity*, 215.

22. This is a history that Porter, *Trust in Numbers*, and Scott, *Seeing Like a State*, have done well to explain.

23. See Lears, *Fables of Abundance*; Leach, *Land of Desire*; Hoganson, *Consumer's Imperium*; Balleisen, *Fraud*; and Rutherford, *Adman's Dilemma*.

24. A series of texts followed Accum in the United Kingdom. For a sampling of the second wave of works, see Pereira, *Treatise on Food and Diet* (1843); Marcet, *On the Composition of Food* (1856); Hassall, *Adulterations Detected* (1857); Moleschott, *Chemistry of Food and Diet* (1860); and Hassall again with *Food* (1876). Texts in the United States soon followed in parallel. For those in just the fifty odd years after Accum, see Cutbush, *Lectures on the Adulteration of Food* (1823); Beck, *Adulteration of Various Substances* (1846); Byrn, *Detection of Fraud and Protection of Health* (1852); Hoskins, *What We Eat* (1861); and Smith, *Foods* (1876). Chapter 2 provides a further explanation and accounting of these texts.

25. Britain passed its first attempt at national regulation with the Food Adulteration Act of 1860. See Teuteberg, "Food Adulteration and the Beginnings of Uniform Food Legislation," and Burnett and Oddy, *Origins and Development of Food Policies in Europe* for further detail about European policy. On Wiley, the first and still main biography is Anderson, *Health of a Nation*. Blum, *Poison Squad*, offers a recent work with Wiley as the central character. Also see Okun, *Fair Play in the*

Marketplace; Young, *Pure Food*; and Coppin and High, *Politics of Purity* for more detail on Wiley.

26. For readers old enough to remember stirring a yellow color pack into their store-bought white margarine—a practice still common into the mid-twentieth century—this was the first broad attempt at coloring laws that required margarine to be sold uncolored. Also, notice the bandage on the neck of the oleomargarine hydra head on the 1887 frontispiece to this book, a feature included by the artist in recognition of the attack made against the oleo offense in 1886.

27. See *Pure Adulteration* Digital Companion Site, https://purefood.lafayette.edu.

28. The statistics on national proposals come from an analysis of records at the NARA Record Group 97, Box 16, College Park, MD. Also see Bailey, "Congressional Opposition to Pure Food Legislation," and Strasser, *Satisfaction Guaranteed*. The statistics on margarine-specific bills comes from an analysis of legislative activity by Matthew Plishka, Lafayette College. See *Pure Adulteration* Digital Companion Site, https://purefood.lafayette.edu, for the legislative maps.

29. See Freidberg, *Fresh*, for more context about how our understanding of freshness and decay was itself undergoing a major historical transition.

30. By the dawn of World War II in 1940, the rural population had dropped from 74 percent to 43 percent; agriculture was 25 percent of the labor force. Figures from the United States Census Bureau and Carter et al., *Historical Statistics of the United States*. Also, see Cohen, "Three Peasants on Their Way to a Meal" for a sketch of the changes from agrarian to consumer society between the 1850s and 1910s.

31. Chandler in United States Congress, *Imitation Dairy Products*, 67–81.

32. *San Francisco Daily Morning Call*, "Oleaginous Paraf," February 11, 1876.

33. *El Comercio* (Lima, Peru), August 20, 1878.

34. *Daily Star and Herald* (Panama City, Panama), December 18, 1879. One account admitted that his "artificial butter" plans were overwrought, as he "only managed to obtain a loss of $17,200." From "The Sage of Barcelona" pamphlet, page 18. Credit and thanks to Ana Luhrs for the translation.

35. *New Orleans Daily Democrat*, December 9, 1877, 4; *Daily Star and Herald* (Panama City, Panama), October 18, 1879.

36. *Daily Star and Herald* (Panama City, Panama), October 18, 1879.

Chapter Two

1. La Rochefoucauld from *Maxims* 83, no. 256, as quoted in Magill, *Sincerity*, 62; Beecher, as quoted in "Adulterations of Coffee and Pepper," *Hunt's Merchant's Magazine and Commercial Review* 24, no. 3 (March 1851): 395.

2. Cook, *Arts of Deception*, 16–17.

3. Beecher, too, was born into the young republic, the son of temperance advocate and Calvinist evangelist Lyman Beecher and the product of a culture where deliberations on public character framed a life's work. His older sister Catharine was a prominent author of books on domestic economy, cooking, and kitchen management, guidebooks intended to uphold the character of the household. (She returns briefly in chap. 3.) His other older sister, Harriet, would write *Uncle Tom's Cabin* and coauthor books with Catherine. Henry's own biography saw him suffering a great public downfall in 1875 for an adultery scandal. After that, a wary public came to understand his public proclamations for purity and sincerity as perhaps

concealing the truth of his inner life. The surface did not match the interior. The Beecher sisters and their brother Henry published their works from the tradition of the protestant virtues of their father, Lyman Beecher, a preacher who brought his principles to New England through years of studying and teaching Scottish philosophical tenets that were at that time but a generation or two removed from the Enlightenment figures David Hume and Adam Smith. The philosopher William James would introduce a distinctly American, nineteenth-century pragmatist tradition to these questions after the Civil War. That American pragmatist tradition, put into practice by the likes of John Dewey, took aim at exactly this assumption, that surface indicates substance or that beliefs are true if they match reality. "Pragmatists," Louis Menand writes, "think that the mistake most people make about beliefs is to think that a belief is true, or justified, only if it mirrors 'the ways things really are.'" See Menand, *Metaphysical Club*, 356.

4. It also followed from the fracturing of older guild-based structures of labor and production. That is, the newer modes of production bringing people to the factories outside the home that Jefferson worried over in the 1780s were indeed causing friction in trust, connection, and familiarity. A British economist writing in 1812 touched on the relationship between industry and food when he foresaw a future of laborers working beyond the home for wages. They would, he wrote, "be converted into a body of men who earn their subsistence by working for others [and will] be under a necessity of going to market for all they want." New workplaces stretched the bounds of prior home-based, cottage industries and challenged the policing of foods within previously circumscribed communities. To the working class, Engels wrote, "fall all the adulterated, poisoned provisions." Richard Price, *Observations on Reversionary Payment* (1812), quoted in Stoll, "Metabolism of Society," 378. A few decades later Friedrich Engels wrote of the class dimension to food frauds, finding that "The lion's share of the evil results of these frauds falls to the workers." In a telling observation that ripples across twenty-first-century debates over food justice, Engels recognized the contrast with the poor, "the working-people, to whom a couple of farthings are important, who must buy many things with little money, who cannot afford to inquire too closely into the quality of their purchases." Engels, *Condition of the Working-Class*, 70.

5. Laudan, *Cuisine and Empire*, 250.

6. In the United States, one in sixteen lived in an urban center in 1800, one in six in 1850, and one in three by 1880, marking a nearly sixfold increase in under a century. United States Bureau of the Census, *Historical Statistics of the United States* (1976), series C89.

7. The nation's population more than doubled in that period, from 38 to 92 million.

8. The observation of the loss of a bounded system over time is a commonplace in studies of food adulteration, especially in the United States. It rests heavily on the physical-distance-leads-to-adulteration equation. For many who have examined pure food debates as exemplars in modern bureaucratic regulation, the shift from an agrarian to a manufacturing culture in the Western world was precipitated by a shift away from local knowledge and control. For similar attributions of distance as the explanation for adulteration, see Okun, *Fair Play in the Marketplace*, ix; Young, *Pure Food*, 35; Orvell, *Real Thing*, xvii; Strasser, *Satisfaction Guaranteed*, 255; Vileisis, *Kitchen Literacy*, 6; and Atkins, *Liquid Materialities*, xvi–xviii.

9. Horowitz, *Kosher USA*; Plato, *Laws*, 11.917; Wiley, LOC archives. Long after, in medieval Europe Charlemagne implemented legal codes in 802 CE to stave off wine adulteration. Eisinger, "Lead and Wine."

10. On French butcher laws, see Ferriéres, *Sacred Cow, Mad Cow*, chap. 2; Plowman as quoted in Wilson, *Swindled*, 85; for the Nuremberg story see "Adulteration of Food in the Middle Ages," *Sanitary News*, November 1, 1884, 11.

11. *Advice to the Unwary, or, An Abstract of Certain Penal Laws Now in Force against Smuggling in General and the Adulteration of Tea* (London: E. Cox, 1780), 2, 18. "Terra Japonica" is an extract from acacia trees brought from Indian lands through trade with the East.

12. Trilling, *Sincerity and Authenticity*, 13.

13. Trilling, *Sincerity and Authenticity*, 13. Kant's work may be the signature example of moving the seat of moral decision making out of the realm of the abstract theological and into the mind of the individual with will and reason. The very invention of human will, autonomous moral agency, and deference to rational decision-making powers of the mind grew from a host of philosophical works, most famously Kant's. See Rohlf, "Immanuel Kant."

14. Magill, *Sincerity*, 108. Shakespeare tackled the expression of earnest expression in his work, as sincerity made its way beyond the church. Even more, as one modern critic has argued, *all* of Shakespeare's "tragic heroes have the same concerns about appearance and reality, performance and inner life." See Tim Parks, "In Search of Authenticity," NYR Daily, *New York Review of Books*, February 4, 2015, http://www.nybooks.com/blogs/nyrblog/2015/feb/04/novels-search-authenticity/.

15. Arnaud, *An alarm to all persons*. A 1756 treatise by a British instrument maker helped readers seek out the "genuine" article by detecting counterfeits. See Bradford, *The gentleman's and trader's guide*. As well, a medical guide by a British pharmacist in 1772 showed readers how to understand the role of adulterated foods as they stood against genuine and true consumables. Hill, *Virtues of Wild Valerian*.

16. A widely reprinted British guide to farriery first published in 1737—this is on horseshoeing and animal care—argued for genuine treatments and against the adulteration of medicines for animals. In this case, what might seem like an aside, the care of horses, was a central feature of community life and agrarian activity. Bracken, *Farriery Improved*; Croker, *Complete Dictionary of Arts and Sciences*.

17. Richard Steele, as quoted in Magill, *Sincerity*, 76–77.

18. Johnson, *Dictionary of the English language*. Another guidebook explained how to gauge drugs and spices from "Asia and the East Indies" and, more to the point, ways to distinguish the "characteristics of those that are genuine, and the arts practised in their adulteration." Steel and Draper, *Portable instructions*, 1.

19. Markham, *Syhoroc*, 41.

20. Not everyone thought artifice was a bad idea, of course, as certainly innovators and factory owners claimed the benefits of artifice. But the opposite, the worry over artifice, was growing to gain a modern foothold. Misa, *Leonardo to the Internet*, provides a primer on the various versions of industrial revolutions.

21. Manufacturing methods for all manner of production increased in scope, scale, and complexity. In the United States, this was a second industrial revolution,

one following on the heels of the earlier versions in Britain that Jefferson responded to. Again, see Misa, *Leonardo to the Internet.*

22. Emerson, "Friendship," 140. Regulating the process of baking and selling bread has for so long shown that worries over food are about power and control, too, not mere bodily nutrition. See Fullilove, *Profit of the Earth,* for the context of a longer historiography about markets, control, and protest; E. P. Thompson, "The Moral Economy of the English Crowd in the Eighteenth Century"; L. Tilly, "The Food Riot as a Form of Political Conflict in France"; and C. Tilly, *The Politics of Collective Violence.* Amy Bentley and Christy Spackman write that public protest over food was "a preindustrial expression of collective action, as a gendered form of collective protest, and as a form of nationalistic display and identity tied to the consumption of material goods." Bentley and Spackman, "Food Riots: Historical Perspectives." Also see Aaron Bobrow-Strain, who writes, "Accused—often with good reason—of false weights, grain hoarding, hunger profiteering, and cutting flour with cheap whiteners like chalk, alum, or borax, bakers earned dubious reputations over the centuries. When things went wrong in town... the baker often got the blame." Bobrow-Strain, *White Bread,* 6.

23. Halttunen, *Confidence Men and Painted Women*; Salazar, *Bodies of Reform,* 1, 193. Also see Orvell, *Real Thing.*

24. The first quote is from McWilliams, *Revolution in Eating,* 3, and the second is from Bushman, *American Farmer in the Eighteenth Century,* x.

25. Vileisis, *Kitchen Literacy,* 31.

26. Simmons, *American Cookery,* 9.

27. See Vileisis, *Kitchen Literacy*; for the European context, see Atkins, Lummel, and Oddy, eds., *Food and the City in Europe since 1800.*

28. An Enemy to Fraud and Villainy, *Deadly Adulteration and Slow Poisoning Unmasked,* 15, 138. Also see Ferrières, *Sacred Cow, Mad Cow,* and, for further detail about the changing concepts of taste and flavor in the early twentieth century, Berenstein, "Flavor Added," and Hisano, "'Eye Appeal Is Buy Appeal.'"

29. Ferrières, *Sacred Cow, Mad Cow,* 71.

30. It means that appearances properly represent reality. The mimetic impulse became a dominant value of knowledge making in the nineteenth century. The inner nature of the individual would be reflected in outer material possessions. As with brands. Or, put the other way around, the outward material expressions of a person were windows onto their inner moral nature. "Material objects not only express the taste of those who own them, but shape the character of their owners and observers." Tonkovich, *Domesticity with a Difference,* xxv. This played out more spuriously at the time with phrenology, the premise of which was "that the outward form of the skull was an index to the individual's mental faculties, social proclivities, and personalities: to the phrenologist, in other words, outer equals inner." Orvell, *Real Thing,* 14.

31. Lears, *Fables of Abundance,* 83–84. In that cultural history of advertising's origins, Lears locates "early forms of 'consumer education'" in the later nineteenth century in the demand wrought by itinerant peddlers to "unmask commercial chicanery" (341). Also see Orvell, *Real Thing* and Vileisis, *Kitchen Literacy,* for further explorations on the same theme.

32. United States Bureau of the Census, *Historical Statistics of the United States* (1976) series C89.

33. Theodore Dreiser's *Sister Carrie* (1900) explored a similar theme. His urban ingenue, Carrie Meeber, made her starry-eyed way in Chicago after a country upbringing.

34. Cook, *Arts of Deception*, 16–17.

35. Reading, *Mark Inside*, 25.

36. Quotes from Melville, *Confidence-Man*, xvi, 8, and 21, respectively. Frank and Charlie discuss Polonius at length in chapter 30, pages 238–55. Although the novel is less read or visible to modern readers, it is a frequent point of discussion for historians of the nineteenth century. See, for example, Mihm, *Nation of Counterfeiters*; Cook, *Arts of Deception*; Salazar, *Bodies of Reform*; and Reading, *Mark Inside*.

37. Not only did the novel from 1851 concern the difficulties of knowledge and understanding and of Captain Ahab's quest for the white whale and for his own identity, it also forced the visual metaphor of surfaces and depths I address in this chapter. Ahab's search for truth required looking below the thin skin of the ocean's surface. What is more, as many a modern reader has come to note, a great bulk of the novel attended not just to whale hunting but to the acquisition of illuminating and edible oil and the prevailing concern for pure, unadulterated oil. For instance, Melville illustrates the regal character of this prized commodity, spermaceti oil. Writing about the coronation process for a king, he has it that the "king's head is solemnly oiled" like a head of salad. "Certainly it cannot be olive oil, nor macassar oil, nor castor oil, nor bear's oil, nor train oil, nor cod-liver oil. What then can it possibly be, but sperm oil in its unmanufactured, unpolluted state, the sweetest of all oils?" (Melville, *Moby-Dick*, 133). Ahab's hunt for oil is one of honest acquisition, "so as to be sure of its freshness and genuineness, even as the traveller on the prairie hunts up his own supper of game." The whaleman burns only "the purest of oil," and the purest oil is that "in its unmanufactured, and, therefore, unvitiated state." It is untarnished by human contrivances, "sweet as early grass butter in April." Ishmael's understanding of purity put an exclamation point on the previous century's transition to defining purity and genuineness in distinction to artifice and manufacturing.

38. The first reference is in chapter 30, and the second one (to cottonseed) is in chapter 39.

39. See Sandage, "The Gilded Age" for an account placing Twain's work in relation to the later historiography of the era. White, *Republic for Which it Stands*, revises the periodization to 1865–1896.

40. It may be overstating it to say that analysis wouldn't have helped detect the true identity of a con man, given that this period also saw the rise of phrenology, a kind of study premised on the notion that the external shape of the skull revealed internal character. See Van Wyhe, *Phrenology and the Origins of Victorian Scientific Naturalism*; Fabian, *Skull Collectors*; and Pearl, *About Faces*.

41. Robertson, "An Essay on Culinary Poisons"; Farley, *London Art of Cookery*, 307. See also Wilson, *Swindled*, 80.

42. Accum's career was nurtured in the decades after a so-called chemical revolution ushered in new notions of atomic theory and laboratory design. It had its basis in instrumental practice and theories of the element; it brought an older tradition of pharmacy and workshop trades to a new academic and disciplinary phase, and those disciplinary and theoretical changes were key considerations for a debate over food identity. It may seem obvious from the vantage point of the twenty-first century, but in the early nineteenth century molecular and elemental theories were

still in the making. Detecting the proper identity of a substance required some idea of what substances were supposed to be there in the first place. Without a chemical theory of substances there could be little to identify. See Brock, *Fontana History of Chemistry*, and Schickore, *Microscope and the Eye*.

43. Filby, *History of Food Adulteration and Analysis*, 17–18.

44. Filby, 19.

45. See Cotsell, *Companion to Our Mutual Friend*.

46. See Ferrières, *Sacred Cow, Mad Cow*; Atkins, Lummel, and Oddy, *Food and the City in Europe since 1800*; Oddy and Drouard, *Food Industries of Europe*, on Brussels, Paris, German cities, and London.

47. As cited in Atkins, Lummel, and Oddy, *Food and the City in Europe since 1800*, 92.

48. See Young, *Pure Food*, for more on the early lineage of these texts in the United States. For France, consult F. Aulangier, *Dictionnaire de substances alimentaires indigenes et exotique et de leur properties* [Dictionary of native and exotic food substances and their properties]; in Prussia, *Der Chemiker fur's Haus* (1830), as cited by Ferrières, *Sacred Cow, Mad Cow*, 276; in England, Pereira, *Treatise on Food and Diet*.

49. Chemist, *Domestic Chemist*.

50. See Cohen, *Notes from the Ground*.

51. Davy was a renowned chemist of the early century, eventually rising to become head of the Royal Society. His contributions were monumental; his work on agricultural chemistry was a pillar of the acclaim. On Davy, see Golinski, *Experimental Self*. Liebig did not compose a work on the adulteration of food directly. Much of his chemistry was intended to explain how to do things rather than how to spot people doing improper things. See Brock, *Justus von Liebig*, for more detail. Quote from Finlay, "Early Marketing of the Theory of Nutrition," 50.

52. His work has been well canvassed by historians of public health and adulteration. The secondary literature helps explain the rise of public analysts, the invention of modern nation-state food governance, and the ways scientific efforts followed from a cultural inertia for locating truth beyond surfaces. Hassall's early work arrived in tandem with publications by the upstart medical journal *Lancet*. For broader histories placing Hassall and the *Lancet* into their nineteenth-century context, consult Rosen, *History of Public Health*; Hamlin, *Public Health and Social Justice*; and Porter, *Greatest Benefit to Mankind*.

53. Hassall's work was not groundbreaking for the fact of microscopy but for his work in connecting the small scale of the unseen with the large scale of public health. Antonie Van Leeuwenhoek was recording observations of "animalcules" in the 1670s. See Lane, "Unseen World," for more on early discussions of "little animals."

54. See Tomes, *Gospel of Germs*; Hamlin, *Cholera*; Whooley, *Knowledge in the Time of Cholera*. Also note that many governments were busy managing the daily realities of market interactions and food identity. Under Napoleon's direction, the French had initiated broad programs for public health in the first half of the century. By midcentury, the British began to build their structure of similar public health governance. German states, not unified until 1871, nonetheless sought similar measures. Belgian, Dutch, and Italian measures worked in kind to more effectively manage the public's trust of food. See Atkins, Lummel, and Oddy, *Food and the City*

in Europe since 1800; Oddy and Drouard, *Food Industries of Europe*; and Burnett and Oddy, *Origins and Development of Food Policies in Europe.*

Chapter Three

1. Richards, *Appleton's Cyclopedia of American Biography* (New York: D. Appleton, 1888–1900), 5:239; Talbot, "Ellen Richards Obituary"; Shapiro, *Perfection Salad*; Biltekoff, *Eating Right in America.*

2. Thus, while I am surprised Richards is not better known in the textbooks and lectures of today's classrooms, I wouldn't suggest her work was without its problems. Hoganson, *Consumers' Imperium*, 122; Shapiro, *Perfection Salad*; and Biltekoff, *Eating Right in America*, provide more context.

3. By the time Frazier and Wealthy were regulars in the 1840s, the store had a few dozen Paiges on its rolls. Lucius Paige, an 1850 Harvard graduate and Frazier's fourth cousin, penned the town's history in 1883 for the Massachusetts Historical Society. He recorded a genealogy of seventy-six Paiges that extended as far back as before the town's official incorporation in the 1740s. Paige, *History of Hardwick*, 433–50. Material about Frazier and Wealthy Paige in this and subsequent pages is drawn from business records of the Mixter family and the Knight family, 1793–1905, Mss: 77 1793–1903, HBS.

4. Paige, *History of Hardwick*, 302.

5. Thoreau, *Walden*, 45.

6. By the mid-1800s, the town developed ties to the textile and paper mill industries that were common to the region. Those new employment opportunities beyond the home helped Hardwick's population double in the years Frazier shopped at the store. Hardwick reached its historical peak of thirty-five hundred people in the early 1900s. Paige, *History of Hardwick*, 306–7.

7. More than mere lined sheets of paper, account books were complicated financial instruments. See Rockman, "What Makes the History of Capitalism Newsworthy?" They tabulated daily market activity in neat columns and punctuated lists while opening a window onto the store manager's control and authority as a crucial node in the food chain. Hardwick's general store joins a host of examples from antebellum Southern merchants to the backcountry of colonial Virginia or frontier Kentucky to inland Pennsylvania and to Upper Canada that show the degree of integration in Atlantic markets throughout the century before manufactured foods stocked country store shelves. See Spellman, *Cornering the Market*, and Byrne, *Becoming Bourgeois*, 3. Other accounts of storekeeping include Atherton, *Southern Country Store*; Carson, *Old Country Store*; Perkins, "Consumer Frontier"; Martin, *Buying Into the World of Goods*; Wenger, "Delivering the Goods," and *Country Storekeeper in Pennsylvania.*

8. Wenger, *Country Storekeeper in Pennsylvania*, 45.

9. Paige, *History of Hardwick*, 151–54.

10. Customers relied on the integrity of the shopkeeper to maintain confidence in the authenticity of the food. But businesspeople were turning at just this time "from personal criteria for trustworthiness to formal, institutionalized criteria." Rather than simple "nostalgic notions of trust" that the hackneyed image from television's version of *Little House on the Prairie* might suggest, modes of exchange backed by brands, endorsements, and certified claims to purity grew to anchor new

distribution networks for food and dry goods. Spellman, "Trust Brokers," 280. Also see Lears, *Fables of Abundance*.

11. Spellman, *Cornering the Market*, chap. 5; Deutsch, *Building a Housewife's Paradise*, chap. 2.

12. Bussing, *Grocerdom*, 3.

13. Simmonds, *Practical Grocer*, 13.

14. Thomas DeVoe, a nineteenth-century chronicler of the history of public markets and butchers, devotes two volumes to his vast survey. See DeVoe, *Market Book*. Also consult Lobel, *Urban Appetites* and McNeur, *Taming Manhattan*.

15. Lobel, *Urban Appetites*; Baics, *Feeding Gotham*.

16. Ernst Haeckel had coined the term *oekology* a quarter century earlier. Richards, who knew German and Greek, wrote the German scientist for permission to use the term and borrowed it anew for work in the 1890s, after which the Boston Globe credited her for the "new science" of ecology. Swallow, *Remarkable Life and Career of Ellen Swallow Richards*, 93. This and subsequent paragraphs are also indebted to MIT's Collection on Ellen Swallow Richards, MC 659, Institute Archives and Special Collections, Massachusetts Institute of Technology, Cambridge, MA.

17. See Swallow, *Remarkable Life and Career of Ellen Swallow Richards*: "Small, compactly built woman" (xiii); "as a small child" (6); chapters 2–4, passim; "right living" (62). See also Tonkovich, *Domesticity with a Difference*.

18. Richards, "Paper on the Adulterations of Groceries," 59.

19. Richards, *Chemistry of Cooking and Cleaning*, viii. Coppin and High, *Politics of Purity*, note that "The seriousness of the servant problem should not be underrated; between the Civil War and World War I, discussions of this problem occupied more space than any other topic in women's magazines. Up to this point, the middle-class wife had done more supervising than preparing of food. As the nineteenth century progressed, the dual problems of a lack of knowledgeable servants and the changing food industry resulted in what middle-class women's magazines called the 'tyranny of the kitchen'" (26–27).

20. "Minor Notices," *Critic* 5 (April 24, 1886): 205.

21. "New Books," *Science* 7 (February 12, 1886): 154.

22. Richards, *Food Materials and Their Adulterations* (1886), 1.

23. Richards, 5. See Vileisis, *Kitchen Literacy*, for further discussion on the theme of lost knowledge.

24. *Testimony in Regard to the Manufacture and Sale of Imitation Dairy Products: Hearings before the United States Senate Committee on Agriculture and Forestry*, 49th Cong., 4910 (1886) (statement of Rep. Lewis Beach). She also wasn't alone in the view that agrarian lives were slipping away, as the Granger movement and populist campaigns were similarly seeking to address challenges to agrarian livelihoods. On the Granger movement, see Woods, *Knights of the Plow*; Summerhill, *Harvest of Dissent*; White, *Republic for Which It Stands*.

25. Young, *Pure Food*, 406. Horton and Company circular from Chandler Papers, MS#0209, July 27, 1878, CUA.

26. Richards, *Food Materials and Their Adulterations* (1886), iv.

27. Shapiro, *Perfection Salad*, provides a fuller exploration of this literature. Laudan, *Cuisine and Empire*, 253–60 provides more global analysis of moral, religious, gendered, and nutritional efforts at household management.

28. Biltekoff, *Eating Right in America*, 27, 21. Also see Strasser, *Never Done*, and *Satisfaction Guaranteed.*

29. Mason hearings.

30. Young, *Pure Food*, 398, 400. R. M. Littler, president of the Chicago Produce Exchange, added to the point: "we men who suffer and endure everything and have endured everything that we may have a country and have a commerce, if we are to be crushed by this hideous fraud, then I say, God bless America"; Miller hearings.

31. Hoganson, *Consumers' Imperium*, notes that "even the writings that celebrated foreign foods fueled anxiety, for they reminded those wary of foreignness just how close the foreign was—across the harbor, down the street, in the pantry, on the table" (122).

32. Soap, Box 6, Warshaw Collection, NMAH. The anthropologist Mary Douglas took the same concept of dirt and boundaries as a means to examine cultural relations in her influential study of taboo in culture, *Purity and Danger*. Her work held purity and danger apart, noting "accusations of causing dirt and defilement are weapons against disorderly behavior" (as quoted in Mody, "Little Dirt," 8). Douglas's argument for the work of cultural demarcation suggested to some an entirely nonmaterial idea, that decisions about food identity are strictly cultural ones. Others have argued that the cultural construct approach of anthropologists like Douglas (and Norbert Elias) limits attention to material circumstances in food choice. I share that critique but don't want to throw the baby out with the bath water. It isn't one or the other—cultural "ideas" or material "stuff"—but a complicated mutual shaping of both. Among others, Nancy Tomes discusses how to avoid that either/or-ness in *Gospel of Germs*.

33. Dreydopple's 1882 ad campaign was "Light and Shade." The text for a rhyme telling of a young black child who tried and tried to become white was short but direct. "A mite of queer humanity, / As dark as a cloudy night, / Was displeased with his complexion, / And wished to change from black to white." He had tried regular soap, medicine, Sulphur Springs, an airtight sweat box, and then finally a box of Dreydopple Soap. That did the trick. From Soap, Box 2, Warshaw Collection, NMAH. Also see Loeb, *Consuming Angels*. Ivory soap wasn't any subtler. Their ad from the 1890s:

> We once were factious, fierce and wild
> To peaceful arts unreconciled
> ....
> But Ivory Soap came like a ray
> of light across our darkened way
> and now we're civil, kind, and good
> And keep the laws as people should
> ....
> And now I take where'er we go
> This cake of Ivory Soap to show
> What civilized my squaw and me
> And made us clean and fair to see.

The food and household purveyor's goal was to separate the clean from the dirty, to maintain the line between civilization and savagery.

34. Stowe, *House and Home Papers*, 233. Hoganson, *Consumers' Imperium*, 126. Bobrow-Strain, *White Bread*, 7. On prison bread, see Fouser, "The Global Staff of Life." For an analysis of the racial and ethnic morality of consumption in the nineteenth century, see Tompkins, *Racial Indigestion*, and for a recent account of the historically consistent association of cleanliness and whiteness, see Zimring, *Clean and White*.

35. The "home front" and metaphors of domesticity extended "female influence outward to civilize the foreign" at the same time they "extend[ed] domesticity outward to the tutelage of the heathens and inward to regulate the threat of foreignness within the boundaries of the home." Kaplan, *Anarchy of Empire*, 32; the home front implied "a line that seals off domestic space from a foreign battlefield, but as a front, [also providing] a formidable line of attack and engagement" (16).

36. Downing is known now as a founder of landscape architecture. In the nineteenth century he was more fully read as a leader in the integrated pursuits of gardening, architecture, and horticulture. Downing, *Architecture of Country Houses*, v, vii.

37. Richards, *Food Materials and Their Adulterations* (1911), 4. Richards, like her contemporaries, offered this as an observation of common sense.

38. Brillat-Savarin, *Handbook of Dining*.

39. Figures compiled from *Boyd's Co-Partnership and Residence Business Directory of Philadelphia City* (Philadelphia: C. E. Howe, 1859–60, 1888); Z. F. Williams. *Baltimore's 400 Business Directory* (Baltimore: Press of the Friedenwald Co., 1887, 1895); *The Combined Business Directory of Philadelphia, Baltimore, and Pittsburgh* (New York: Curtin, 1893–1894); *Wilson's Business Directory of New York City* (New York: H. Wilson, 1869–1870); *Gopsill's Philadelphia Business Directory* (Philadelphia: J. Gopsill, 1869); Richard Edwards, *Chicago Census Report and Statistical Review, Embracing a Complete Directory of the City* (Chicago: R. Edwards, 1871); Thomas Hutchinson, comp., *Lakeside Annual Business Directory of the City of Chicago* (1881). In addition, Baics, *Feeding Gotham*, includes a wealth of data on bakers, butchers, and grocers graphed, charted, and mapped for the period immediately preceding the Civil War.

40. Richards, *Food Materials and Their Adulterations* (1886), iv.

41. See Baics, *Feeding Gotham*, for more on the trend toward private sellers over public markets.

42. "The Cornerstone Laid: New Home of the New-York Retail Grocers' Union," *New York Times*, February 3, 1893, 9; and Bussing, *Grocerdom*. Love of coffee seems to be the only remaining point of continuity with the building. As of 2019, a Juan Valdez Café occupies the ground floor at 138 East 57th.

43. Ward, *Grocer's Companion and Merchant's Handbook*, 6.

44. For a broader assessment of the farm-to-fork life cycle that was implicated in the grocers' blame game, see Belasco and Horowitz, *Food Chains*.

45. Ward, *Grocer's Companion and Merchant's Handbook*, 5–6.

46. Ward, 6.

47. Felker, *What the Grocers Sell Us*. A Chicago manual compiled in 1888 certainly confused the profession's broader claims by including instructions to its readers on *how to* adulterate. "To adulterate or cheapen cost" of baking powder, they advised, use Terra Alba "by mixing it in place of starch or flour filler." The entry on chocolate explained that grocers should "use plenty powdered sugar and

corn starch" to make sweet chocolate out of their stock of French chocolate. *Grocers' Manual*, 12, 90.

48. Thurber was half of Thurber and Whyland, a wholesaling giant in Manhattan. He had accrued enough business to concern himself with regional and federal politics. He wrote and spoke on foreign trade, monopolistic practices, tariffs, colonial missions, and import treaties.

49. The *National Provisioner*, a rival paper in Chicago, was less friendly with Thurber.

50. Bussing, *Grocerdom*, 12.

51. McNeur, *Taming Manhattan*, 136, 150. "Swill milk" controversies flared up in the 1850s in Manhattan when *Frank Leslie's Illustrated Paper* exposed corruption in the dairy trade linked to waste practices from butchers and meat-packers. "Startling Exposure of the Milk Trade of New York and Brooklyn," *Frank Leslie's Illustrated Newspaper*, May 8, 1858; McNeur, 153.

52. The cheap milk was a bargain at 6 cents a quart, and thus the problems of low quality fell mostly to the poorer classes. Yet as McNeur summarizes it, "Even the rich were swindled into purchasing swill or adulterated milk that was deceptively advertised as pure country milk. It was difficult for New Yorkers to tell if they were buying swill milk or something more pure." McNeur, *Taming Manhattan*, 153.

53. Berghaus, "Hydra-Headed Adulteration," *Rural New Yorker*, May 14, 1887, 8.

54. The sixth entry in the series portrayed stock traders on a platform held up by four pillars of dubious validity, one being bribery and corruption, another flattery, a third "seed distribution," and the fourth "oleomargarine." A farmer wielding a *Rural New Yorker*–brand axe has chopped down the margarine pillar, presumably on its way to upholding the platform with honest dairy butter. The seventh entry showed a Gulliver-sized giant farmer strapped to a tree, bound by the ties of Lilliputian-like grain and land speculators, lawyers, insurance agents, and "Dr. Quack" as thieves scurry away wearing sashes labeled "humbug" and "swindle."

55. "Inspection of Food and Drugs," *New York Times*, December 21, 1884, 8.

56. "A Grocery Order of the Future," *New York Times*, October 29, 1905, SM3.

Chapter Four

1. In its guise as most famous, most enduring, and most legislated adulterant of the era, margarine has also been one of the most written about since: Burns, *Bogus Butter*; Dupre, "'If It's Yellow, It Must Be Butter'"; Riepma, *Story of Margarine*; Snodgrass, *Margarine as a Butter Substitute*; Van Stuyvenberg, *Margarine*; Wiest, *Butter Industry in the United States*; Wilson, *Swindled*.

2. U.S. Department of State, *Reports from the Consuls of the United States*, no. 7 (Washington, DC: Government Printing Office, 1881), 695 (Britain and Holland); Deutscher Landwirtschaftsrat, *Archiv des Deutscher Landwirtschaftsrat* 1/2 (1876–1878): 129 (Germany); Lanzillotti-Buonsanti, *Dizionario delle science mediche e veterinarie* (Milan, 1875), 955 (Italy); Canadian House of Commons, *Official reports of the debates of the House of Commons* 21 (Toronto: R. Briggs, 1886), 547–45 (Canada); *Révue d'hygiène: organe du Ministère de l'hygiène et de l'assistance sociale* [The journal of hygiene: Organ of the Ministry of Health and Social Assistance] (Istanbul: M. Sadık Kâğıfçı, 1882) (Turkey); Dehra Dun and R. P. Sharma, *The Indian Forester* (1886), 569 (India); *Laws of Barbados, 1888–1889* (Barbados: Barclay and

Fraser, 1889), 40 (Barbados). With thanks to Josh Sanborn for the Russian "zemstvos" reference from Stolypin, *Search for Stability in Late Imperial Russia*, 365.

3. The consequences were notable not just for policy but for the politicians broaching the subject. Miller lost his congressional seat the next year, chased out of office by urban interests who considered his championing of the anti-oleo laws evidence that he was in the pocket of grangers.

4. That was a priority in the political agenda that frustrated others. That same week, the Office of Indian Rights Association (OIRA) was hoping to hear congressmen speak on the floor about railroad rights across Indian lands, a Sioux bill, and missionary relief in California. Instead they sat in the halls waiting as a "hobgoblin" took over. "Oleomargarine was its name." With its mere presence, "an affrighted House [would be thrown] into a most demoralized panic, in which it lost its head and all the reckoning as to its course." For a full week, the officers of the OIRA complained, "the whole House was cowered and shuddered as the finger of the granger has pointed to this ghastly phantom, and all Indian business has been forgotten." C. C. Painter, Indian Rights Association, *Oleomargarine versus the Indian* (Philadelphia: Indian Rights Association, 1886). Letter of June 3, 1886, MHS archives, Boston, MA.

5. "Oleomargarine and Counter Legislations," *AA* 4 (May 15, 1888): 206; "blessing" from J. S. W. Arnold, as quoted in Mason hearings, viii.

6. Miller hearings, 137.

7. The president of Baltimore's produce exchange was similarly drawn to the monster metaphor, considering that when Mège received his patent, "no one supposed it would assume the hydra-headed proportions it has developed into." Miller hearings, 10.

8. Miller hearings, 4868.

9. Smith-Howard, *Pure and Modern Milk*, 58.

10. Skillman, *Biography of a College*, 292–303.

11. McNeur, *Taming Manhattan*; Steinberg, *Down to Earth*, 157; and Atkins, *Animal Cities*.

12. There are a number of excellent cultural and environmental histories of milk and cheese. See DuPuis, *Nature's Perfect Food*; Goodchild, "Building "A Natural Industry of This Country'"; Paxson, *Life of Cheese*; Smith-Howard, *Pure and Modern Milk*; Valenze, *Milk*.

13. Smiley, *New Federal Calculator*, 75.

14. Seaman in Miller hearings, 4.

15. Most milk, cheese, and butter production derived from on-farm sites until a shift to local factories and creameries began in the last third of the century. "Although some butter was made in early cheese plants, the first commercial creamery was not established until 1861." United States Bureau of the Census, *Historical Statistics of the United States, 1789–1945*, 85.

16. Based on poundage per square mile, the rankings had Vermont first, followed by New York, Connecticut, Pennsylvania, and Ohio.

17. "As late as 1877 butter factories in Minnesota were considered something new." Jarchow, "The Beginnings of Minnesota Dairying," 112, and Seim, "The Butter-Margarine," 3.

18. Calvin Peck, US Patent 96,477, issued November 2, 1869; George Kirchhoffer, US Patent 139,796, issued June 10, 1873. The US Patent Office classification is

426/530, a section under "Food or Edible Material" (426) for "Renovated Butter" (530).

19. Paraf discussed safflower as a dye in his early meeting with Columbia's Charles Chandler. It was the point of conversation that brought the two together to discuss animal chemistry, organic dyes, and industrial novelties. Safflower alone often gave a pinkish or red hue to fabrics; new aniline dyes pioneered near Mulhouse were at the time displacing the earlier botanical dyes for textiles and edibles alike. Chandler, "Safflower," *American Chemist* 3 (November 1872): 170. Incidentally, there were suspicions that Paraf was swindling people with his aniline dye patents as well. Depositions in an 1876 case where a Swiss textile maker brought suit against the Merrimack Manufacturing Company of Lowell, Massachusetts, spoke to the difficulty in replicating Paraf's aniline black process. S. Dana Hays, "Aniline Black in Court," *American Chemist* 7 (July 1876): 6.

20. Garret Cosine, US Patent 173,591, issued February 15, 1876.

21. Van Norman, *First Lessons in Dairying*.

22. "Color: The general market requires that butter be as nearly the June color as possible throughout the year. If necessary to secure this, butter color may be used. The standard butter color is harm less and cannot be detected if used only in such an amount as is required for cream. The standard coloring matter is prepared from the coating of the annatto seed combined with a neutral oil. The color unites only with the butter fat, and more color will be required with rich than with thin cream. It should be added to the cream just before starting to churn. Twelve to fifteen drops of color for each gallon of cream that will churn out two and a half pounds of butter will be about right in the fall and winter, while less may be required during spring and early summer." Van Norman, *First Lessons in Dairying*, 47.

23. Morton in Miller hearings, 51.

24. Frederick in Cong. Rec., 49th Cong., 1st Sess., 4901 ("growing of grain, raising of cattle").

25. Seaman in Miller hearings, 4.

26. Young, *Pure Food*, 398.

27. Seaman in Miller hearings, 4. "And let me say here, gentlemen, that in Central Minnesota, which has been producing the best wheats that we have had for many years, who find there that the wheats are running out, as they did in the Genesee Valley some years ago, and farmers there are learning that those lands are as well adapted to dairying as they have been in past times to wheat growing."

28. From charts included in Miller hearings, diagrams 6 and 8, after p. 184. A light recovery by the time of the margarine hearing of 1886 had U.S. exports at about twenty million pounds.

29. Collier in Miller hearings, 185.

30. See Finger, "Harvesting Power."

31. Cong. Rec., 49th Cong., 1st Sess., 4867 (1886) (statement of Rep. Scott). The Suez Canal opened under French control in 1869 with little British support, but within the decade it had lured the interest of Great Britain's investment.

32. Quote from *Memphis Daily Appeal*, December 23, 1877; Chicago Board of Trade speculation from the *Sun*, July 26, 1883.

33. Details on company organization from the Commissioner of New Jersey, *Report of the Dairy Commissioner of the State of New Jersey*, 18; French statistic from Commercial Manufacturing Company, *A Brief History of the Mége Discovery:*

Oleo-Margarine Butter or Butterine; Microscopically and Chemically Analyzed [...] *Demonstrating Its Purity* (New York: Commercial Manufacturing Company, 1881), 18. Other details from UK Parliament, House of Commons, *Correspondence Respecting the Manufacture of Oleomargarine in the United States* (London, 1880–), 70–72.

34. See the *Oxford English Dictionary* (3rd ed., 2004), s.v. "oleomargarine." Also see Burns, *Bogus Butter*, 16–18. The Wisconsin Dairy Commission made that oleo-butterine distinction in its 1895 report.

35. International Library of Technology, *Packinghouse Industries* (Scranton, PA: International Textbook), sec. 41, 5.

36. See Cong. Rec., 49th Cong., 1st Sess., 4932 (statement of Rep. Grout) for "bastard butter."

37. Statistics from Stanziani, "Municipal Laboratories and the Analysis of Foodstuffs in France under the Third Republic," 108 (Paris); Commissioner of New Jersey, *Report of the Dairy Commissioner of the State of New Jersey*, 17; "Western Bogus Butter in Eastern Markets," *New York Tribune* March 13, 1886, 5; "Manufacture of Artificial Butter in the Netherlands," *Journal of the Royal Agricultural Society of England* 42 (1881): 435 (Vienna).

38. Not to mention the auxiliary industry of mechanical equipment used to mince, chop, heat, crush, and separate the fat from cows and the lard of hogs.

39. L.S.H., "Caul Fat Butter," *Wallace's Monthly*, July 1879, 432.

40. See Commissioner of New Jersey, *Report of the Dairy Commissioner of the State of New Jersey*, 20. During the 1905 Supreme Court case, the "Big Six" were Swift, Armour, Morris, Cudahy, Wilson, and Schwartzchild. The "Big Four" were the Chicago-based Swift, Armour, Wilson, and Cudahy. See Swift & Co. v. United States, 196 U.S. 375 (1905); Gordon, "Swift & Co. v. United States"; Young, *Pure Food*, 226; Chandler, *Visible Hand*, 295–302; Santlofer, *Food City*, 320–21.

41. Cronon, *Nature's Metropolis*, 212.

42. Cronon, *Nature's Metropolis*, 212, 445; Armour, *Packers and the People*, 154–58. By the early 1900s, there were 921 meatpacking establishments across forty-two states and the District of Columbia. Chicago may have been "the great bovine city" to some observers, but it was also home to "the biggest pig-killing concerns in God's Creation" to at least one Scottish visitor (David Macrae as quoted in Wade, *Chicago's Pride*, 98).

43. Young, "'This Greasy Counterfeit,'" 394.

44. US Patent 146,012 A, issued 1873.

45. Armour, *Packers and the People*, 206.

46. Hibbard in Miller hearings, 31. The expanding tentacles of the meat-packers led to more multiheaded imagery. Tentacles were a common metaphor at the time. Critics called the railroad trust an octopus for its growing and suctioning tentacles. Chicago novelist Frank Norris intended a trilogy about the furious convergence of industry and agriculture as tentacles in the later nineteenth century, managing *McTeague* and *The Octopus* before he passed away in 1902. As Norris put it, "If it is not a railroad trust, it is a sugar trust, or an oil trust, or an industrial trust, that exploits the People, because the people allow it." "The Octopus," a close cousin to the many-headed hydra, was an often-used reference to the Southern Pacific Railway of the early 1880s with whom farmers fought for land rights. Norris, *Octopus*, 221.

47. "Oleomargarine," *Chicago Daily Tribune*, February 21, 1877, 7.

48. United States Treasury Department, *Annual Report and Statements of the*

Chief of the Bureau of Statistics. Also see https://purefood.lafayette.edu for the mapped version of export flows.

49. "Manufacture of Artificial Butter in the Netherlands," *Journal of the Royal Agricultural Society of England* 42 (1881): 443. Jurgens had merged with another continental oil processing behemoth, the Van den Bergh company in 1927. Three years later they joined up with the Lever brothers to form Unilever.

50. W. J. Reader, *Unilever: A Short History* (London: Information Division, Unilever House, 1960), 15. Also see the *Quarterly Journal of the Agriculture Society of England* 17 (1881): 434; UK Parliament, House of Commons, *Correspondence Respecting the Manufacture of Oleomargarine in the United States* (London, 1880).

51. See "The War on Oleomargarine," *SM* 2 (January 1978): 2; 4 (July 1880): 200.

52. *Boston Evening Transcript*, January 1, 1898, 17; "War on Oleomargarine Vendors," *New York Times*, December 24, 1894; *OPD* 50 (July 19, 1896): 135. See also "Pushing the War on Oleomargarine," *Philadelphia Inquirer*, December 2, 1891, 1, and "The War on Oleomargarine: An Effort to Be Made to Repeal the Present Law," *Philadelphia Inquirer*, January 24, 1893, 7. The United States waged the war with particular commitment, but it wasn't an American phenomenon alone. The German Reichstag and the British House of Commons were also debating bills to "destroy the industry" in the 1890s.

53. Mason hearings, ix.

54. Miller hearings, 90. Babcock was famous at the time for the Babcock Tester used to verify the purity of milk, and he testified that the new product was safe and healthy.

55. Miller hearings, 105.

56. Commercial Manufacturing Company, Consolidated, "A Great Wrong," *New York Times*, March 22, 1881.

57. Miller hearings, 76–81.

58. Miller hearings, 56–57.

59. See Specht, "Failure to Prohibit"; Spiekermann, "Dangerous Meat?"

60. Mason hearings, ix. Harvey Wiley confirmed the tenor of the Mason report that legitimate adulteration was a healthy and benign modification of normal agricultural processes. "These mixtures of animal fats and vegetable oils, in my opinion, are perfectly wholesome and good in nutritious food." He made the common pitch to lower classes, too, pointing out that "persons in straitened circumstances prefer to use them." Ibid., 14.

61. "Oleomargarine," *Chicago Tribune*, February 21, 1877, 7.

62. *New York Tribune*, March 13, 1886, 5.

63. Miller hearings, 14 (Piollet); 40 (Washington, DC, spokesman); 250 (merchant); 62 (Blair).

64. "Waging War on Oleomargarine," *New York Tribune*, April 14, 1880, 8.

65. This is a point DuPuis, *Nature's Perfect Food*, and Smith-Howard, *Pure and Modern Milk*, explore in greater depth.

66. Miller hearings, 14 and 79.

67. Armour, *Packers and the People*, 208; The Beef Trust had Armour's back, stating that "oleomargarine butter is an industry valuable to the public in furnishing a pure and nutritious food product at comparatively low cost to the consumer," as quoted in "A Great Wrong," *New York Times*, March 22, 1881.

68. *Puck* was a satirical newspaper much like the *Onion* of today. The weekly

paper initially took its cues from its older sibling, the English paper *Punch*. *Puck* began in the 1860s as a German-language paper in New York, publishing its first several volumes solely in German before expanding to an English version in the late 1870s. To an audience of immigrant craftspeople, of butchers and bakers and shopkeepers and grocers, *Puck* spoke with a heavily illustrated paper to make its points clear and accessible to a semiliterate population.

69. O'Ferrall, Cong. Rec., 49th Cong., 1st Sess., 4921 (1886).

70. It is possible to look back to the Granger Movement and Progressive Era populism as broad social responses to economic and demographic change and see responses to oleomargarine fitting within those movements. But just as evident is to see those movements as bolstered and given specificity by the very example of cases like oleomargarine. Populism may not explain margarine, that is, but margarine helps explain populism. With bogus butter, the laboring agrarian class would take a hit in health, in wages, and in labor opportunities.

71. From the Capital City Dairy Company of Columbus, Ohio. See Box 1, Dairy, Warshaw Collection, NMAH.

72. Canola was not really a thing until the later twentieth century. It is a name invented in the 1970s by the Rapeseed Association of Canada to stand either for "Canadian Oil" or "Canadian Oil Low Acid," and it was "used by the Manitoba government to label the seed during its experimental stages." It is a variant of the rapeseed family of plants, a term less common now but referring to the family including turnips, rutabagas, and cabbages. See "Canola oil," Wikipedia, https://en.wikipedia.org/wiki/Canola_oil.

73. Miller hearings, 60–63. Charles Chandler was also in the hot seat as a witness for Blair's questions, at one point arguing over the validity of "red" oleomargarine. Blair wanted to press the contrast between "natural butter" and its yellow tint to the oleo. Chandler thought the oleo should be colored yellow so people would enjoy it. "It would be a great hardship," he replied, "to compel people who want to eat oleomargarine to have it red when they want it yellow." Ibid., 77.

74. See Dupré, "'If It's Yellow, It Must Be Butter,'" 355. *Puck* responded to the *Sun* sarcastically: "We are ardent admirers of our contemporary, *The Sun*... but our happiness [with them] would be complete if our esteemed contemporary would mention, incidentally, that [their margarine proposals] were PUCK's ideas.... We refer to an article which appeared in *The Sun* on December 24th, 1881, advocating a law compelling Oilymargarine to take a distinctive color." See "Oilymargarine," *Puck* 10 (January 4, 1882): 283.

75. See Wiest, *Butter Industry in the United States*; Snodgrass, *Margarine as a Butter Substitute*; Dupré, "'If It's Yellow, It Must Be Butter'"; "Oleomargarine in Minnesota," *New York Times*, December 22, 1897, 1.

76. Collins v. New Hampshire, 171 U.S. 30 (1898). "Oleomargarine Laws Invalid," *New York Times*, May 24, 1898, 5.

Chapter Five

1. Gantz and Company, *Manufacture of Cottonseed Oil*, 1; National Provisioner, *Directory and Handbook of the Meat and Provision Trades*, 381; Wrenn, *Cinderella of the New South*; Wesson, "Contributions of the Chemist to the Cottonseed Industry."

2. Atkinson, *Cheap Cotton by Free Labor*, 18. Twain, *Life on the Mississippi*, 306.

3. *Correspondence Scientifique* quote as reprinted in "Falsifications of Olive Oil," *OPD* 16 (October 29, 1879): 421; *Morning Call, Prosperous California*, 15–16; Bailey, *Holy Earth*, 101–102. See also Tompkins, *Cotton and Cotton oil*, for an early account of the industry.

4. Clemen, *By-Products in the Packing Industry*, 159.

5. *Scientific American*, January 10, 1891, 20; Charles Marie, "The Adulteration of Olive Oil," *American Journal of Pharmacy* 55 (January 1883): 25; Frank Moerk, "Lard Adulteration with Cotton-Seed Oil," *American Journal of Pharmacy*, 60 (November 1888): 573; Frederic Mather, "Waste Products: Cottonseed Oil," *Popular Science* 45 (May 1894): 104–8.

6. Wiley in Mason hearings, 16; Wrenn, *Cinderella of the New South*, 83; Weber and Alsberg, *American Vegetable-Shortening Industry*, 94; List and Jackson, "Battle over Hydrogenation."

7. See Sterne in Miller hearings, 256, for testimony on cottonseed oil. The industry of oil crushing, in retrospect, sounds quite like that for petroleum processors. Indeed, cottonseed or cotton oil predated petroleum and in many respects provided the infrastructural basis for the later petroleum oil industry. It was cottonseed crushers who developed not just an edible liquid product but a subsidiary industry of processing equipment, small-scale refineries, delivery pipelines, and shipping barrels to bring oil from southern cotton fields to the world. John D. Rockefeller did not abscond with cotton crushing plans to build his petroleum empire, but the later petroleum industry derives from patterns of oil delivery already in place with organic, edible oils.

8. This was in the Antilles, specifically. Jean Baptise Du Tertre, *Historie generale des Antilles habitués par les Francois* (Paris: Thomas Jolly, 1665).

9. J. Leander Bishop, Edwin T. Freedley, and Edward Young, *History of American Manufactures from 1608–1860* (Philadelphia: E. Young, 1861). The Moravians are a mechanically inclined Reformation sect from Czech Bohemia who settled in America in the 1740s. They built a trading post near the Lehigh Valley of Pennsylvania, about fifty miles north of Philadelphia. Existing records do not explain how they happened upon this new industry.

10. See Shields, *Southern Provisions*, 300–302.

11. "Cottonseed Oil," *Pee Dee Gazette*, December 1, 1820, 470.

12. J. Hamilton Couper, "Cotton Seed Oil," *Southern Agriculturist, Horticulturist, and Register of Rural Affairs* 6 (January 1846): 13. Couper continues: "The seed of the Sea-Island or long staple cotton weighs about 40 lbs. to the bushel. As it is less coated with fibre, the yield of oil to the bushel will be considerably greater than that of the upland [variety of cotton fiber].... The processes employed were such as were used in the Netherlands, France, and in America." David Shields notes that Couper had begun his experiments in the 1820s and that by the end of the 1840s he was out of the business.

13. Shields, *Southern Provisions*, 294.

14. Shields, 302–5. See Zallen, "American Lucifers," 2014, for further exploration of preelectric lighting oils.

15. These quantities of cotton came from over nine hundred counties in seventeen states and eight "Indian Territories" (in 1900). See University of Minnesota, Minnesota Population Center, *National Historical Geographic Information System*,

for derivation of data sets. Some records consider bales of four hundred pounds, others note five-hundred-pound bales. See Beckert, *Empire of Cotton*, for more details about the culture of the global cotton trade.

16. In the 1880s, the *Oil, Paint, and Drug Reporter* related industry-best yields at about thirty-seven gallons per ton, a figure that increased as the industry designed more efficient processing equipment in the coming years. American cotton producers would extract oil in the hundreds of millions of gallons by the turn of the century. "Close of the Cotton Oil Season," *OPD* 34 (September 5, 1888): 5. Couper's earlier measures were confident overestimates. He considered in antebellum Georgia that a bushel of seed (which is about thirty-two pounds) would yield a gallon of crude oil when pressed. That would be a yield of about sixty-two gallons per ton of seed. A gallon of cottonseed oil weighs about the same as a gallon of water (8.3 pounds).

17. Curtis, *Cottonseed Meal, Origin, History, Research*, 29.

18. Wrenn, *Cinderella of the New South*, 15, 19.

19. The growth of cottonseed crushing mills in figures 5.1 and 5.2 map neatly onto a specific soil region in central Texas. Geologically, zone 21 represents the western edge of the same coastal shelf that also defines the edge of the crescent-shaped black belt cutting across South Carolina, Alabama, and Mississippi. Zone 21 is the Black Waxy Prairie region, an area that alone is more than half the size of Mississippi. It has the soil conditions of "Black clay and clay loams from marly limestone" particularly favorable to cotton cultivation. "The Cottonseed Oil Business in Texas," *OPD* 19 (May 4, 1881): 591.

20. "Cottonseed Annual Report," *OPD* 12 (October 10, 1877): 12; Daniel C. Roper, "Cottonseed Products," in United States Census Office, *Bulletins of the Twelfth Census of the United States Published between November 1, 1901 and April 20, 1902* (Washington, DC: Census Office), 3. National Provisioner, *Directory and Handbook of the Meat and Provision Trades*, 183–87.

21. Labor questions by western investors and southern cultivators also helped lead this push into Texas. The abolitionist Edward Atkinson of Massachusetts offered a complicated take on the issues with *Cheap Cotton by Free Labor*. He ran textile mills, owned cotton fields, and managed labor pools with an interesting view of diet, funding, and working with his friend Ellen Richards to make nutrition scientific. Before that, he contrived a system of land cultivation that addressed the purported fear that emancipation would destroy the labor force—freed slaves would not work, so the argument went, and "without his cheap labor American cotton would be an impossibility." Atkinson thought that free white labor offered one solution; he also saw the German immigrant population of Texas as a good source of labor, lessening the worries over freed slaves. Furthermore, he thought that maximizing cottonseed oil from the cotton fields would help the economics of the labor arguments. With Atkinson's help, Texas became the leading cottonseed oil producer in the world by the early twentieth century. Williamson, *Edward Atkinson*, 7, quoting the London *Spectator*'s review of Atkinson's *Cheap Cotton by Free Labor*. See also Atkinson et al., *Science of Nutrition*, his book coauthored with Richards and part of the New England kitchen's history.

22. Mueller, "Slippery Business"; also see Mueller, *Extra Virginity*.

23. Shields, "Prospecting for Oil."

24. England's olive oil was officially exported from England to the United States, but it had derived from colonial holdings in the Middle East.

25. United States Treasury Department, *Annual Report and Statements of the Chief of the Bureau of Statistics.*

26. Importers paid between just under two dollars per gallon for salad oil in the 1870s; the price dropped toward the end of the century to about one dollar per gallon. Customers paid less for oil used in "other" applications, ranging from seventy cents to eighty-seven cents per gallon. Data from range of *OPD* volumes (1875–1900).

27. Reprinted in J. Burns, "Cottonseed Oil Manufacture," *SM* 5 (February 1882): 47.

28. Reported in "The Chemistry of Cotton," *New England Farmer* 62, no. 46 (November 17, 1883): 2.

29. "Do Plants Think?," *Puck* 14 (January 16, 1884): 314. Elsewhere, *Puck* took aim at cottonseed oil's incursion into the butter industry. While the Cotton Oil Product Company proudly advertised its Manteca de Semilla de Algodon (butter of cotton oil) in the 1890s, *Puck* thought the posturing as improvement was one of deception in its column on "The March of Progress." The cottonseed oil butter was only the latest false innovation unleashed on "unfortunate citizens." "The March of Progress," *Puck* 10 (September 21, 1881): 37.

30. The industry was seasonal, divided into summer and winter. Winter lard yielded higher returns. In a given year, 1886 for example, the US meatpacking industry processed 173 million pounds of lard from summer hogs and 215 million pounds from winter hogs. In the summer, they got a 13 percent yield from every one hundred pounds of pig—thirteen pounds per hundred. In the winter, they got closer to 33 percent—thirty-three pounds of lard for every one hundred pounds of processed pig. At the stock market in Chicago, traders sold pork by the barrel and lard in one-hundred-pound lots. Lard prices fluctuated over the latter decades of the century from a low of three dollars per one hundred pounds to a high of thirteen dollars. Osman, *The Price Current-Grain Reporter Year Book*, 50–57.

31. As quoted in Cronon, *Nature's Metropolis*, 226.

32. Wrenn, *Cinderella of the New South*, 28.

33. Wrenn, 15.

34. C. Wood Davis, "Corn and Cotton Seed: Why the Price for Corn Is So Low," *Forum* 24 (1898): 730–31.

35. United States Tariff Commission, *Survey of the American Cottonseed Oil Industry*, 7–8. Whereas the 1884 Encyclopedia Britannica did not include cottonseed among the top nineteen oils used for food, in its 1903 edition they listed cottonseed oil as the most valuable of all cooking oils. *OPD* 65 (September 9, 1904): 42.

36. "Cottonseed Oil Annual Report," *OPD* 16 (October 29, 1879): 419.

37. *OPD* 18 (August 4, 1880): 131.

38. *OPD* 19 (February 2, 1881): 140.

39. Twain, *Life on the Mississippi*, 244. Tariffs across Europe pushed shipments of cottonseed oil from one part of the continent to another. Italy had already increased their import tariff in 1881. That led to increased trade from the US South to France and Austria instead of Italy. But by the early 1900s, Germany and the Austro-Hungarian governments had imposed new import tariffs of their own. In Germany's

case, the trade barriers were motivated by their African colonial holdings. They sought to ramp up cotton production of their own. Raising the tariff on American imports encouraged their colonists to compete in the oil market. United States Congress, *Foreign Tariffs*, 202 (on Italy); True, *Cotton Plant*, 372 (on Germany); "Austria," *National Provisioner* 53 (September 18, 1915): 26 (on Austria).

40. In early volumes in the 1870s, the *OPD*'s editor, W. O. Allison, pitched the paper as "representing the Oil, Paint, Drug, Dyestuffs, and Chemical Trades." By the end of the 1880s, the list of audiences had expanded to include related fats and oils industries, including the *Oil and Paint Review, Weekly Drug News*, and *Soap-Makers' Journal*. See *OPD*'s front covers for January 1879 (vol. 15) and July 1888 (vol. 34). The similarly named *Paint, Oil, and Drug Review* (*PODR*), from Chicago, had comparable ambitions to become an organ for the new market. Their masthead included the common appeal that "Westward the course of Empire takes its way." The *PODR* was no friend of Allison or his *Oil, Paint, and Drug Reporter*, noting that while he was "a failure" as a broker in American Cotton Oil Trust Certificates, as "a renegade publisher of a renegade trade paper [he] was a first-rate success." *PODR* 5 (May 15, 1887): 9.

41. Part of it followed from competition from John D. Rockefeller's consolidated Standard Oil. To the Cotton Trust, it was "that monster fraud, the Standard Oil Co," a constant source of frustration and anger as the petroleum giant threatened cotton-based trade. *OPD* 21 (March 22, 1882): 565. Rockefeller was clever, though, as he placed some of his own men on the board of the Cotton Trust. In efforts to build cottonseed oil into a stable commodity and combat Standard Oil, the crushing-mill operators also joined forces to gain strength in numbers. With committees on general business, statistics, cooperage, finance, and arbitration, the crushers stood together for the foundations of a new industry. *OPD*, 17 (April 21, 1880): 470, 497–98, 502.

42. Fairbank had help from chemists in this regard. One of them, David Wesson, would go on to pioneer a process that removed the odor and colored tint of the raw oil, eventually founding his own company in 1899, Wesson Oil, with the Cotton Oil Trust's backing. Wesson then developed Snowdrift, a lard substitute made from cottonseed oil. It went on the market in the early 1900s, to be eclipsed a few years later by Crisco. Wesson Oil as a vegetable oil is still sold today, a subsidiary of Con Agra; cf. Olsen, "American Contemporaries—David Wesson" and Shields, "Prospecting for Oil."

43. Strasser, *Satisfaction Guaranteed*, 5.

44. Fairbank, *Cottolene*, 1892.

45. Cottonseed had a dual role to play with respect to soap in the nineteenth century. Newer, cheaper cotton textiles were actually washable, unlike wool, so that the same cotton that made soap possible and necessary provided the oil that was used to make the soap.

46. "Annual Report of the Chief Inspector of Milk and of the Chemist of the Board of Health of the City and Port of Philadelphia" for the year 1897 (Philadelphia: Dunlap, 1898).

47. *Grocers' Manual*, 46, 97.

48. "The Adulteration of Olive Oil," *SM* 3 (March 1880): 80–81; 5 (1882): 47–48.

49. "The Adulteration of Olive Oil," *SM* 3 (March 1880): 45.

50. "Lard and Leviticus," *AA* 5 (January 31, 1889): 50.

51. A proposal had sought to "establish in the Department of the Treasury a Bureau on Adulteration . . . and to regulate or prohibit the importation, manufacture, and sale of adulterated articles of food and drugs." But as a representative of one member of the National Board of Trade, the Chicago Board of Trade, put it, regulating lard to prevent "spurious lard" would also cut into the margarine business. National Board of Trade, *Proceedings of the Eighteenth Annual Meeting of the National Board of Trade: Held in Washington, January 1888* (Boston: George E. Crosby, 1888), 20–21.

52. Wiley in Mason hearings, 16.

53. Mrs. J. P. Walker, "Cottonseed Oil for Cooking Purposes," *OPD* 28 (December 2, 1885): 11.

54. Rorer, *How to Use Olive Butter*, 20–23.

55. Wrenn, *Cinderella of the New South*, xv.

56. As reported in "Lard and Cotton Oil," *OPD* 31 (March 9, 1887): 6.

57. La Compania de Productos de Aceite de Algodon, New York.

58. Companies such as the Union Oil Company of Providence, Rhode Island, had a similar point to make. They argued that they offered "Pure Salad Oil" as a "pure, highly refined Cottonseed Oil" because it was "identical with much of the so-called 'Pure Salad Oil' [olive oil] and superior to most of it." "Providence Pure Salad Oil Trade Card," 92. 229, Box 10: Foods, Pictorial Collections Dept. Advertising Cards, Hagley Museum, Wilmington, DE. David Wesson also worked on distinguishing pure and spurious lard. See Wesson, "Examination of Lard for Impurities."

59. In 1886 adulteration hearings, George Sterne described the deodorizing process for his congressional audience in his capacity as representative of the Chicago packinghouses. The process used chemical filtering equipment to remove the odor from animal fats and to neutralize the color of the oil. Senator Miller, the chair of the meeting, pressed him on the legitimacy of filtering, but Sterne explained that it was all quite normal. When the packers got the oil it still had an unrefined reddish tint, even after they pressed the dense cottonseed cakes. So their process filtered the oil "through bone dust and charcoal," put it under "high heat," and charged the presses "with superheated steam." They then mixed the new oil into other oils, like lard and margarine. An adversarial Senator Miller took issue with this, arguing that food and charcoal should not go together. The corruption, to his mind, was that the bone dust, charcoal, and superheated metalwork presses made the new spread unsafe and inappropriate. Sterne in Miller hearings, 256.

60. As quoted in "Adulteration of Lard," *AA* 5 (May 9, 1889): 220.

61. The Italian government knew adulteration was rampant. They advocated for olive oil as one of its biggest exports, seeking to preserve its reputation as pure and authentic. They wanted to show that they were better at catching the adulterators before they left the harbor than anyone. The vice president of the Italian Senate and the analytical chemist of the Italian customs coauthored "An Examination as to the Reliability of Certain Tests for Determining the Purity of Olive Oil," reviewing 70 oils with six separate tests. By the 1890s, the question was not whether or not growers mixed olive oil with other oils or detecting the contamination. So rampant was the adulteration that the customs agents had moved to the next level of deciding *which* of the many tests was most effective. See testimony of Anthonio Zucca in Mason hearings, 484–86.

62. Mueller, "Truth in Olive Oil."

63. For more detail, see Fairbank, *Cottolene*. See also Schleifer, "Perfect Solution," and List and Jackson, "Battle over Hydrogenation" for more context on oils and fats.

64. Clemen, *By-Products in the Packing Industry*, 159.

Chapter Six

1. "The True Business," *Wasp*, November 11, 1881, 306.

2. "What Fools These Mortals Be," *Brooklyn Daily Eagle*, January 9, 1889, 4. The same line adorned *Puck*'s masthead each issue.

3. "A 'Sweet' Affair: The President of the Grape Sugar Refinery Arrested," *Chicago Tribune*, June 21, 1882, 3; "Friend's Chicago Deal," *Chicago Tribune*, January 6, 1889, 6.

4. "A Mysterious Machine," *Macon Telegraph*, January 13, 1887, 5.

5. "Sieve" quote from "Keely Motor Sugar!," *Sun*, January 5, 1889.

6. "Hard Labor," *Daily Herald*, June 21, 1889.

7. "Sterling Syrup Works to Edison," December 23, 1888, from TEP, with thanks to David Singerman and Paul Israel.

8. "Claus Spreckels Declined," *Philadelphia Inquirer*, January 23, 1889.

9. See Beckert, *Empire of Cotton* on global cotton. Mintz, *Sweetness and Power*, remains a necessary starting point for studies of sugar, food, and culture. In recent years, a number of other studies use sugar as their anchor for examining empire, power, science, race, gender, and more. See, for example, Merlaux, *Sugar and Civilization*; Singerman, "Inventing Purity in the Atlantic Sugar World"; Mapes, *Sweet Tyranny*; Warner, *Sweet Stuff*; De la Pena, *Empty Pleasures*; Woloson, *Refined Tastes*; Rogers, *Deepest Wounds*; Coppin and High, *Politics of Purity*, chap. 9.

10. Even sucrose, the supposed "pure" sugar against which adulterated sugars were measured, was the result of construction. As David Singerman, "Inventing Purity in the Atlantic Sugar World," notes, "Sugar has not just been sucrose all along: it had to be made that way" (12). This chapter focuses mostly on the late nineteenth century with the example of glucose; Carolyn De la Pena's *Empty Pleasures* provides an account encompassing the full twentieth century and a number of cases.

11. Warner, *Sweet Stuff*, 109.

12. Frankel, *Practical Treatise on the Manufacture of Starch, Glucose, Starch-Sugar, and Dextrine*, 196.

13. American Grocer, as quoted in Wells, *Sugar Industry of the United States*, 17–19.

14. USDA reference is from Wiley, August 3, 1881, Box 1, RG 97, NARA,; "Inspection of Food and Drugs," *New York Times*, December 21, 1884, 8.

15. "Sham Lard and Other Shams," *New York Times*, March 2, 1887, 4.

16. "A Consummate Strategy," *Wasp*, September 16, 1881, 180.

17. De la Pena, *Empty Pleasures*, writes, "because sugar [from cane] contained calories and saccharin did not and sugar had its origins in plantation fields while saccharin had its origins in coal tar, individuals saw the former as natural, or connected to nature, and the latter as artificial, a product of the chemist's lab. . . . In the early twentieth century, sugar was as much the product of the field as it was of the factory" (14).

18. Richards, *Food Materials and Their Adulterations* (1886), 92. Wiley, August 3, 1881, Box 1, RG 97, NARA.

19. See Bobrow-Strain, *White Bread*; Fouser, "The Global Staff of Life."

20. See Mintz, *Sweetness and Power*; Merlaux, *Sugar and Civilization.*

21. Singerman, "Inventing Purity in the Atlantic Sugar World," 13–14, and Deerr, *History of Sugar*, 2:284.

22. To quote an observer from the 1860s, "The work of sugar making from cane juice is most anxious.... It admits of no irregularity, no laziness." William Reed, *The History of Sugar and Sugar-Yielding Plants* (London: Longmans, Green, 1866), 53. "For each kilogram of sugar produced," writes Thomas Rogers about the seventeenth-century process, "about fifteen kilograms of wood were burned, leading to the logging of 210,000 tons of wood each year." Hoeing a "cane field [was] a most laborious operation." So, too, was the harvesting that followed, usually in March and April. The cane stalks were cut close to the ground, where sugar concentration was highest. Once cut, the stalks were sorted and sent to boilers for processing. Operators filtered the sucrose from the broth to dry and separate it in pans. See Rogers, *Deepest Wounds*, 30–32.

23. See Woloson, *Refined Tastes*, 24–25. Also see Glickman, "'Buy for the Sake of the Slave'"; Fox, *Address to the People of Great Britain*, 4; and Hoganson, *Consumers' Imperium*, 121, for commentary on a 1909 effort by the National Consumer's League to campaign against slave-grown cocoa from the Portuguese islands of San Thome and Principe.

24. Faulkner, "Root of the Evil." Also see Schoolman, "Building Community, Benefiting Neighbors."

25. See Schwartz, *Tropical Babylons*; Rodrigue, *Reconstruction in the Cane Fields*; Follett, *Sugar Masters.*

26. Warner, *Sweet Stuff*, 70.

27. Mapes, *Sweet Tyranny*, 3. Also see Cram, *Statistical Diagram*, 1893. It might be nice to say the motivation for noncane sugar first took root to avoid the moral problems of slave-produced sugar. Alas, that was not the case. While free produce advocates eventually encouraged beet cultivation, the more pressing original motivation was European wars. In the early half of the nineteenth century during the Napoleonic Wars, blockades and dangerous seas limited shipment of Caribbean sugar cane into Europe, while Britain struggled to keep its East Indies trade route secure. After scattered fits and starts, government-supported chemists began tinkering with sugar derived from more locally sourced starches such as beets, sorghum, potatoes, and corn. A German chemist, Andreas Marggraf, had identified the process of sugar extraction from beets in the mid-1700s. In the early nineteenth century, Prussian state scientists further developed the earlier workshop practice.

28. Warner, "How Sweet It Is," 162. Wiley was initially sanguine in his 1885 USDA report (pp. 13 and 69) about the possibility that beet sugar could find a footing in the United States. He wrote that domestic activity paled in comparison to the "signal success" of European efforts. Yet he was also encouraged that "the ingenuity and enterprise of the present generation discovered, utilized, and developed the saccharine properties of the beet root." He thought the United States had a decided advantage over the too heavily taxed states of Europe. The "skill and patience" of American growers would help underwrite beet cultivation. And what is more, "wise legislation and generous bounties... encouraged the new industry."

29. Spiekermann, "Claus Spreckels." Spreckels was a German immigrant who began his career as a grocer in San Francisco. Soil conditions, rainfall, mean annual temperatures, and available fuel to run equipment and boilers all made it possible for him to envision his West Coast beet industry. He built his empire with connections to Pacific cane growers and a beet-sugar company that still operates today.

30. Warner, *Sweet Stuff*, 89, 94.

31. Mapes, *Sweet Tyranny*.

32. See Singerman, "Inventing Purity in the Atlantic Sugar World," 17–20 (quote on 19). See also Jones, "Making Chemistry the 'Science' of Agriculture."

33. Nieuwland, *Grangers' Glucose Co.*

34. Nieuwland, 1.

35. Patent Office of Great Britain, *Patents for Inventions*, xv.

36. Zea Mays, "Corn-Sugar," *Chicago Tribune*, March 3, 1877, 3.

37. Warner, *Sweet Stuff*, 112, 241. Warner's work provides the most developed summary of glucose operations at the time. This paragraph and the next are indebted to her work.

38. "Chicago Sugar Refinery," *Daily Inter Ocean*, (December 25, 1880), 3, and "Grape Sugar from Corn," *New York Sun*, July 21, 1880. On water-to-starch rations, 3 to 1 is the average of the five processes summarized by Frenkel, where some used 4 to 1 and some 2 to 1. For reference, the Buffalo Grape Sugar Company handled five thousand bushels of corn a day in 1880. See Cioc, *Rhine*, 112–24, for further details on chemical companies and sulfuric acid production.

39. W. H. Howell Company v. the Charles Pope Glucose Company, Court of Appeals of Illinois, Second District 61, Ill. App. 593.

40. The State of Iowa, Appellee, v. W. S. Smith, Appellant, 82 Iowa 423; 48 N.W. 727; 1891 Iowa Sup. May 1891. Also see Warner, *Sweet Stuff*, 113.

41. Hassall, *Food*, 220, 274.

42. Nichols quoted in *New England Journal of Agriculture*, April 1, 1882; the rural press is "The Glucose Fate," *Colman's Rural World*, May 26, 1881.

43. "Adulterations of Food: Glucose in Sugar and in Sirups," *Scientific American*, December 22, 1883; "Glucose Honey," *Scientific American*, February 1, 1879.

44. Candymakers were mad too. Confectioners were not happy that "the dangerous article [was] extensively used in the adulteration of candies." Blake, *Candy Making at Home*, 3. Quotes from "Grape-Sugar from Corn: Important Facts Brought Out in a Law Suit," *Chicago Daily Tribune*, July 26, 1880, 2, and *Sugar Cane* 10 (1878): 372.

45. These figures are from 1890. See *Gentistoria from Government Records and Official Sources* (Washington, DC: American Society, 1913), 893, 1255.

46. Warner, "How Sweet It Is," 2.

47. Wells, *Sugar Industry of the United States*, 52.

48. Ibid.

49. *BKM* 7 (January 1879): 7.

50. Ibid., 4–6.

51. "Glucose," *BKM* 7 (February 1879): 36. "About that honey confiscation," *American Bee Journal* 15 (February 1879): 51.

52. "Glucose Dealers," *BKM* 7 (February 1879): 43. "D" was a reference to D. Quimby, who had recently been accused of adulteration.

53. There have been a number of significant battles over sugar, health, marketing,

and trust in the past ten years alone, with public health officials leading a debate about sugar, obesity, diabetes, and liver damage. See Lustig, *Fat Chance*; Moss, *Salt, Sugar, Fat*. For the recent corn syrup court cases, see Salisbury, "Sugar Growers' Lawsuit over 'Sweet Surprise,'" and Fuhrmeister, "Bitter Sugar vs. Corn Syrup Lawsuit Ends in Settlement." Also see Western Sugar Coop., et al., Plaintiffs, v. Archer-Daniels-Midland Co., et al., Defendants, No. CV 11-3473 CBM, LexisNexis, https://www.americanbar.org/content/dam/aba/publications/litigation_news/archer-daniels-midland.pdf.

54. Warner, *Sweet Stuff*, 116–21. Also see *Interstate Commerce Commission Reports*, 675. A century later, they tried to change the name again to "corn *sugar*," but in the case of *Western Sugar Coop.*, they lost in a confidential February 2015 settlement.

Chapter Seven

1. *Wahpeton Gazette*, January 15, 1904, as quoted in Kane, "Populism, Progressivism, and Pure Food," 163.

2. Those were long-term and diffuse precedents, from Plato to Accum. In the near term, the chemical analyst in America followed from a vibrant tradition in England that shadowed Accum. Their chemical analysis trade grew from a community of rising professionals interested in developing communication networks and debating new techniques and methods. After adulteration studies took on their modern chemical dimension with Accum, they came to the public stage of governmental authority in the 1850s with Arthur Hassall's microscopy work uncovering theretofore unseen components of water. British advocates soon after founded the Society for Public Analysts (1874). Its members worked as paid analysts for the government. They began publishing their journal's first volume, the *Analyst*, in 1877. Also see Hamlin, *A Science of Impurity*; Sargent, "Scientific Experiment and Legal Expertise."

3. *Meadville Messenger*, February 11, 1895; "The Oleomargarine Law," *Philadelphia Inquirer*, February 15, 1895, 1.

4. "Oleomargarine in the House," *New York Times*, March 2, 1895, 13.

5. Dairy, Box 4, Warshaw Collection of Business Americana, NMAH.

6. "Oleomargarine in the House," *New York Times*, March 2, 1895, 13; "Pennsylvania Dairy Association," *Ohio Farmer*, February 28, 1895, 166; 'Butter not Butterine," *Ohio Farmer*, March 21, 1895, 226; "An Oleo Fraud Shown Up," *Maine Farmer*, April 4, 1895, 1.

7. Pickstone, *Ways of Knowing*; Steere-Williams, "Conflict of Analysis"; Russell, Coley, and Roberts, *Chemists by Profession*, all provide helpful discussions about the new analysts.

8. "Pure Food," *AA* 3 (April 1, 1887): 167.

9. William Atwater, [untitled], *AA* 4 (May 15, 1888): 206.

10. In the 1870s, by comparison, there were twenty-four in Boston, twenty-eight in Philadelphia, and eleven in Chicago. *Boston Almanack and City Directory* (1874, 1884); *Boyd's City Directory* (1874, 1888); *Chicago, Illinois, City Directory* (1874, 1885).

11. Procter & Gamble, *Into a Second Century with Procter & Gamble*, 14. Also cf. Richard Powers, *Gain* (New York: Farrar, Straus, and Giroux, 1998).

12. Armour, *Packers and the People,* 188.

13. See Commercial Manufacturing Company, *A Brief History of the Mege Discovery.*

14. Okun, *Fair Play in the Marketplace,* 293.

15. These efforts also came through publishing and scientific organizations. Charles Chandler, for instance, founded the journal *American Chemist* in 1870 with a focus on the activities of chemical analysts and microscopists and their forays into food and drug investigations. He ceded that work to the new *Journal of the American Chemical Society* (*JACS*) in 1877, when chemists organized to form the American Chemical Society (ACS; Chandler would be their president from 1881 to 1889). In its original framing, the journal was meant to provide a forum for analytical activity assessing new urban problems—such as food, water, and waste. The ACS still publishes *JACS,* albeit with a broader chemical mission, in its 141st volume in 2019. See Bogert, *Biographical Memoir of Charles Frederick Chandler.*

16. That meant about one in forty thousand residents in the 1870s, one in thirty-six thousand in the 1880s, and one in thirty-three thousand before consolidation of the boroughs in the 1890s.

17. That was one per eleven thousand residents by the 1890s, triple the rate of New York City.

18. Moore, *Price List Analytical Laboratory and Assay Office.*

19. Okun, *Fair Play in the Marketplace,* 95–96; Spellman, *Cornering the Market,* 13–14.

20. *AG* 29 (February 8, 1883): 281–82.

21. *AG* 29 (February 22, 1883).

22. *AG* 29 (May 3, 1883): 957 (on the UK), and (May 24, 1883): 1163 (on the Germans). When Darby left the paper, Thurber hired two grocers to edit it. The paper then brought its own chemist on staff. No longer would readers need to hire out a consultant; the grocer could manage analyses in house. They hired Joseph Geisler—he'd advertised early in the paper's run—who had built a career analyzing samples just as Darby and Gideon Moore had.

23. *SM* 5 (February 1882): 47–48.

24. "What the American Analyst Is Doing," *AA* 3 (July 1, 1887): 310.

25. The paper's constantly changing titles and subtitles told their own story. After launching as the *New York Analyst* in 1885, Lassig ambitiously renamed it *American Analyst* in its second volume. It changed again the next year to *American Analyst and Practical Cook and Housekeeper.* The subtitle began as *A Popular Semi-monthly Review Devoted to Industrial Progress, Sanitation, and the Chemistry of Commercial Products.* It then morphed into *A Semi-monthly Journal of Pure Food, the Suppression of Adulteration and Everyday Advice to the Physician, Druggist, and Dentist.* Lassig was more concise by year four with *A Semi-monthly Advocate of Pure Food and the Suppression of Adulteration.* What began as a forum for analyzing agricultural goods as they appeared in urban storefronts shifted to a more focused view of the household by the later 1880s. From the early 1890s until the paper closed its doors in the second half of the decade, Lassig tried to make his way as a publisher by appealing to an audience agitating for analytical clarity.

26. "The Anti-Adulteration War," *AA* 3 (June 1, 1887): 256.

27. "Adulterations," *AA* 3 (April 15, 1887): 383; *AA* 3 (November 1, 1887): 498. Also see Singerman, "Science, Commodities, and Corruption in the Gilded Age."

28. "Bogus Butter," *AA* 4 (January 15, 1888): 45.

29. *AA* 3 (September 15, 1887): 433.

30. Ibid.

31. Rueber, "Is the Milk the Measure of All Things?," 93.

32. Goodwin, *Pure Food, Drink, and Drug Crusaders*, 65–73.

33. "American Society of Analysts," *Sanitary News* 5 (December 1, 1884): 34. See also "Minutes and Reports from the American Society of Public Analysts, 1884–1885," SHI archives.

34. "Minutes and Reports from the American Society of Public Analysts, 1884–1885," October 31, 1884, SHI archives.

35. "Minutes and Reports from the American Society of Public Analysts, 1884–1885," December 8, 1884, SHI archives.

36. *New York Times*, December 27, 1876, 3.

37. One of the glass instruments was a lactometer. Schrumpf's lawyer grilled Waller about it, casting it as suspicious and unreliable. The lactometer had been calibrated to determine specific gravity. When Charles Chandler was called as a witness, he explained that a French chemist developed the lactometer's metric after substantive studies of the cows of France. Waller noted that 1.029 specific gravity was the "lowest quality" minimum for "genuine milk" measured at 60° Fahrenheit (the French standard). Schrumpf's milk was .85. Asked how he knew his instrument was valid, Waller explained that he had "visited the country and tested milk direct from cows." He had, that is, combined the new methods (instrumental) with the old ones (unmediated sensory tests) to confirm their veracity. New York (State), Court of General Sessions (New York County), *People vs. Daniel Schrumpf*, 14. See also Atkins, *Liquid Materialities*; Smith-Howard, *Pure and Modern Milk*; Steere-Williams, "A Conflict of Analysis"; Orland, "Enlightened Milk."

38. *New York Times*, January 29, 1876, 4; "Professor Doremus on Milk," *New York Times*, January 27, 1876; *Chemical News*, January 14, 1876, 20. See Atkins, *Liquid Materialities*, 41–45 for further discussion of the matter.

39. Charles Cameron, Read Committee (1874), Q.4955, as quoted in Atkins, *Liquid Materialities*, 65.

40. In the end, Schrumpf's case was too easily prosecuted even after his lawyers had called the lactometer's value into question. The onslaught of chemical tools on display proved overwhelming. The judge sided with the prosecution. Schrumpf would serve ten days in jail and pay a $250 fine. The expert witnesses would continue arguing over how best to elevate their public standing as adulteration police.

41. *AG* 32 (November 13, 1884): 8.

42. "People v Daniel Schrumpf," 126, 179; "The Adulteration of Milk," *American Chemist*, 7 (1877): 365–87. As Mitchell Okun summarized it, "Chandler claimed that alum was dangerous; Dr. Love's report for Chandler's Sanitary Committee denied its danger. By 1884 Love himself condemned it. Yet no chemist had ever conducted an objective, controlled experiment on the use of alum. Chandler claimed the lactometer was useless; when he realized that his enforcement of the milk laws necessitated its use, he argued that he now trusted the instrument implicitly. Moreover, for every chemist who defended a product or a process, another could be found who condemned it." Okun, *Fair Play in the Marketplace*, 293.

43. Lassig, as quoted in Burns, *SM* 9 (November 1886): 331.

44. See Atkins, *Liquid Materialities*, 58.

45. Richards, *Food Materials and Their Adulterations* (1886), 58–59.

46. "They Mixed Those Babies Up," *AA* 3 (February 15, 1887): 77.

47. One of the judges, McClintock, was there because he had represented the county with distinction at the Columbian Exposition (the World's Fair) in Chicago just over a year before; he stood for the combination of local pride and expertise vetted beyond those local confines. Another judge, A. L. Wales, was on the panel because of his years of experience in the local creamery, an elder statesman knowledgeable and respected by his peers.

48. See, for example, W. Frear and G. Holter, "Simple Methods of Determining Milk Fat," *Pennsylvania State College Agricultural Experiment Station Bulletin* 12 (July 1890): 2. Frear was thus an agent of the state experiment station, a benefactor of the USDA's federal structures that mandated those stations, and a chemistry professor with his own lab and classroom.

49. "Butter vs. Butterine," *Agricola*, March 14, 1895, 31; "Oleomargarine in the House," *New York Times*, March 2, 1895, 13.

Chapter Eight

1. Letter of August 26, 1905, Entry 4 ("Series 4"), Box 45, RG 97, NARA; Sinclair, *The Jungle*, 254.

2. "Song Book of the AOAC," 4, Wiley Papers, LOC.

3. Ibid., 12.

4. Wines, *Fertilizer in America*; McKinley, *Stinking Stones and Rocks of Gold*; Cohen, *Notes from the Ground*; True, *History of Agricultural Experimentation*.

5. United States Food and Drug Administration, *A Guide to Resources on the History of the Food and Drug Administration* (Rockville, MD: Food and Drug Administration History Office, 1995), discusses the name change.

6. Wilson, *Swindled*, 189. Also see Blum, *Poison Squad*.

7. Anderson, *Health of a Nation*, 4.

8. Anderson, 20; Wiley, *Autobiography*, 147.

9. Wiley, 139.

10. Deborah Warner, in *Sweet Stuff*, shows that Wiley was only the latest to be enamored with French and German polariscopes when he encountered them in the 1870s. As early as the 1840s, the US Senate had asked the Department of the Treasury to consider the French physicist Jean-Baptiste Biot's polariscope. Customs houses, sugar refiners, and government agencies tangled for decades over the validity of the instruments.

11. Technically speaking, the instrument was made up of polarized glass lenses and a light source. Testers put the sample of sugar between the lenses to find how much the sample rotated the polarization of light. They then set standards for "pure" sugar that could be gauged against new samples. See Warner, *Sweet Stuff*, 162 and H. Wiley, "The Effect of Heating with Dilute Acids and Treating with Animal Charcoal, on the Rotary Power of Glucose; with Notes on the Estimation of Cane Sugar and Glucose in Mixture," *Journal of the American Chemical Society* 2 (1880): 395–402; "The Rotary Power of Commercial Glucose," *Science* 2 (1881): 53–54; "The Rotary Power of Commercial Glucose and Grape Sugar: A Method of Determining the Amount of Reducing Substance Present, by the Polariscope," *Journal of the American Chemical Society* 2 (1880): 387–95; reprinted in *Science* 2 (1881), 393–95.

12. Wiley, *Autobiography*, 151.

13. Anderson, *Health of a Nation*, 26–27.

14. Wiley, *Autobiography*, 150.

15. Anderson, *Health of a Nation*, 28.

16. Anderson, 28.

17. Anderson, 22–23.

18. Wiley, *Autobiography*, 156–58.

19. Wiley, 157–58. Note that Butler was known as North Western Christian University at the time, having opened in 1855. The school changed its name to Butler in 1877.

20. Anderson, *Health of a Nation*, 69.

21. Anderson, 68. Throughout the era and with his stewardship, the USDA continued its patronage by providing the chemists meeting spaces and offices to record deliberations. It also printed the society's proceedings and official methods as USDA publications.

22. It was also the same point that British, French, Belgian, and German principles had made in their own governmental offices against adulteration. See Atkins, Lummel, and Oddy, eds., *Food and the City in Europe since 1800*, and Stanziani, "Food and Expertise" and "Negotiating Innovation in a Market Economy."

23. So active were efforts to address purity that one history of the pure food crusades covers only the years between the 1870s and 1890s. Okun, *Fair Play in the Marketplace*.

24. RG 97, Series 4, Box 45, File #89, Records of the Bureau of Agricultural and Industrial Chemistry, NARA.

25. See letters in RG 97, Series 4, Files 1–19 of Special Files, NARA.

26. RG 97, Series 5, Box 4, Letter of April 20, 1887.

27. Historians of science have done well to explain the rise of quantitative and instrumental thinking; Wiley's work provides a vision of the process in the making. The larger set of issues of relevance here includes metrology, calibration, gauges, and standardization. For more on the virtues of quantification in scientific practice that those subjects touch on, see Crosby, *Measure of Reality*; Gooday, *Morals of Measurement*; Hacking, *Taming of Chance*; Mudry, *Measured Meals*; Lampland and Star, eds., *Standards and Their Stories*; O'Connell, "Metrology"; Porter, *Trust in Numbers*; and Scott, *Seeing Like a State*.

28. Another technique involved refractive index. The tests could be fickle and the equipment finicky, requiring a deft hand and proper training. Wiley was still smitten with the more established instruments he found in his European travels the decade before. He used an imported German refractometer for his work even though the sensitivity of the instrument required a room temperature above 40°C. This was hard to achieve even for the USDA, let alone your average grocer's chemist. Wiley had a work-around. "In the absence of any such room in our laboratory," he reported, he "used the hot-rooms of the Turkish bath house to good advantage." He admitted difficulty while assuring credibility. Even if instrumental tests were finicky, specialists used trained judgment to draw conclusions. USDA, Bulletin 13, pt. 1 (1887), 420, 442. Also see Daston and Galison, *Objectivity*, for a fuller explication of the ways the scientists' commitment to instrumental analysis framed their views of objectivity.

29. USDA, Bulletin 13, pt. 4 (1889), 424.

30. USDA, Bulletin 13, pt. 6 (1892), 742–809.

31. Ibid., 747.

32. Mudry, *Measured Meals*, 29.

33. See Finlay, "Transnational Exchange of Agricultural Scientific Thought," 35–39. The German precedents influenced Wiley's work in the Division of Chemistry toward the same ends of consistency in methods just as Wiley's influence as the head of the division would then influence the ways his bosses at the USDA imagined the new OES.

34. The Morrill Land Grant Act of 1862 created a system whereby each state would provide agricultural and mechanical education for its citizens. Passed the same year as the USDA, it has a long-shared history with agricultural research.

35. Atwater earned his doctorate in 1869 under Samuel Johnson's supervision at Yale. He then taught in Tennessee and Maine before returning to his undergraduate alma mater Wesleyan University in 1873. Atwater was also a member of the AOAC. See Mudry, *Measured Meals*; Coppin and High, *Politics of Purity*, 39; and Jou, "Controlling Consumption," 11–59.

36. Coppin and High, *Politics of Purity*, 39, offer more comment on disputes between Wiley and Atwater.

37. See Rosenberg, "Wilbur Olin Atwater"; Carpenter, "Short History of Nutritional Science: Part 2"; Mudry, *Measured Meals*.

38. Cullather, "Foreign Policy of the Calorie," 342. Also see Whorten, *Crusaders for Fitness*. In her discussion of colonial productions of bodies and food knowledge in the Spanish Americas, Rebecca Earle provides another view of how people understood nutrition before the twentieth century. Earle, "'If You Eat Their Food.'"

39. Liebig's meat extract was a novel commercial enterprise that anticipated the later integration of science and commerce in a consumer world of prepared foods. It was the result of bringing the chemistry of food to the marketplace, and it was a good example of the kind of thinking that understood food as a measurable quantity. See Finlay, "Early Marketing of the Theory of Nutrition."

40. Atkinson et al., *Science of Nutrition*.

41. Cullather, "Foreign Policy of the Calorie," 340, and *Hungry World*, chap. 1. Also see Susan Lederer, *Subjected to Science: Human Experimentation in America before the Second World War* (Baltimore: Johns Hopkins University Press, 1995), for more insight on the use of human subjects in research at the time and Simmons, *Vital Minimum*. Atwater was not the first to conduct these kinds of input-output studies of food and health. Earlier in the century, in 1823, the British physician Edward Smith conducted experiments on prisoners to measure how much urea they excreted after walking on a treadmill for hours. See Carpenter, "Short History of Nutritional Science: Part 1."

42. Cullather, "Foreign Policy of the Calorie," 340.

43. Cullather, 345. At the OES, the concept giving substance to Atwater's approach was captured in the German term *Kraftwechsel*. This referred to "a mathematically reliable system of equivalence between the amount of potential energy ingested in the form of nutriments and the amount of energy produced." Mudry, *Measured Meals*, 32.

44. The Poison Squad has since been a staple of food adulteration history. I draw on it as an example of the ways Wiley sought to quantify and control the purity debate through his office and with his instruments, but it has appeared in basically

every subsequent written account as evidence of the march to the 1906 law. See, for example, Wiley, *Autobiography*; Anderson, *Health of a Nation*; Wilson, *Swindled*; Harvey Levenstein, *Fear of Food: A History of Why We Worry about What We Eat* (Chicago: University of Chicago Press, 2012); Vileisis, *Kitchen Literacy*; Satin, *Death in the Pot*; Thomas, *In Food We Trust*; Blum, *Poison Squad*.

45. Levenstein, *Fear of Food*, 63.

46. Thomas, *In Food We Trust*, 18.

47. Wiley, *Autobiography*, 216.

48. "Poison Squad to Resume Work," *Brooklyn Eagle*, March 14, 1905, 5.

49. Wiley, *Autobiography*, 216.

50. "Vacation for Poison Squad," *Brooklyn Daily Eagle*, May 22, 1904, 9; "Scientific Martyrdom of Poison Squad Over," *Washington Times*, May 21, 1904, 1; "pink cheeks" in "Points about People," *Brooklyn Daily Eagle*, July 9, 1904, 4.

51. Ceylon placed the ad all over New York City. See *New York Tribune*, December 8, 1902, 2; *Brooklyn Daily Eagle*, December 8, 1902, 9.

52. "Gridiron Club's Annual Dinner," *San Francisco Call*, February 1, 1903, 33.

53. On embalmed beef, see Olmstead, *Arresting Contagion*, 198–200. Also see Spiekermann, "Dangerous Meat?," for a discussion of broader and more lingering meat scandals. And if the borax joke landed in 1903 because of the prior scandal, the rhymes and doggerel landed because chemists still loved a good song. It was AOAC banquet days all over again. The Gridiron Club dinner was punctuated with "jingling rhymes" and burlesque songs. Public accounts of Wiley's experiments took it further, with reporters crafting poetic couplets in their stories. Writer S. W. Gillian penned the "Song of the Pizen Squad." Wiley quoted it in his memoir. "We are the pizen squad / On prussic acid we break our fast; / We lunch on morphine stew; / We dine with a matchhead consommé / Drink carbolic acid brew." This wasn't high art. The journalists seemed to relish using chemical terms. One of the more widely circulating of the tunes was "They'll Never Look the Same":

> If ever you should visit the Smithsonian Institute,
> Look out that Professor Wiley doesn't make you a recruit.
> He's got a lot of fellows there that tell him how they feel,
> They take a batch of poison every time they eat a meal.
> For breakfast they get cyanide of liver, coffin shaped,
> For dinner, undertaker's pie, all trimmed with crepe;
> For supper, arsenic fritters, fried in appetizing shade,
> And late at night they get a prussic acid lemonade.

Lew Dockstade, "They'll Never Look the Same" (see Wilson, *Swindled*, 188).

54. *Washington Bee*, January 10, 1903, 7.

55. Vileisis, *Kitchen Literacy*, 127.

56. "Borax Begins to Tell," *Washington Post*, December 26, 1902.

57. He followed Atwater's precedent with this work, but he also took note of German experiments that did much the same thing as the Poison Squad. Between 1900 and 1902, German chemists, under the auspices of government oversight, fed the four men borax. They found that a "single dose remains in the body for eight days," that the borax sometimes affected "a threatening aspect" to their health, and that "excessive loss of liquids and decrease of weight" were some of the problems of the preservative. "Effect of Using Borax," *Washington Post*, June 2, 1902, 3.

58. Quoted in "Dr. Wiley's Experiments," *Washington Post*, December 24, 1902.

59. Wilson, *Swindled*, 188.

60. Finck, *Food and Flavor*, 25–26.

61. H. Finck, *New York Evening Post*, April 8, 1903.

62. Also see Balleisen, *Fraud*, 120–22.

63. "All Are Eager to Drink Wine," *Saint Paul Globe*, October 18, 1903, 13.

64. Pirog et al., *Local Food Movement*.

65. Mason hearings, 16. Wiley was nuanced. "If food looks appetizing it will render the flow of digestive juices stronger.... These things appeal to the taste in its figurative sense, and the question, it seems to me, of aesthetics should be considered in the process of digestion" (Mason hearings, 42).

66. Wiley in Mason hearings, 20.

67. Rep. Lewis Beach of New York, in Cong. Rec., 49th Cong., 1st Sess., 4910.

68. Wiley in Mason hearings, 54.

69. United States Congress, *Hearings before the Committee on Interstate and Foreign Commerce of the House of Representatives on Pure-Food Bills H.R. 3109, 12348, 9352, 276, and 4342*, 10–12 and 177–276.

70. United States Congress, *Hearings before the Committee on Interstate and Foreign Commerce of the House of Representatives on Pure-Food Bills H. R. 5077 and 6294*.

71. United States Congress, *Hearings before the Committee on Interstate and Foreign Commerce of the House of Representatives on the Pure-Food Bills H.R. 3044, 4527, 7018, 12071, 13086, 13853, and 13859*.

72. Young, *Pure Food*, 230.

73. Sinclair, *The Jungle*, 194, 254, and 355. This was a view Sinclair carried forward in years to come. As he ran for governor of California in 1934 on the socialist ticket, he extended ideological positions published in *The Book of Life*, what historian Doug Sackman calls a nonfictional sequel to *The Jungle*. In it, Sinclair was still railing against the transitions Wiley and friends had enunciated in the previous half century that "the modern intensive gardener, by use of glass and chemical test-tube, has developed an entirely new science of plant raising. He is independent of climate, he makes his own climate." Sinclair, *The Book of Life* (New York: Macmillan, 1921), 26. Sinclair wanted the fight for pure foods to be a fight about an eater's place in nature, not the vouching that comes from a chemist's tool kit. See Sackman, *Orange Empire*, 189–98, for a fuller assessment of Sinclair's dual political and food-health commitments.

74. United States Congress, *Hearings before the Committee on Interstate and Foreign Commerce of the House of Representatives on the Pure-Food Bills H.R. 3044, 4527, 7018, 12071, 13086, 13853, and 13859*, 238.

75. In many ways, the experience of chemists at the turn of the century (Wiley included) anticipated a phenomenon identified much later in the century by the sociologist Ulrich Beck. Beck, *Risk Society: Towards a New Modernity* (London: Sage, 1992), argues that by the later twentieth century, the challenges of scientific knowledge in the public sphere were less about assessing whether a given public problem was addressed scientifically and more about whose science was addressing it. Pure food chemists experienced a similar dynamic when their work was challenged as either good or bad based on who employed them rather than whether or not the purity question was being addressed by chemists at all.

76. USDA, Bulletin 13, pt. 4, 546.

77. This sense was also captured by some of the more influential later historians of the Progressive Era, including Wiebe, *Search for Order*, and Kolko, *Triumph of Conservatism*. For a more recent analysis of the dynamic, see Loner, "Alchemy in Eden."

78. See Petrick, "Arbiters of Taste"; "Feeding the Masses"; and Domosh, "Pickles and Purity."

79. Francis Barrett chastised the USDA's pretensions in handling the larger problem and rebuked Wiley for his comments about the various pure food bills under debate. Barrett feared authoritative overreach with a bill that vested too much power in one bureau chief. Letter to Wiley from Barrett, November 8, 1901, Wiley archives, NARA. Also see Letters from Barrett to Wiley, June 3, 1902, and February 17, 1902, NARA.

80. Letter from Barrett to Wiley, February 7, 1900, NARA.

81. Cong. Rec., 59 Cong., 1st Sess., 8955-56. Also see Young, *Pure Food*, 255.

82. Bobrow-Strain, "White Bread"; Smith-Howard, *Pure and Modern Milk*; Rappaport, "Packaging China"; Rappaport, *A Thirst for Empire*.

83. We see shades of this today with raw milk debates. Raw milk is banned in many states because of its danger to health. Advocates argue that they are avoiding the problems created by a process of a concentrated commercial dairy that endangers cow health in the first place—requiring pasteurization to overcome the problems of an overcentralized (mechanized, industrialized) dairy process. The raw milk supporters make a case about process over product. They don't need to pasteurize milk because it does not have the elevated levels of salmonella, *E. coli*, and listeria pasteurization knocks out. See United States Food and Drug Administration, "Dangers of Raw Milk," and Weston A. Price Foundation, "Campaign for Real Milk."

84. Senator Blair in Miller hearings, 62. Also see Cohen, "Three Peasants on Their Way to a Meal."

85. D. E. L., "Letter to the Editor," *Brooklyn Eagle*, November 11, 1900, 23.

86. Bailey, *Holy Earth*.

Epilogue

1. Solnit, "Mysteries of Thoreau Unsolved."

2. Victor, "ABC Settles with Meat Producer."

3. One of the chocolate episodes on my list struck readers as a quintessential Brooklyn story for our day, though for me it felt so Gilded Age. The Mast Brothers sell "bean-to-bar" chocolate, meaning they process judiciously chosen cocoa beans into artisanal candy bars. As one writer summarized their new and unwanted public repute, chocolate aficionados had discovered something afoot in the Mast Brothers operations. "At some point early in their operations," a reporter noted, the two brothers used commercial chocolate "that they melted, reformed, and sold under their label." That is, for a time they masqueraded a cheaper chocolate with their fancy label. The saga, the reporter thought, exposed "dark truths about capitalism." Andries, "Colorful Honey"; Mulvaney, "The Parmesan Cheese You Sprinkle on Your Penne Could Be Wood"; Engber, "Rump Faker"; Rosner, "What the Mast Brothers Scandal Tells Us about Ourselves."

4. Virilio, *Paul Virilio Reader*; Balleisen, *Fraud*.

5. Consider just a sampling: chromatography, mass spectrometry, and hyperspectral imaging are some of the more technically sophisticated, but we've also had tenderometers, olfactometers, and even something called a "stinkometer." See Lahne, "Sensory Science, the Food Industry, and the Objectification of Taste"; Shapiro, *Something from the Oven*.

6. See United States Census figures at https://www.census.gov; Stanley Lebergott, "Labor Force and Employment, 1800–1960" in *Output, Employment, and Productivity in the United States after 1800*, ed. Dorothy S. Brady (New York: National Bureau of Economic Research, 1966), http://www.nber.org/chapters/c1567.pdf.

7. The term *adulteration* might not dominate GMO debates, but the validity of GMO foods as "pure" is very much part of them. See Bain and Selfa, "Non-GMO vs organic labels."

8. Vogel, *Thinking Like a Mall*, 16–27, provides a useful review of the different ways ethicists have framed arguments over "nature." Pollan, "Why 'Natural' Doesn't Mean Anything Anymore," takes a more ahistorical approach to suggest the term has lost meaning over time rather than never having a solid basis. See also Berenstein, "Making a Global Sensation."

9. See Lustig et al., "Roundtable"; Jasanoff, *Designs on Nature*; Kleinman, ed., *Science and Technology in Society*; Johnson, "What I Learned from Six Months of GM Research."

10. Thompson, *From Field to Fork*, 205.

11. Thompson, 189–90.

Bibliography

Archives

CUA Columbia University Archives, New York, NY
HBS Harvard Business School Baker Library, Cambridge, MA
HM Hagley Museum, Newark, DE
LC Library Company, Philadelphia, PA
LOC Library of Congress, Washington, DC
MHS Massachusetts Historical Society, Boston, MA
MIT Massachusetts Institute of Technology Archives, Cambridge, MA
NARA National Archives and Records Administration, College Park, MD
NMAH National Museum of American History, Smithsonian Institute, Washington, DC
NYHS New-York Historical Society, New York, NY
SHI Science History Institute, Philadelphia, PA
TEP Thomas Edison Papers, Rutgers University, Piscataway, NJ
UCB University of California Bancroft Library, Berkeley, CA
UVA University of Virginia Special Collections, Charlottesville, VA

Periodicals

AA *American Analyst*
AG *American Grocer*
BKM *Beekeepers' Magazine*
CG *Chicago Grocer*
GH *Good Housekeeping*
Godey's
LHJ *Ladies Home Journal*
OPD *Oil, Paint, and Drug Reporter*
PODR *Paint, Oil, and Drug Review*
SM *The Spice Mill*

Primary Sources

Accum, Fredric. *A Treatise on Adulterations of Food, and Culinary Poisons*. London: Longman, Hurst, 1820.

Alsberg, C. L., and A. E Taylor. *The Fats and Oils: A General View*. Fats and Oils

Studies of the Food Research Institute, no. 1. Palo Alto, CA: Stanford University Press, 1928.

An Enemy to Fraud and Villainy. *Deadly Adulteration and Slow Poisoning Unmasked; or, Disease and Death in the Pot and the Bottle*. London: Sherwood, Gilbert, and Piper, 1832.

Armour, J. Ogden. *The Packers and the People*. London: T. W. Laurie, 1906.

Arnaud, Jasper. *An Alarm to All Persons Touching Their Health and Lives*. London: Printed for T. Payne, 1740.

Atkinson, Edward. *Cheap Cotton by Free Labor*. Boston, MA: A. Williams, 1861.

Atkinson, Edward, Ellen Henrietta Richards, Mary W. Abel, Maria Daniell, and Wilbur Olin Atwater. *The Science of Nutrition*. Springfield, MA: Bryan, 1891.

Atwater, W. O., and A. P. Bryant. "Availability and Fuel Value of Food Materials." In *Twelfth Annual Report of the Storrs Agricultural Experiment Station, Storrs, Conn., 1899–1900*, 73–100. Storrs, CT: Storrs Agricultural Experiment Station, 1900.

Aulangier, F. *Dictionnaire de substances alimentaires indigenes et exotique et de leur properties* [Dictionary of native and exotic food substances and their properties]. Paris: Pillet aine, 1830.

Beck, Lewis Caleb. *Adulteration of Various Substances Used in Medicine and the Arts*. New York: S. S. and W. Wood, 1846.

Blake, Isabel. *Candy Making at Home*. Boston: F. R. Everston, 1884.

Bracken, Henry. *Farriery Improve; or, A Compleat Treatise upon the Art of Farriery*. Dublin: R. Reilly, G. Ewing.

Bradford, William. *The Gentleman's and Trader's Guide*. Stratford: J. Keating, 1756.

Brillat-Savarin, *Handbook of Dining*. Translated by L. Simpson. London: Longman, Green, Longman, Roberts, and Green, 1864.

Brooks, R. O., "Our Future 'Public Analysts.'" *Science* 19, no. 481 (March 18, 1904): 465–67.

Bussing, C. F. *Grocerdom: A History of the New York and Brooklyn Grocers' Associations*. New York: Retail Grocers Publishing, 1892.

Byrn, Marcus. *Detection of Fraud and Protection of Health*. Philadelphia: Lippincott, Grambo, 1852.

Carter, Oscar C. S., "On the Detection of Adulterations in Oils." *Proceedings of the American Philosophical Society* 22, no. 120, pt. 4 (October 1885): 296–299.

Casamajor, P. "Detection of Oleomargarine." *Science* 2, no. 75 (December 3, 1881): 576–77.

Chemist. *The Domestic Chemist: Comprising Instructions for the Detection of Adulteration in Numerous Articles Employed in Domestic Economy, Medicine, and the Arts*. London: Bumpus & Griffin, 1831.

Commercial Manufacturing Company. *A Brief History of the Mege Discovery: Oleo-Margarine Butter or Butterine; Microscopically and Chemically Analyzed* [...] *Demonstrating Its Purity*. New York: Commercial Manufacturing Company, 1881.

Commissioner of New Jersey. *Report of the Dairy Commissioner of the State of New Jersey*. Trenton, NJ: The Commissioner, 1887.

Cram, George. *Statistical Diagram: Sugar and Coffee Production*. Chicago: George F. Cram, 1893.

Croker, Temple Henry, Thomas Williams, and Samuel Clark. 3 vols. *The Complete Dictionary of Arts and Sciences*. London, 1764–1766.

Cutbush, James. *Lectures on the Adulteration of Food and Culinary Poisons*. Newburgh, NY: Ward M. Gazlay, 1823.

Denyer, C. H. "The Consumption of Tea and Other Staple Drinks." *Economic Journal*, no. 3 (March 1893): 33–51.

DeVoe, Thomas F. *The Market Book: Containing a Historical Account of the Public Markets in the Cities of New York, Boston, Philadelphia, and Brooklyn* [...]. 1862. Reprint, New York: Burt Franklin, 1969.

Downing, Andrew. *The Architecture of Country Houses: Including Designs for Cottages, Farm Houses, and Villas* [...]. New York: D. Appleton, 1851.

Ebert, Albert E., "Glucose." *Science* 2, no. 74 (November 26, 1881): 567–68.

Electric Sugar Refining Company Stock Certificate, https://commons.wikimedia.org/wiki/File:Stock_certificate_-_Electric_Sugar_Refining_Co.jpg#/media/File:Stock_certificate_-_Electric_Sugar_Refining_Co.jpg

Emerson, Ralph Waldo. "Friendship." In *Select Writings of Ralph Waldo Emerson*, 134–47. Toronto, Canada: W. J. Gage, 1888.

Engels, Friedrich. *The Condition of the Working-Class in England in 1844*. London: G. Allen and Unwin, 1844.

Fairbank, N. K. *Cottolene: The New Shortening*. Chicago: N. K. Fairbank, 1892.

Farley, John. *The London Art of Cookery, and Housekeeper's Complete Assistant*. 1783. 7th ed., London, 1792.

Felker, Peter. *What the Grocers Sell Us*. New York: O. Judd, 1880.

Filby, F. *A History of Food Adulteration and Analysis*. London: George Allen and Unwin, 1934.

Fox, William. *An Address to the People of Great Britain, on the Propriety of Abstaining from West India Sugar and Rum*. Sunderland, England: T. Reed, 1791.

Frankel, Julius. *A Practical Treatise on the Manufacture of Starch, Glucose, Starch-Sugar, and Dextrine*. Philadelphia: H. C. Baird, 1881.

Gantz and Company. *The Manufacture of Cottonseed Oil and Allied Products*. New York: National Provisioner Publishing, 1897.

Grocers' Manual: containing recipes, formulas and instructions for the manufacture of baking powders, flavoring extracts, essences, condiments, etc., in their purity, also their imitations and adulterations. Chicago: Grocers' Manual Publishing, 1888.

Hassall, Arthur. *Adulterations Detected; or, Plain Instructions for the Discovery of Frauds in Food and Medicine*. London: Longman, Brown, Green, Longmans, and Roberts, 1857.

Hassall, Arthur. *Food: Its Adulterations, and the Methods for Their Detection*. London: Longmans Green, 1876.

Hill, J. *The Virtues of Wild Valerian in Nervous Disorders*. 12th ed. London: R. Baldwin and E. and C. Dilly, 1772.

Hoskins, Thomas. *What We Eat, and an Account of the Most Common Adulterations of Food and Drink*. Boston: TOHP Burnham, 1861.

International Library of Technology. *Packinghouse Industries, Cottonseed Oil and Products, Manufacture of Leather, Manufacture of Soap*. Scranton, PA: International Textbook, 1901.

Interstate Commerce Commission Reports: Decisions of the Interstate Commerce Commission of the United States, 28. Washington, DC: Government Printing Office, 1914.

Johnson, Samuel. *A Dictionary of the English Language*. London: W. Strahan, 1755.

La Compañía de Productos de Aceite de Algodón. *Cotton Oil Product Co.* New York, ca. 1895.

Letheby, H. *Lectures on Food*. Rev. 2nd ed. New York: William Wood, 1872.

Liebig, Justus. *Researches on the Chemistry of Food*. London: Taylor and Walton, 1847.

Marcet, William. *On the Composition of Food and How It Is Adulterated*. London: J. Churchill, 1856.

Markham, Peter. *Syhoroc: or, Considerations on the Ten Ingredients Used in the Adulteration of Bread-Flour, and Bread*. London: M. Cooper, 1758

Marx, Karl. *Capital: A Critical Analysis of Capitalist Production*. Translated by Samuel Moore and Edward Aveling, edited by Frederick Engels. New York: Appleton, 1889.

[Mason hearings]. United States Congress, Senate Committee on Manufacture. *Adulteration of Food Products: Hearings before the United States Senate Committee on Manufactures, Fifty-Sixth Congress, First Session and Fifty-Fifth Congress, Third Session, on Mar. 7, May 3–5, 8, 9, 11, 12, June 5–9, Oct. 20, Nov. 11, 13–17, 20, 22, 27, Dec. 21, 1899, Jan. 13, 17, 20, 1900*. Washington, DC: Government Printing Office, 1899.

Melville, Herman. *Moby-Dick*. Everyman's Library. 1851. New York: Knopf, 1988.

Melville, Herman. *The Confidence-Man: His Masquerade*. 1857. Champaign, IL: Dalkey Archive Press, 2006.

[Miller hearings]. United States Congress, Senate Committee on Agriculture and Forestry. *Testimony in Regard to the Manufacture and Sale of Imitation Dairy Products: Hearings before the United States Senate Committee on Agriculture and Forestry, Forty-Ninth Congress, first session, on Apr. 28, 29, June 15–18, 1886*. Washington, DC: Government Printing Office, 1886.

Minnesota Population Center. *National Historical Geographic Information System: Version 2.0*. Minneapolis: University of Minnesota, 2011.

Moleschott, Jacob. *The Chemistry of Food and Diet with a Chapter on Food Adulterations*. Translated by Edward Bronner. London: Richard Griffin, 1860.

Moore, Gideon. *Price List Analytical Laboratory and Assay Office*. New York: John Rankin, 1888.

Morning Call. *Prosperous California: A Land of Money, Progress and Content*. San Francisco, CA: Morning Call, 1883.

Morris, Walter. "On the Adulteration of Food, Principally with a View to Its Detection by the Microscope." *Chemical News and Journal of Physical Science* 24 (1871): 197–98.

National Provisioner. *Directory and Handbook of the Meat and Provision Trades*. Chicago: National Provisioner Publishing, 1895.

New York (State), Court of General Sessions (New York County). *The People vs. Daniel Schrumpf: Misdemeanor, Adulteration of Milk*. New York: Martin B. Brown, 1881.

Nieuwland, Edward. *The Grangers' Glucose Co.: A Very Safe and Most Profitable Investment for Farmers*. New York, 1874.

Norris, Frank. *The Octopus: The Epic of the Wheat*. New York: Collier, 1901.

Osman, Eaton. *The Price Current-Grain Reporter Year Book for 1917*. Chicago: Price Current-Grain Reporter, 1917.

Paige, Lucius. *History of Hardwick, Massachusetts with a Genealogical Register.* Boston: Houghton, Mifflin, 1883.

Patent Office of Great Britain. *Patents for Inventions: Abridgements of Specifications Relating to Sugar; A.D. 1663–1866.* London: G. E. Eyre and W. Spottiswoode, 1871.

Pereira, Jonathan. *A Treatise on Food and Diet.* London: Fowler and Wells, 1843.

Procter & Gamble. *Into a Second Century with Procter & Gamble.* Cincinnati, OH: Procter & Gamble, 1944.

Richards, Edgar, "Butter and Oleomargarine," *Science* 16, no. 392 (August 8, 1890): 71–75.

Richards, Edgar, "Butter and Oleomargarine," *Science* 16, no. 393 (August 15, 1890): 88–90.

Richards, Edgar, "The Use of the Microscope as a Practical Test for Oleomargarine," *Science* 12, no. 287 (August 3, 1888): 59–60.

Richards, Ellen Henrietta. "Paper on the Adulterations of Groceries." *Annual Report of the State Board of Health, Lunacy, and Charity of Massachusetts: Supplement Containing the Report and Papers on Public Health* 1 (1880): 55–56.

Richards, Ellen Henrietta. *The Chemistry of Cooking and Cleaning: A Manual for Housekeepers.* Bedford, MA: Applewood Books, 1882.

Richards, Ellen Henrietta. *Food Materials and Their Adulterations.* Boston: Estes and Lauriat, 1886.

Richards, Ellen Henrietta. *Food Materials and Their Adulterations.* 3rd ed. Boston: Whitcomb and Barrows, 1911.

Robertson, Joseph. *An Essay on Culinary Poisons* [. . .]. London: G. Kearsly, 1781.

Rorer, Sarah. *How to Use Olive Butter: A Collection of Valuable Cooking Recipes.* Philadelphia: Washington Butcher's Sons, 1883.

Scoffern John. "Food Adulteration." In Jacob Moleschott, *The Chemistry of Food and Diet with a Chapter on Food Adulterations*, translated by Edward Bronner, 91–124. London: Richard Griffin. London: Houlston and Stonement, 1860.

Simmonds, W. H. *The Practical Grocer: A Manual and Guide for the Grocer, the Provision Merchant, and Allied Trades.* London: Gresham, 1906.

Simmons, Amelia. *American Cookery.* 1796.

Sinclair, Upton. *The Jungle.* 1906. New York: Bedford/St. Martin's, 2005.

Smiley, Thomas. *The New Federal Calculator, or Scholar's Assistant.* Philadelphia: Grigg and Elliot, 1846.

Smith, Edward. *Foods.* New York: D. Appleton, 1876.

Steel, H. D., and Henry Draper. *Portable Instructions for Purchasing the Drugs and Spices of Asia and the East-Indies.* London: D. Steel, 1779.

Sterling Syrup Works. "Sterling Syrup Works to Edison." December 23, 1888. Thomas A. Edison Papers, Rutgers School of Arts and Sciences, Piscataway, NJ.

Stowe, Harriet [as Christopher Crowfield]. *House and Home Papers.* Boston: Tichnor and Fields, 1865.

Swift and Company. *The Meat Packing Industry in America: A Brief History of the Development of the World's Greatest Food Industry, Together with a Description of the Service Which It Performs.* Chicago: Swift, 1931.

Talbot, H. "Ellen Richards Obituary." *Technology Review* 13 (1911): 365–73.

Taylor, Thomas. "Microscopic Investigations Relating to Tea and Its Adulterations." *Proceedings of the American Society of Microscopists* 11 (1889): 46–52.

Thoreau, Henry. *Walden and Other Writings*. Modern Library Classics. 1854. New York: Modern Library, 2000.

Tompkins, Daniel Augustus. *Cotton and Cotton Oil: Cotton, Cotton Seed Oil Mills, Cattle Feeding, Fertilizers; Full Information for Investor, Student and Practical Mechanic*. Charlotte, NC: D. A. Tompkins, 1901.

True, Alfred. *The Cotton Plant: Its History, Botany, Chemistry, Culture, Enemies, and Uses*. Washington, DC: Government Printing Office, 1896.

Twain, Mark. *The Writings of Mark Twain*, vol. 9, *Life on the Mississippi*. Hartford, CT: American, 1899.

UK Parliament. *Adulteration of Food, Drink, and Drugs, the Evidence Taken before the Parliamentary Committee*. London: David Bryce, 1855.

UK Parliament, House of Commons. *Correspondence Respecting the Manufacture of Oleomargarine in the United States*. London, 1880.

United States Census Office. *Bulletins of the Twelfth Census of the United States Published between November 1, 1901 and April 20, 1902; Numbers 107 to 163 by the Department of the Interior Census Office*. Washington, DC: Government Printing Office, 1902.

United States Congress. *Foreign Tariffs: Discriminations against the Importation of American Products*. Washington, DC: Government Printing Office, 1884.

United States Congress. *Hearings before the Committee on Interstate and Foreign Commerce of the House of Representatives on Pure-Food Bills H.R. 3109, 12348, 9352, 276, and 4342*. Washington, DC: Government Printing Office, 1902.

United States Congress. *Hearings before the Committee on Interstate and Foreign Commerce of the House of Representatives on Pure-Food Bills H. R. 5077 and 6294* [. . .]. Washington, DC: Government Printing Office, 1904.

United States Congress. *Hearings before the Committee on Interstate and Foreign Commerce of the House of Representatives on the Pure-Food Bills H.R. 3044, 4527, 7018, 12071, 13086, 13853, and 13859*. Washington, DC: Government Printing Office, 1906.

United States Tariff Commission. *Survey of the American Cottonseed Oil Industry*. Washington, DC: Government Printing Office, 1920.

United State Treasury Department. *Annual Report and Statements of the Chief of the Bureau of Statistics on the Foreign Commerce and Navigation, Immigration, and Tonnage of the United States*. Washington, DC: Government Printing Office, 1885–1909.

Vander Weyde, P. H., "The Mege Discovery." *New York Times*, March 19, 1880.

Van Norman, Hubert. *First Lessons in Dairying*. New York: O. Judd, 1908.

Ward, Artemus. *The Grocer's Companion and Merchant's Handbook*. Boston: New England Grocer, 1883.

Wiest, Edward. *The Butter Industry in the United States: An Economic Study of Butter and Oleomargerine*. 1916. Reprint, London: Forgotten Books, 2016.

Wells, David. *The Sugar Industry of the United States, and the Tariff: Report on the Assessment and Collection of Duties on Imported Sugars*. New York: Evening Post Press, 1878.

Wesson, David. "Contributions of the Chemist to the Cottonseed Industry." *Journal of Industrial and Engineering Chemistry* 7 (1915): 276–77.

Wesson, David. "The Examination of Lard for Impurities." *Journal of the American Chemical Society* 17 (1895): 723–35.

Wiley, Harvey. *Foods and Their Adulteration*. 2nd ed. Philadelphia: P. Blakiston Sons, 1907.
Wiley, Harvey. *Harvey W. Wiley: An Autobiography*. Indianapolis, IN: Bobbs-Merrill, 1930.

Secondary Sources

Alsberg, Carl. "Economic Aspects of Adulteration and Imitation." *Quarterly Journal of Economics* 46 (December 1931): 1–33.
Anderson, Oscar. *The Health of a Nation: Harvey W. Wiley and the Fight for Pure Food*. Chicago: University of Chicago Press, 1958.
Anderson, Oscar. "The Pure-Food Issue: A Republican Dilemma, 1906–1912." *The American Historical Review* 61 (April 1956): 550–73.
Andries, Kate. "Colorful Honey." *National Geographic*, October 13, 2012. https://news.nationalgeographic.com/news/2012/10/pictures/121011-blue-honey-honeybees-animals-science/.
Atherton, Lewis E. *The Southern Country Store, 1800–1860*. Baton Rouge: Louisiana State University Press, 1949.
Atkins, Peter, ed. *Animal Cities: Beastly Urban Histories*. Burlington, VT: Ashgate, 2012.
Atkins, Peter. *Liquid Materialities: A History of Milk, Science, and the Law*. Burlington, VT: Ashgate, 2010.
Atkins Peter. "Sophistication Detected: or, the Adulteration of the Milk Supply, 1850–1914." *Social History* 16 (1991): 317–339.
Atkins, Peter, and Ian Bowler. *Food in Society: Economy, Culture, Geography*. New York: Oxford University Press, 2001.
Atkins, Peter, Peter Lummel, and Derek Oddy, eds. *Food and the City in Europe since 1800*. Burlington, VT: Ashgate, 2012.
Baics, Georg. *Feeding Gotham: The Political Economy and Geography of Food in New York, 1790–1860*. Princeton, NJ: Princeton University Press, 2016.
Bailey, Liberty Hyde. *The Holy Earth*. New York: C. Scribner's Sons, 1915.
Bailey, Thomas. "Congressional Opposition to Pure Food Legislation, 1879–1906." *American Journal of Sociology* 36 (1930): 52–64.
Bain, Carmen, and Theresa Selfa. "Non-GMO vs Organic Labels: Purity or Process Guarantees in a GMO Contaminated Landscape." *Agriculture and Human Values* 34 (December 2017): 805–18.
Ball, Richard and J. Robert Lilly. "The Menace of Margarine: The Rise and Fall of a Social Problem." *Social Problems* 29, (1982): 488–498.
Balleisen, Edward J. *Fraud: An American History from Barnum to Madoff*. Princeton, NJ: Princeton University Press, 2017.
Beckert, Sven. *Empire of Cotton: A Global History*. New York: Alfred A. Knopf, 2015.
Belasco, Warren, and Roger Horowitz, eds. *Food Chains: From Farmyard to Shopping Cart*. Philadelphia: University of Pennsylvania Press, 2009.
Bentley, Amy, and Christy Spackman. "Food Riots: Historical Perspectives." In *Encyclopedia of Food and Agricultural Ethics* ed. P. B. Thompson and D. M. Kaplan. Dordrecht: Springer Reference, 2014.
Berenstein, Nadia. "Flavor Added: The Sciences of Flavor and the Industrialization of Taste in America." PhD diss., University of Pennsylvania, 2018.

Berenstein, Nadia. "Making a Global Sensation: Vanilla Flavor, Synthetic Chemistry, and the Meanings of Purity." *History of Science* 54 (2016): 399–424.

Biltekoff, Charlotte. *Eating Right in America: The Cultural Politics of Food and Health*. Durham, NC: Duke University Press, 2013.

Block, Daniel. "Protecting and Connecting: Separation, Connection, and the U.S. Dairy Economy, 1840–2002," *Journal for the Study of Food and Society* 6 (2002): 22–30.

Blum, Deborah. *The Poison Squad: One Chemist's Single-Minded Crusade for Food Safety at the Turn of the Twentieth Century*. New York: Penguin, 2018.

Bobrow-Strain, Aaron. *White Bread: A Social History of the Store Bought Loaf*. Boston: Beacon, 2012.

Bobrow-Strain, Aaron. "White Bread Bio-politics: Purity, Health, and the Triumph of Industrial Baking." *Cultural Geographies* 15 (2008): 19–40.

Bogert, Marston. *Biographical Memoir of Charles Frederick Chandler*. Washington, DC: National Academy of Sciences, 1931.

Boyer, Paul. *Urban Masses and Moral Order in America, 1820–1920*. Cambridge, MA: Harvard University Press, 1978.

Brock, William. *The Fontana History of Chemistry*. London: Harper Press, 2010.

Brock, William. *Justus von Liebig: The Chemical Gatekeeper*. Cambridge: Cambridge University Press, 1997.

Brown, Kathleen. *Foul Bodies: Cleanliness in Early America*. New Haven, CT: Yale University Press, 2009.

Burnett, John. *Plenty and Want: A Social History of Food in England from 1815 to the Present Day*. 2nd ed. London: Routledge, 1977.

Burnett, John, and Derek J. Oddy, eds. *The Origins and Development of Food Policies in Europe*. London: Leicester University Press, 1994.

Burns, Chris. *Bogus Butter: An Analysis of the 1886 Congressional Debates on Oleomargarine Legislation*. Master's thesis, University of Vermont, 2009.

Bushman, Richard. *The American Farmer in the Eighteenth Century: A Social and Cultural History*. New Haven, CT: Yale University Press, 2018.

Byrne, Frank J. *Becoming Bourgeois: Merchant Culture in the South, 1820–1865*. Lexington: University Press of Kentucky, 2006.

Carpenter, Daniel. *Reputation and Power: Organizational Image and Pharmaceutical Regulation at the FDA*. Princeton, NJ: Princeton University Press, 2010.

Carpenter, Kevin. "A Short History of Nutritional Science: Part 1 (1785–1885)." *Journal of Nutritional Science* 133 (March 2003): 638–45.

Carpenter, Kevin. "A Short History of Nutritional Science: Part 2 (1885–1912)." *Journal of Nutritional Science* 133 (April 2003): 975–84.

Carson, Gerald. *The Old Country Store*. New York: E.P. Dutton, 1965.

Carter, Susan, Scott Sigmund Gartner, Michael R. Haines, Alan L. Olmstead, Richard Sutch, and Gavin Wright, eds. *Historical Statistics of the United States: Earliest Times to the Present*. New York: Cambridge University Press, 2006.

Chan, E., S. Griffiths, C. Chan. "Public-Health Risks of Melamine in Milk Products." *Lancet* 372 (2008): 1444–45.

Chandler, Alfred D., Jr. *The Visible Hand: The Managerial Revolution in American Business*. Cambridge, MA: Belknap Press of Harvard University Press, 1977.

Charnley, Berris. "Arguing over Adulteration: The Success of the Analytical Sanitary Commission." *Endeavour* 32 (2005): 129–33.

Cioc, Mark. *The Rhine: An Eco-Biography, 1815–2000*. Seattle: University of Washington Press, 2002.

Clarke, Robert. *Ellen Swallow: The Woman Who Founded Ecology*. Chicago: Follett, 1973.

Clemen, Rudolf. *By-Products in the Packing Industry*. Chicago: University of Chicago Press, 1927.

Cohen, Benjamin. "Analysis as Border Patrol: Chemists along the Boundary between Pure Food and Real Adulteration." *Endeavour* 35 (2011): 66–73.

Cohen, Benjamin. *Notes from the Ground: Science, Soil, and Society in the American Countryside*. New Haven, CT: Yale University Press, 2009.

Cohen, Benjamin. "Three Peasants on Their Way to a Meal: Millet's *The Gleaners*, Macaroni, and Human Intervention in Nature." *Environmental History* 14 (October 2009): 744–52.

Cohen, William, ed. *Filth: Dirt, Disgust, and Modern Life*. Minneapolis: University of Minnesota Press, 2005.

Collins, E. J. T. "Food Adulteration and Food Safety in Britain in the 19th and 20th Centuries," *Food Policy* 18 (1993): 95–109.

Cook, James. *The Arts of Deception: Playing with Fraud in the Age of Barnum*. Cambridge, MA: Harvard University Press, 2008.

Coppin, Clayton, and Jack High. *The Politics of Purity: Harvey Washington Wiley and the Origins of the Federal Food Policy*. Ann Arbor: University of Michigan Press, 1999.

Cotsell, Michael. *The Companion to Our Mutual Friend*. London: Routledge, 2014.

Cronon, William. *Nature's Metropolis: Chicago and the Great West*. New York: W. W. Norton, 1991.

Crosby, Alfred. *The Measure of Reality: Quantification and Western Society, 1250–1600*. New York: Cambridge University Press, 1997.

Cullather, Nick. "The Foreign Policy of the Calorie." *American Historical Review* 112 (April 2007): 337–64.

Cullather, Nick. *The Hungry World: America's Cold War Battle against Poverty in Asia*. Cambridge, MA: Harvard University Press, 2010.

Curtis, Robert. *Cottonseed Meal, Origin, History, Research: For Farmers, Stockmen, College Students and Research Workers*. Raleigh, NC: Robert S. Curtis, 1938.

Deerr, Noel. *History of Sugar*. 2 vols. London: Chapman and Hall, 1949.

De la Pena, Carolyn. *Empty Pleasures: The Story of Artificial Sweeteners from Saccharin to Splenda*. Chapel Hill: University of North Carolina Press, 2010.

Deutsch, Tracey. *Building a Housewife's Paradise: Gender, Politics, and American Grocery Stores in the Twentieth Century*. Chapel Hill: University of North Carolina Press, 2010.

Domosh, M. "Pickles and Purity: Discourses of Food, Empire and Work in Turn-of-the-Century USA." *Social and Cultural Geography* 4, no. 1, (2003): 7–26.

Douglas, Mary. "Deciphering a Meal." *Daedalus* 101, no. 1 (Winter, 1972): 61–81.

Douglas, Mary. *Purity and Danger: An Analysis of the Concepts of Pollution and Taboo*. 1966. Reprint, New York: Routledge, 2003.

Dupré, Ruth. "'If It's Yellow, It Must Be Butter': Margarine Regulation in North America since 1886." *Journal of Economic History* 59, no. 2 (June 1999): 353–71.

DuPuis, E. Melanie. *Nature's Perfect Food: How Milk Became America's Drink*. New York: New York University Press, 2002.

Earle, Rebecca. "'If You Eat Their Food . . .': Diets and Bodies in Early Colonial Spanish America." *American Historical Review* 115 (2010): 688–713.

Eisinger, Josef. "Lead and Wine: Eberhard Gockel and the *Colica Pictonum*." *Medical History* 26 (1982): 279–302.

Engber, Dan. "Rump Faker." *Slate*, January 18, 2013. http://www.slate.com/articles/life/food/2013/01/calamari_made_of_pig_rectum_the_this_american_life_rumor_isn_t_true_but.html.

Fabian, Ann. *Card Sharps, Dream Books, and Bucket Shops: Gambling in Nineteenth-Century America.* Ithaca, NY: Cornell University Press, 1991.

Fabian, Ann. *The Skull Collectors: Race, Science, and America's Unburied Dead.* Chicago: University of Chicago Press, 2010.

Faulkner, Carol. "The Root of the Evil: Free Produce and Radical Antislavery, 1820–1860." *Journal of the Early Republic* 27 (Fall 2007): 377–405.

Ferrières, Madeleine. *Sacred Cow, Mad Cow: A History of Food Fears.* Translated by Jody Gladding. New York: Columbia University Press, 2006.

Finck, Henry. *Food and Flavor: A Gastronomic Guide to Health and Good Living.* 1913. Reprint, Applewood Books, 2010.

Finger, Thomas. "Harvesting Power: Transatlantic Merchants and the Anglo-American Grain Trade, 1795–1890." PhD diss., University of Virginia, 2015. http://libra.virginia.edu/catalog/libra-oa:8343

Finlay, Mark. "Early Marketing of the Theory of Nutrition: The Science and Culture of Liebig's Extract of Meat." In *The Science and Culture of Nutrition, 1840–1940*, edited by H. Kamminga and A. Cunnigham, 48–74. Amsterdam: Editions Rodopi, 1995.

Finlay, Mark. "Transnational Exchange of Agricultural Scientific Thought from the Morrill Act through the Hatch Act." In *Science as Service: Establishing and Reformulating the Land-Grant Universities, 1865–1930*, edited by Alan Marcus, 33–60. Tuscaloosa: University of Alabama Press, 2015.

Follett, Richard. *The Sugar Masters: Planters and Slaves in Louisiana's Cane World, 1820–1860.* Baton Rouge: Louisiana State University Press, 2005.

Fordyce, Eleanor. 1987. "Cookbooks of the 1800s." In *Dining in America, 1850–1950*, edited by Kathryn Grover, 85–113. Amherst: University of Massachusetts Press, 1987.

Fouser, David. "The Global Staff of Life: Wheat, Flour, and Bread in Britain, 1846–1914." PhD diss., University of California, Irvine, 2016.

Freidberg, Susanne. *Fresh: A Perishable History.* Cambridge, MA. Harvard University Press, 2009.

French, Michael, and Jim Philips. *Cheated Not Poisoned? Food Regulation in the United Kingdom, 1875–1938.* Manchester: Manchester University Press, 2000.

Fuhrmeister, Chris. "Bitter Sugar vs. Corn Syrup Lawsuit Ends in Settlement," Eater, November 20, 2015. https://www.eater.com/2015/11/20/9772470/sugar-corn-syrup-lawsuit-settlement.

Fullilove, Courtney. *The Profit of the Earth: The Global Seeds of American Agriculture.* Chicago: University of Chicago Press, 2017.

Genosko, G. "Better than Butter: Margarine and Simulation." In *Jean Baudrillard: Fatal Theories*, edited by D. B. Clarke, M. A. Doel, W. Merrin, and R. G. Smith, 83–90. Abingdon: Routledge, 2009.

Gieryn, Thomas. *Cultural Boundaries of Science: Credibility on the Line.* Chicago: University of Chicago Press, 1999.

Glickman, Lawrence B. "'Buy for the Sake of the Slave': Abolitionism and the Origins of American Consumer Activism." *American Quarterly* 56, no. 4 (December 2004): 889–912.

Golinski, Jan. *The Experimental Self: Humphry Davy and the Making of a Man of Science.* Chicago: University of Chicago Press, 2016.

Gooday, Graeme. *The Morals of Measurement: Accuracy, Irony, and Trust in Late Victorian Electrical Practice.* New York: Cambridge University Press, 2004.

Goodchild, Hayley. "Building 'A Natural Industry of This Country': An Environmental History of the Ontario Cheese Industry from the 1860s to the 1930s." PhD diss., McMaster University, 2017.

Goodwin, Loraine. *The Pure Food, Drink, and Drug Crusaders, 1879–1914.* Jefferson, NC: McFarland, 1999.

Gordon, David. "Swift & Co. v. United States: The Beef Trust and the Stream of Commerce Doctrine." *American Journal of Legal History* 28 (1984): 244–79.

Hacking, Ian. *The Taming of Chance.* Cambridge: Cambridge University Press, 1990.

Haltunnen, Karen. *Confidence Men and Painted Women.* New Haven, CT: Yale University Press, 1982.

Hamilton, Shane. "Analyzing Commodity Chains: Linkages or Restraints?" In *Food Chains: From Farmyard to Shopping Cart,* edited by W. Belasco and R. Horowitz, 16–25. Philadelphia: University of Pennsylvania Press, 2009.

Hamlin, Christopher. *Cholera: The Biography.* Oxford: Oxford University Press, 2009

Hamlin, Christopher. *Public Health and Social Justice in the Time of Chadwick.* Cambridge: Cambridge University Press, 2008.

Hamlin, Christopher. *A Science of Impurity: Water Analysis in Nineteenth Century Britain.* Berkeley: University of California Press, 1990.

Harding, T. Swann. "False and Fraudulent," *North American Review,* 236 (November 1933): 439–47.

Henke, Chris. *Cultivating Science, Harvesting Power: Science and Industrial Agriculture in California.* Cambridge, MA: MIT Press, 2008.

Higgins, David, and Dev Gangjee. "'Trick or Treat?' The Misrepresentation of American Beef Exports in Britain during the Late Nineteenth Century." *Enterprise and Society* 4 (2010): 203–41.

Hisano, Ai. "'Eye Appeal Is Buy Appeal': Business Creates the Color Foods." PhD diss., University of Delaware, 2016.

Hoganson, Kristin. *Consumers' Imperium: The Global Production of American Domesticity, 1865–1920.* Chapel Hill, NC: University of North Carolina Press, 2007.

Horowitz, Roger. *Kosher USA: How Coke Became Kosher and Other Tales of Modern Food.* New York: Columbia University Press, 2016.

Horowitz, Roger. *Putting Meat on the American Table: Taste, Technology, Transformation.* Baltimore: Johns Hopkins University Press, 2004.

Hosking, Richard, ed. *Authenticity in the Kitchen: Proceedings of the Oxford Symposium on Food and Cookery 2005.* Totnes: Prospect Books, 2006.

Hunter, Lynette. "Nineteenth- and Twentieth-Century Trends in Food Preserving: Frugality, Nutrition and Luxury." In *"Waste Not, Want Not" Food Preservation from Early Times to the Present,* edited by Anne Wilson, 134–58. Edinburgh: Edinburgh University Press, 1991.

Jacobsen, Jan. "Adulteration as Part of Authenticity." In *Authenticity in the Kitchen: Proceedings of the Oxford Symposium on Food and Cookery 2005*, edited by Richard Hoskin, 252–61. Totnes: Prospect Books, 2006.

Jarchow, Merrill. "The Beginnings of Minnesota Dairying." *Minnesota History* 27 (1946): 107–21.

Jasanoff, Sheila. *Designs on Nature: Science and Democracy in Europe and the United States*. Princeton, NJ: Princeton University Press, 2007.

Johnson, Nathanael. "What I Learned from Six Months of GM Research." Grist.org, January 9, 2014. http://grist.org/food/what-i-learned-from-six-months-of-gmo-research-none-of-it-matters/.

Jones, Peter. "Making Chemistry the 'Science' of Agriculture, c. 1760–1840." *History of Science* 54 (2016): 169–94.

Jou, Chin. "Controlling Consumption: The Origins of Modern American Ideas about Food, Eating, and Fat, 1886–1930." PhD diss., Princeton University, 2009.

Kamminga, Harmke, and Andrew Cunningham, eds. *The Science and Culture of Nutrition, 1840–1940*. Amsterdam: Ropoli, 1995.

Kane, James. "Populism, Progressivism, and Pure Food," *Agricultural History* 38, no. 3 (July 1964): 161–66.

Kaplan, Amy. *The Anarchy of Empire in the Making of U.S. Culture*. Cambridge, MA: Harvard University Press, 2005.

Kassim, Lozah. "The Cooperative Movement and Food Adulteration in the Nineteenth Century." *Manchester Region History Review* 15 (2001): 1–10.

Kleinman, Daniel, ed. *Science and Technology in Society: From Biotechnology to the Internet*. New York: Wiley and Sons, 2009.

Kolko, Gabriel. *The Triumph of Conservatism: A Reinterpretation of American History, 1900–1916*. New York: Free Press, 1963.

LaBerge, Ann. *Mission and Method: The Early Nineteenth-Century French Public Health Movement*. Cambridge: Cambridge University Press, 1992.

Lahne, Jake. "Sensory Science, the Food Industry, and the Objectification of Taste." *Anthropology of Food* 10 (2016). https://journals.openedition.org/aof/7956.

Lampland, Martha, and Susan Leigh Star, eds. *Standards and Their Stories: How Quantifying, Classifying, and Formalizing Practices Shape Everyday Life*. Ithaca, NY: Cornell University Press, 2008.

Lane, Nick. "The Unseen World: Reflections on Leeuwenhoek (1677) 'Concerning Little Animals.'" *Philosophical Transactions of the Royal Society B* 370 (2015): 1–10.

Laudan, Rachel. *Cuisine and Empire: Cooking in World History*. Berkeley: University of California Press, 2013.

Leach, William R. *Land of Desire: Merchants, Power, and the Rise of a New American Culture*. New York: Pantheon, 1993.

Lears, Jackson. *Fables of Abundance: A Cultural History of Advertising in America*. New York: Basic Books, 1994.

Lears, Jackson. *No Place of Grace*. Chicago: University of Chicago Press, 1980.

Lears, Jackson. *Rebirth of a Nation: The Making of Modern America, 1877–1920*. New York: HarperCollins, 2009.

List, Gary, and Michael A. Jackson, "The Battle over Hydrogenation (1903–1920) Part II: Litigation," *Inform* 20 (2009): 395–97.

Lobel, Cindy R. *Urban Appetites: Food and Culture in Nineteenth-Century New York*. Chicago: University of Chicago Press, 2015.

Loeb, Lori. *Consuming Angels: Advertising and Victorian Women.* Oxford: Oxford University Press, 1994.

Loner, Teri. "Alchemy in Eden: Entrepreneurialism, Branding, and Food Marketing in the United States, 1880–1920." PhD diss., New York University, 2009.

Lustig, Robert H. *Fat Chance: Beating the Odds against Sugar, Processed Food, Obesity, and Disease.* New York: Hudson Street Press, 2013.

Lustig, Robert, Marsha Cohen, Urvashi Rangan, Vani Hari, Richard Williams, Doug Van Hoewyk, and Benjamin Cohen. "Roundtable: Should the F.D.A. Regulate the Use of 'Natural' on Food Products?" *New York Times,* November 11, 2014. https://www.nytimes.com/roomfordebate/2014/11/10/should-the-fda-regulate-the-use-of-natural-on-food-products-15.

Magill, R. Jay. *Sincerity: How a Moral Ideal Born Five Hundred Years Ago Inspired Religious Wars, Modern Art, Hipster Chic, and the Curious Notion That We All Have Something to Say (No Matter How Dull).* New York: W. W. Norton, 2012.

Mapes, Kathleen. *Sweet Tyranny: Migrant Labor, Industrial Agriculture, and Imperial Politics.* Chicago: University of Illinois Press, 2009.

Marine, Gene, and Judith Van Allen. *Food Pollution: The Violation of Our Inner Ecology.* Holt, Rinehart and Winston, 1972.

Martin, Ann Smart. *Buying Into the World of Goods: Early Consumers in Backcountry Virginia.* Baltimore: Johns Hopkins Press, 2008.

McCalla, Douglas. "Retailing in the Countryside: Upper Canadian General Stores in the Mid-19th Century." *Business and Economic History* 26, no. 2 (Winter 1997): 393–403.

McCalla, Douglas. "Textile Purchases by Some Ordinary Upper Canadians, 1808–1861." *Material History Review* 53 (Spring/Summer 2001): 4–27.

McCalla, Douglas. "Upper Canadians and Their Guns: An Exploration via Country Store Accounts (1808–61)." *Ontario History* 97 (2005): 121–37.

McKinley, Shepherd W. *Stinking Stones and Rocks of Gold: Phosphate, Fertilizer, and Industrialization in Postbellum South Carolina.* Gainesville: University Press of Florida, 2014.

McNeur, Catherine. *Taming Manhattan: Environmental Battles in the Antebellum City.* Cambridge, MA: Harvard University Press, 2014.

McWilliams, James. *Revolution in Eating: How the Quest for Food Shaped America.* New York: Columbia University Press, 2005.

Menand, Louis. *The Metaphysical Club: A Story of Ideas in America.* New York: Farrar, Straus, and Giroux, 2002.

Merlaux, April. *Sugar and Civilization: American Empire and the Cultural Politics of Sweetness.* Chapel Hill: University of North Carolina Press, 2015.

Mihm, Stephen. *A Nation of Counterfeiters: Capitalists, Con Men, and the Making of the United States.* Cambridge, MA: Harvard University Press, 2007.

Mintz, Sidney W. *Sweetness and Power: The Place of Sugar in Modern History.* New York: Viking, 1985.

Mintz, Sidney W., and Christine M. Du Bois. "The Anthropology of Food and Eating." *Annual Review of Anthropology* 31 (2002): 99–119.

Misa, Thomas. *Leonardo to the Internet: Technology and Culture from the Renaissance to the Present.* Baltimore: Johns Hopkins University Press, 2004.

Mody, Cyrus. "A Little Dirt Never Hurt Anyone: Knowledge-Making and Contamination in Materials Science." *Social Studies of Science* 31 (2002): 7–36.

Moss, Michael. *Salt, Sugar, Fat: How the Food Giants Hooked Us*. New York: Random House, 2014.

Mudry, Jessica. *Measured Meals: Nutrition in America*. Albany, NY: SUNY Press, 2009.

Mudry, Jessica. "Quantifying an American Eater: Early USDA Food Guidance, and a Language of Numbers." *Food, Culture and Society* 9 (2006): 49–67.

Mueller, Thomas. *Extra Virginity: The Sublime and Scandalous World of Olive Oil*. New York: W. W. Norton, 2012.

Mueller, Thomas. "Slippery Business." *New Yorker*, August 13, 2007. http://www.newyorker.com/magazine/2007/08/13/slippery-business.

Mueller, Thomas. *Truth in Olive Oil*, (blog). http://www.truthinoliveoil.com/about-truth-olive-oil.

Mulvaney, Lydia. "The Parmesan Cheese You Sprinkle on Your Penne Could Be Wood." *Bloomberg*, February 16, 2016. Retrieved from https://www.bloomberg.com/news/articles/2016-02-16/the-parmesan-cheese-you-sprinkle-on-your-penne-could-be-wood.

Nixon, H. C., "The Rise of the American Cottonseed Oil Industry." *Journal of Political Economy* 38 (February 1930): 73–85.

O'Connell, J. "Metrology: The Creation of Universality by the Circulation of Particulars." *Social Studies of Science* 23 (1993): 129–73.

Oddy, Derek J., and Alain Drouard, eds. *The Food Industries of Europe of the Nineteenth and Twentieth Centuries*. London: Routledge, 2014.

Okun, Mitchell. *Fair Play in the Marketplace: The First Battle for Pure Food and Drugs*. DeKalb: Norther Illinois University Press, 1986.

Olmstead, Alan. *Arresting Contagion: Science, Policy, and Conflicts over Animal Disease Control*. Cambridge, MA: Harvard University Press, 2015.

Olmsted, Larry. *Real Food/Fake Food: Why You Don't Know What You're Eating and What You Can Do About It*. New York: Algonquin Books, 2016.

Olsen, J. C. "American Contemporaries—David Wesson," *Industrial and Chemical Engineering* 21 (1929): 290–91.

Orland, Barbara. "Enlightened Milk: Reshaping a Bodily Substance into a Chemical Object." In *Materials and Expertise in Early Modern Europe*, edited by Ursula Klein and E. C. Spary, 163–97. Chicago: University of Chicago Press, 2010.

Orvell, Miles. *The Real Thing: Imitation and Authenticity in American Culture, 1880–1940*. Chapel Hill: University of North Carolina Press, 1989.

Passariello, Phyllis, "Anomalies, Analogies, and Sacred Profanities: Mary Douglas on Food and Culture, 1957–1989." *Food and Foodways* 4 (Spring 1990): 53–71.

Paxson, Heather. *The Life of Cheese: Crafting Food and Value in America*. Berkeley: University of California Press, 2013.

Pearl, Sharrona. *About Faces: Physiognomy in Nineteenth-Century Britain*. Cambridge: Harvard University Press, 2010.

Pearson, Andrea. "*Frank Leslie's Illustrated Newspaper* and *Harper's Weekly*: Innovation and Imitation in Nineteenth-Century American Pictorial Reporting." *Journal of Popular Culture* 23 (1990): 81–111.

Perkins, Elizabeth. "The Consumer Frontier: Household Consumption in Early Kentucky," *Journal of American History* 78 (1991): 486–510.

Petrick, Gabriella. "The Arbiters of Taste: Producers, Consumers, and the Industrialization of Taste in America, 1900–1960." PhD diss., University of Delaware, 2006.

Petrick, Gabriella. "Feeding the Masses: H.J. Heinz and the Creation of Industrial Food." *Endeavor* 33 (2009): 29–34.

Pickstone, John V. *Ways of Knowing: A New History of Science, Technology, and Medicine*. Chicago: University of Chicago Press, 2004.

Pirog, R., C. Miller, L. Way, C. Hazekamp, and E. Kim. *The Local Food Movement: Setting the Stage for Good Food*. East Lansing, MI: MSU Center for Regional Food Systems, 2014. https://www.canr.msu.edu/foodsystems/uploads/files/Local_Food_Movement.pdf.

Pollan, Michael. "Why 'Natural' Doesn't Mean Anything Anymore." *New York Times*, April 28, 2015. https://www.nytimes.com/2015/05/03/magazine/why-natural-doesnt-mean-anything-anymore.html?_r=0.

Poppe, C., and U. Kjaernes. *Trust in Food in Europe: A Comparative Analysis*. Oslo: National Institute for Consumer Research, 2003.

Porter, Roy. *The Greatest Benefit to Mankind: A Medical History of Mankind from Antiquity to the Present*. New York: W. W. Norton, 1997.

Porter, Roy. *Health for Sale: Quackery in England, 1660–1850*. Manchester: Manchester University Press, 1989.

Porter, Ted. *Trust in Numbers*. Princeton, NJ: Princeton University Press, 1995.

Rappaport, Erika. "Packaging China: Foreign Articles and Dangerous Tastes in the Mid-Victorian Tea Party." In *The Making of the Consumer: Knowledge, Power and Identity in the Modern World*, edited by Frank Trentmann, 125–46. Oxford: Berg, 2006.

Rappaport, Erika. *A Thirst for Empire: How Tea Shaped the Modern World*. Princeton, NJ: Princeton University Press, 2017.

Reading, Amy. *The Mark Inside: A Perfect Swindle, a Cunning Revenge, and a Small History of the Big Con*. New York: Vintage, 2013.

Reinhardt, Carston, and Anthony Travis. *Heinrich Caro and the Creation of Modern Chemical Industry*. Dordrecht, The Netherlands: Kluwer, 2000.

Riepma, S. F. *The Story of Margarine*. Washington, DC: Public Affairs Press, 1970.

Rockman, Seth. "What Makes the History of Capitalism Newsworthy?" *Journal of the Early Republic* 34, no 3 (Fall 2014): 439–66.

Rodrigue, John C. *Reconstruction in the Cane Fields: From Slavery to Free Labor in Louisiana's Sugar Parishes, 1862–1880*. Baton Rouge: Louisiana State University Press, 2001.

Rogers, Thomas. *The Deepest Wounds: A Labor and Environmental History of Sugar in Northeast Brazil*. Chapel Hill: University of North Carolina Press, 2010.

Rohlf, Michael. "Immanuel Kant." Revised January 25, 2016. *Stanford Encyclopedia of Philosophy Archive*. https://plato.stanford.edu/archives/spr2016/entries/kant/.

Rosen, George. *A History of Public Health*. Baltimore: Johns Hopkins University Press, 2015. First published 1958 by MD Publications (New York).

Rosenberg, Charles. "Wilbur Olin Atwater." In *Dictionary of Scientific Biography* edited by Charles Coulson Gillispie. New York: Charles Scribner's Sons, 1970–1980.

Rosner, Helen. "What the Mast Brothers Scandal Tells Us about Ourselves." Eater,

December 23, 2015. https://www.eater.com/2015/12/23/10657022/what-the-mast-brothers-scandal-tells-us-about-ourselves.

Rueber, M. "Is the Milk the Measure of All Things? Babcock Tests, Breed Associations, and Land-Grant Scientists, 1890–1920." In *Science as Service: Establishing and Reformulating American Land-Grant Universities, 1865–1930*, edited by Alan Marcus, 93–114. Tuscaloosa: University of Alabama Press, 2015.

Russell, Colin A., Noel G. Coley, and Gerrylynn K. Roberts. *Chemists by Profession: The Origins and Rise of the Royal Institute of Chemistry.* Milton Keynes: Open University Press, 1977.

Rutherford, Paul. *The Adman's Dilemma: From Barnum to Trump*. Toronto: University of Toronto Press, 2018.

Sackman, Douglas. *Orange Empire: California and the Fruits of Eden*. Berkeley: University of California Press, 2005.

Salazar, James. *Bodies of Reform: The Rhetoric of Character in Gilded Age America.* New York: New York University Press, 2010.

Salisbury, Susan. "Sugar Growers' Lawsuit over 'Sweet Surprise' Advertising Goes to Trial." *Palm Beach Post*, November 2, 2015. http://protectingyourpocket.blog.palmbeachpost.com/2015/11/02/sugar-growers-lawsuit-over-sweet-surprise-advertising-goes-to-trial/.

Sandage, Scott, "The Gilded Age." In *A Companion to American Cultural History*, edited by Karen Halttunen, 139–153. Malden, MA: Blackwell, 2008.

Santlofer, Joy. *Food City: Four Centuries of Food-Making in New York.* New York: W. W. Norton, 2017.

Sargent, Rose-Mary. "Scientific Experiment and Legal Expertise: The Way of Experience in Seventeenth-Century England." *Studies in the History and Philosophy of Science A* 20, no. 1 (1989): 19–45.

Satin, Morton. *Death in the Pot: The Impact of Food Poisoning on History.* New York: Prometheus Books, 2007.

Schickore, Jutta. *The Microscope and the Eye: A History of Reflections, 1740–1870.* Chicago: University of Chicago Press, 2007.

Schleifer, David. "The Perfect Solution: How Trans Fats Became the Healthy Replacement for Saturated Fats." *Technology and Culture* 53 (January 2012): 94–119.

Schoolman, Ethan. "Building community, benefiting neighbors: 'Buying local' by people who do not fit the mold for 'ethical consumers.'" *Journal of Consumer Culture* (August 4, 2017). https://doi.org/10.1177/1469540517717776.

Schwartz, Stuart, ed. *Tropical Babylons: Sugar and the Making of the Atlantic World, 1450–1680.* Chapel Hill: University of North Carolina Press, 2004.

Scott, James. *Seeing Like a State: How Certain Schemes to Improve the Human Condition Have Failed.* New Haven, CT: Yale University Press, 1998.

Seim, Dan. "The Butter-Margarine Controversy and 'Two Cultures' at Iowa State College." *Annals of Iowa* 67 (2008): 1–50.

Shapiro, Laura. *Perfection Salad: Woman and Cooking at the Turn of the Century.* New York: FSG, 1986.

Shapiro, Laura. *Something from the Oven: Reinventing Dinner in 1950s America.* New York: Penguin Books, 2005.

Shields, David. "Prospecting for Oil." *Gastronomica* 10 (2010): 25–34.

Shields, David. *Southern Provisions: The Creation and Revival of a Cuisine.* Chicago: University of Chicago Press, 2015.

Simmons, Dana. *Vital Minimum: Need, Science, and Politics in Modern France.* Chicago: University of Chicago Press, 2015.

Singerman, David. "'A Doubt Is at Best an Unsafe Standard': Measuring Sugar in the Early Bureau of Standards." *Journal of Research of the National Institute of Standards and Technology* 112 (2007): 53–66.

Singerman, David. "Inventing Purity in the Atlantic Sugar World." PhD diss., MIT, 2014.

Singerman, David. "Science, Commodities, and Corruption in the Gilded Age." *Journal of the Gilded Age and Progressive Era* 15, no. 3 (July 2016): 278–93.

Sivulka, Juliann. *Stronger than Dirt: A Cultural History of Advertising Personal Hygiene in America, 1875–1940.* New York: Humanity Books, 2001.

Skillman, David. *Biography of a College: Being the History of the First Century of the Life of Lafayette College.* Easton, PA: Lafayette College, 1932.

Smith, S. D. "Coffee, Microscopy, and the Lancet's Analytical Sanitary Commission." *Social History of Medicine* 14 (2001): 171–97.

Smith-Howard, Kendra. *Pure and Modern Milk: An Environmental History since 1900.* Oxford: Oxford University Press, 2014.

Snodgrass, Katherine. *Margarine as a Butter Substitute.* Stanford, CA: Stanford University Food Research Institute, 1930.

Solnit, Rebecca. "Mysteries of Thoreau Unsolved." *Orion* (May/June 2013): 19.

Specht, Joshua "A Failure to Prohibit: New York City's Underground Bob Veal Trade." *Journal of the Gilded Age and Progressive Era* 12 (October 2013): 475–501.

Spellman, Susan. *Cornering the Market: Independent Grocers and Innovation in American Small Business.* Oxford: Oxford University Press, 2016.

Spellman, Susan. "Trust Brokers: Traveling Grocery Salesman and Confidence in Nineteenth Century Trade." *Enterprise and Society* 13 (2012): 276–312.

Spiekermann, Uwe. "Claus Spreckels: A Biographical Case Study of Nineteenth-Century American Immigrant Entrepreneurship." *Business and Economic History On-Line* 8 (2010): 1–21.

Spiekermann, Uwe. "Dangerous Meat? German-American Quarrels over Pork and Beef, 1870–1900." *Bulletin of the German Historical Institute* 46 (2010): 93–109.

Spiekermann, Uwe. "Redefining Food: The Standardization of Products and Production in Europe and the United States, 1880–1914." *History and Technology* 27 (March 2011): 11–36.

Stacey, Michelle. *Consumed: Why Americans Love, Hate and Fear Food.* New York: Touchstone, 1994.

Stanziani, Alesandro. "Defining 'Natural Product' between Public Health and Business, 17th to 21st Centuries." *Appetite* 51 (2008): 15–17.

Stanziani, Alesandro. "Food and Expertise: The Trichinosis Epidemic in France, 1878–1891." *Food and Foodways* 10 (2003): 209–37.

Stanziani, Alessandro. "Municipal Laboratories and the Analysis of Foodstuffs in France under the Third Republic: A Case Study of the Parisian Municipal Laboratory, 1787–1907." In *Food and the City in Europe since 1800*, edited by L. Atkins, D. Lummel, and D. Oddy, 105–16. Burlington, VT: Ashgate, 2012.

Stanziani, Alessandro. "Negotiating Innovation in a Market Economy: Foodstuffs and Beverages Adulteration in Nineteenth-Century France." *Enterprise and Society* 8 (2007): 375–412.

Steere-Williams, Jacob. "A Conflict of Analysis: Analytical Chemistry and Milk Adulteration in Victorian Britain." *Ambix* 61, no. 3 (2014): 279–98.

Steinberg, Theodore. *Down to Earth: Nature's Role in American History.* New York: Oxford University Press, 2000.

Stoll, Steven. "A Metabolism of Society." In *The Oxford Handbook of Environmental History,* edited by Andrew Isenberg, 378. Oxford: Oxford University Press, 2014.

Stolypin, P. A. *The Search for Stability in Late Imperial Russia.* Palo Alto, CA: Stanford University Press, 2001.

Strasser, Susan. *Never Done: A History of American Housework.* New York: Holt, 1982.

Strasser, Susan. *Satisfaction Guaranteed: The Making of the American Mass Market.* Washington, DC: Smithsonian Institution Press, 1989.

Strasser, Susan. *Waste and Want: A Social History of Trash.* New York: Metropolitan Books, 1999.

Summerhill, Thomas. *Harvest of Dissent: Agrarianism in Nineteenth-Century New York.* Urbana: University of Illinois Press, 2005.

Swallow, Pamela. *The Remarkable Life and Career of Ellen Swallow Richards: Pioneer in Science and Technology.* Hoboken, NJ: Wiley, 2014.

Swift, Louis F., and Arthur Van Vlissinger Jr. *The Yankee of the Yards: The Biography of Gustavus Franklin Swift.* Chicago: A. W. Shaw, 1927.

Teuteberg, Hans Jürgen. "Food Adulteration and the Beginnings of Uniform Food Legislation in Late Nineteenth-Century Germany." In *The Origins and Development of Food Policies in Europe,* edited by John Burnett and Derek J. Oddy, 146–60. London: Leicester University Press, 1994.

Thomas, Courtney. *In Food We Trust: The Politics of Purity in American Food Regulation.* Lincoln: University of Nebraska Press, 2014.

Thompson, E.P. "The Moral Economy of the English Crowd in the Eighteenth Century." *Past & Present* 50 (1971): 76–136.

Thompson, Paul. *From Field to Fork: Food Ethics for Everyone.* Oxford: Oxford University Press, 2015.

Tilly, Charles. *The Politics of Collective Violence.* Cambridge: Cambridge University Press, 2003.

Tilly, Louise. "The Food Riot as a Form of Political Conflict in France." *Journal of Interdisciplinary History* 2 (1971): 23–57.

Tomes, Nancy. *The Gospel of Germs: Men, Women, and the Microbe in American Life.* Cambridge, MA: Harvard University Press, 1998.

Tompkins, Kyla Wazan. *Racial Indigestion: Eating Bodies in the Nineteenth Century* New York: New York University Press, 2012.

Tonkovich, Nicole. *Domesticity with a Difference: The Nonfiction of Catharine Beecher, Sarah J. Hale, Fanny Fern, and Margaret Fuller.* Jackson: University Press of Mississippi, 1997.

Travis, Anthony. "From Manchester to Massachusetts via Mulhouse: The Transatlantic Voyage of Aniline Black." *Technology and Culture* 35 (1994): 70–99.

Trilling, Lionel. *Sincerity and Authenticity.* Cambridge, MA: Harvard University Press, 1971.

True, Alfred. *A History of Agricultural Experimentation and Research in the United*

States, 1607–1925: Including a History of the United States Department of Agriculture. Washington, DC: Government Printing Office, 1937.

United States Bureau of the Census, *Historical Statistics of the United States* series C89. Washington, DC: Government Printing Office, 1976.

United States Food and Drug Administration. "The Dangers of Raw Milk." September 2018. https://www.fda.gov/downloads/Food/FoodborneIllnessContaminants/UCM239493.pdf.

Valenze, Deborah. *Milk: A Local and Global History.* New Haven, CT: Yale University Press, 2011.

Van Stuyvenberg, J. H., ed. *Margarine: An Economic, Social and Scientific History, 1869–1969.* Toronto: University of Toronto Press, 1969.

Van Whye, John. *Phrenology and the Origins of Victorian Scientific Naturalism.* London: Routledge, 2004.

Victor, Daniel. "ABC Settles with Meat Producer in 'Pink Slime' Defamation Case." *New York Times,* June 28, 2017. https://www.nytimes.com/2017/06/28/business/media/pink-slime-abc-lawsuit-settlement.html.

Victor, Daniel. "Butter or Margarine? In Dunkin' Donuts Lawsuit, Man Accepts No Substitutes." *New York Times,* April 4, 2017. https://www.nytimes.com/2017/04/04/business/dunkin-donuts-butter-lawsuit.html?_r=0.

Vileisis, Ann. *Kitchen Literacy: How We Lost Knowledge of Where Food Comes From and Why We Need to Get It Back.* Washington, DC: Island Press, 2008.

Virilio, Paul. *The Paul Virilio Reader.* Edited by Steve Redhead. New York: Columbia University Press, 2004.

Vogel, Steven. *Thinking Like a Mall: Environmental Philosophy after the End of Nature.* Cambridge, MA: MIT Press, 2015.

Wade, Louise Carroll. *Chicago's Pride: The Stockyards, Packingtown, and Environs in the Nineteenth Century.* Urbana: University of Illinois Press, 1987.

Warner, Deborah. "How Sweet It Is: Sugar, Science, and the State." *Annals of Science* 64 (April 2007): 147–70.

Warner, Deborah. *Sweet Stuff: An American History of Sweeteners from Sugar to Sucralose.* Washington, DC: Smithsonian Institution Scholarly Press, 2011.

Webb, Sidney, and Beatrice Webb. "The Assize of Bread." *Economic Journal* 14, no. 54, (June 1904): 196–218.

Weber, C. M., and C. L. Alsberg. *The American Vegetable-Shortening Industry: Its Origin and Development.* Fats and Oils Studies 5. Palo Alto, CA: Stanford, Food Research Institute, 1934.

Weston A. Price Foundation. "A Campaign for Real Milk." Washington, DC: Weston A. Price Foundation, 2006. http://www.realmilk.com/wp-content/uploads/2012/11/RealMilkTrifold.pdf.

Wenger, Diane. "Delivering the Goods: The Country Storekeeper and Inland Commerce in the Mid-Atlantic." *Pennsylvania Magazine of History and Biography* 129 (January 2005): 45–72.

Wenger, Diane. *A Country Storekeeper in Pennsylvania: Creating Economic Networks in Early America, 1790–1807.* University Park: Pennsylvania State University Press, 2008.

White, Richard. "Information, Markets, and Corruption: Transcontinental Railroads in the Gilded Age." *Journal of American History* 90 (June 2003): 19–43.

White, Richard. *The Republic for Which It Stands: The United States during Recon-*

struction and the Gilded Age, 1865–1896. New York: Oxford University Press, 2017.

Whooley, Owen. *Knowledge in the Time of Cholera: The Struggle over American Medicine in the Nineteenth Century.* Chicago: University of Chicago Press, 2013.

Whorton, James. *Crusaders for Fitness: The History of American Health Reformers.* Princeton, NJ: Princeton University Press, 1982.

Wiebe, Robert. *The Search for Order, 1877–1920*. New York: Hill and Wang, 1967.

Williamson, Harold F. *Edward Atkinson: The Biography of an America Liberal 1827–1905*. Boston: Old Corner Book Store, 1934.

Wilson, Bee. *Swindled: The Dark History of Food Fraud, from Poisoned Candy to Counterfeit Coffee.* Princeton, NJ: Princeton University Press, 2008.

Wines, Richard. *Fertilizer in America: From Waste Recycling to Resource Exploitation* Philadelphia: Temple University Press, 1985.

Woloson, Wendy. *Refined Tastes: Sugar, Confectionery, and Consumers in Nineteenth-Century America.* Baltimore, MD: Johns Hopkins University Press, 2002.

Woods, Thomas. *Knights of the Plow: Oliver H. Kelley and the Origins of the Grange in Republican Ideology.* West Lafayette, IN: Purdue University Press, 1991.

Worster, Donald. *Nature's Economy: A History of Ecological Ideas.* Cambridge: Cambridge University Press, 2011. 2nd ed.

Wrenn, Lynette B. *Cinderella of the New South: A History of the Cottonseed Industry, 1895–1955*. Knoxville: University of Tennessee Press, 1995.

Young, James Harvey. "'This Greasy Counterfeit': Butter versus Oleomargarine in the United States Congress, 1886." *Bulletin of the History of Medicine* 53 (1979): 392–414.

Young, James Harvey. *Pure Food: Securing the Food and Drug Act of 1906*. Princeton, NJ: Princeton University Press, 1989.

Young, James Harvey. "The Science and Morals of Metabolism: Catsup and Benzoate of Soda." *Journal of the History of Medicine and Applied Sciences* 23 (1968): 86–104.

Zallen, Jeremy. "American Lucifers: Makers and Masters of the Means of Light 1750–1900." PhD diss., Harvard University, 2014.

Zimring, Carl. *Clean and White: A History of Environmental Racism in the United States.* New York: New York University Press, 2016.

Index

Page numbers in italics refer to illustrations.